The Earth's Atmosphere

Kshudiram Saha

The Earth's Atmosphere

Its Physics and Dynamics

Dr. Kshudiram Saha
4008 Beechwood Road
University Park
MD 20782
USA
krsaha@comcast.net

ISBN: 978-3-540-78426-5   e-ISBN: 978-3-540-78427-2

Library of Congress Control Number: 2008925553

© 2008 Springer-Verlag Berlin Heidelberg

This work is subject to copyright. All rights are reserved, whether the whole or part of the material is concerned, specifically the rights of translation, reprinting, reuse of illustrations, recitation, broadcasting, reproduction on microfilm or in any other way, and storage in data banks. Duplication of this publication or parts thereof is permitted only under the provisions of the German Copyright Law of September 9, 1965, in its current version, and permission for use must always be obtained from Springer. Violations are liable to prosecution under the German Copyright Law.

The use of general descriptive names, registered names, trademarks, etc. in this publication does not imply, even in the absence of a specific statement, that such names are exempt from the relevant protective laws and regulations and therefore free for general use.

Cover design: deblik, Berlin

Printed on acid-free paper

9 8 7 6 5 4 3 2 1

springer.com

Dedicated to

Professor Meghnad Saha, D.Sc. F.N.I., F.Inst.P., F.R.S.
The distinguished astrophysicist and my revered teacher who wisely advised me to opt for a career in meteorology
   In honoring him, I honor all my teachers, advisors and benefactors

Preface

The author has sought to incorporate in the book some of the fundamental concepts and principles of the physics and dynamics of the atmosphere, a knowledge and understanding of which should help an average student of science to comprehend some of the great complexities of the earth-atmosphere system, in which a three-way interaction between the atmosphere, the land and the ocean tends to maintain an overall mass and energy balance in the system through physical and dynamical processes.

The book, divided into two parts and consisting of 19 chapters, introduces only those aspects of the subject that, according to the author, are deemed essential to meet the objective in view. The emphasis is more on clarity and understanding of physical and dynamical principles than on details of complex theories and mathematics. Attempt is made to treat each subject from first principles and trace its development to present state, as far as possible. However, a knowledge of basic calculus and differential equations is *sine qua non* especially for some of the chapters which appear later in the book.

In Part-I (the physics part), Chap. 1, after introductory remarks about the place of the earth in the solar system, stresses the importance of solar radiation and gravitation in atmospheric physics and dynamics. Chap. 2 describes the origin, composition, structure and properties of the atmosphere. Heat and thermodynamics of a dry and moist atmosphere and the physics of formation of cloud and rain are discussed in Chaps. 3–5. Laws of radiation in general are reviewed in Chap. 6. A brief account is given of our current knowledge of the sun as a source of radiation in Chap. 7. Chapter 8 describes the passage of solar radiation through the different layers of the earth's atmosphere and the thermodynamical effects it produces in each layer. Physical processes leading to the warming of the earth's surface by the incoming shortwave solar radiation and its subsequent emission of longwave radiation to produce greenhouse effect, the heat balance of the earth-atmosphere system and formation of heat sources and sinks in the earth-atmosphere system are discussed in Chaps. 9 and 10.

In Part-II (the dynamics part), the first two chapters are devoted to derivation of the fundamental equations of atmospheric motion in different co-ordinate systems and their simplification in order to derive some types of balanced winds. Some essential properties of air flow, such as divergence, vorticity, vertical motion and circulation, involved in the formation of weather and climate, are discussed in Chap. 13. Effects of friction on flow in the boundary layers of the atmosphere and the ocean are discussed in Chap. 14. Chapters 15 and 16 discuss waves and oscillations that are excited in the atmosphere by fluctuations in atmospheric pressure, temperature and wind, including those at or near the equator. Some aspects of dynamical weather prediction by numerical methods and dynamical instability of the atmospheric flows are discussed in Chaps. 17 and 18. The concluding chapter summarizes our current knowledge of the general circulation of the atmosphere derived from observations as well as results of laboratory experiments and numerical simulation studies.

The book is primarily aimed at meeting the needs of students at undergraduate level pursuing courses in earth and atmospheric sciences, but could be used as a reference book by graduate students as well as scientists working in other fields of science, desirous of learning more about the earth-atmosphere system. Inspite of the care taken in the preparation of the book, it is likely that there have been errors and omissions. The author will be thankful if cases of such lapses are brought to his notice.

The author is extremely grateful to his family, especially his daughters, Manjushri and Suranjana, who supported this work from the very beginning. Manjushri helped actively during preparation of the draft manuscript with library and referencing work. Suranjana along with her husband, Professor Dr. Huug van den Dool, provided all the logistic support and helped the author in completing all the technical aspects of the book. Suranjana handled all the diagrams and helped with their insertion in the book. Huug's comments on the draft chapters were immensely helpful in improving the manuscript. Without their help, it would have been well-nigh impossible to complete the work. He also received encouragement from his eldest daughter Jayshri while writing this book. His special thanks are due to the National Centers for Environmental Prediction (NCEP) of the National Weather Service (NWS) of the United States of America for several of their analysis products incorporated in the book. He expresses his indebtedness to the numerous authors, publishers and learned Societies who permitted him to reproduce diagrams and excerpts from their published work.

University Park, U.S.A. *Kshudiram Saha*
January 11, 2008

Contents

Part I Physics of the Earth's Atmosphere

1 The Sun and the Earth – The Solar System and the Earth's Gravitation 3
   1.1   Introduction 3
   1.2   Earth's Gravitational Force – Gravity 4
   1.3   Geopotential Surfaces 6
   1.4   Motion in the Earth's Gravitational Field – The Law of Central Forces 6

2 The Earth's Atmosphere – Its Origin, Composition and Properties .. 9
   2.1   Introduction: Origin of the Earth's Atmosphere 9
   2.2   Composition of the Atmosphere 10
   2.3   Properties and Variables of the Atmosphere 12
       2.3.1   Pressure 12
       2.3.2   Temperature 15
       2.3.3   Density 21
       2.3.4   Other Variables of the Atmosphere 22
       2.3.5   Observing the Atmosphere 22
   2.4   Gas Laws – Equations of State 23
       2.4.1   The Equation of State – General 23
       2.4.2   The Equation of State of an Ideal Gas 24
       2.4.3   The Equation of State of a Mixture of Gases 26
       2.4.4   The Equation of State of a Real Gas 26

3 Heat and Thermodynamics of the Atmosphere 27
   3.1   Introduction. The Nature of Heat and Kinetic Theory 27
   3.2   The First Law of Thermodynamics 27
   3.3   Specific Heats of Gases 28
   3.4   Adiabatic Changes in the Atmosphere 30

		3.4.1	Adiabatic Relationship Between Pressure, Temperature and Volume	30
		3.4.2	Potential Temperature	30
		3.4.3	Dry Adiabatic Lapse Rate of Temperature with Height	31
		3.4.4	Static Stability of Dry Air – Buoyancy Oscillations	31
		3.4.5	Adiabatic Propagation of Sound Waves	33
	3.5	The Concept of Entropy		33
	3.6	The Second Law of Thermodynamics		36
		3.6.1	Carnot Engine	36
		3.6.2	Statement of the Second Law of Thermodynamics	38
	3.7	Thermodynamic Equilibrium of Systems: Thermodynamic Potentials		40
		3.7.1	Free Energy or Helmholtz Potential	40
		3.7.2	Free Enthalpy, or Gibbs' Potential, or Gibbs Free Energy	41
	3.8	The Third Law of Thermodynamics		41
	3.9	The Atmosphere as a Heat Engine		42
4	**Water Vapour in the Atmosphere: Thermodynamics of Moist Air**			**43**
	4.1	Introduction		43
	4.2	Humidity of the Air – Definitions		44
	4.3	Density of Moist Air – Virtual Temperature		45
	4.4	Measurement of Humidity – Hygrometers/Psychrometers		46
	4.5	Ascent of Moist Air in the Atmosphere – Pseudo-Adiabatic Process		48
	4.6	Saturated Adiabatic Lapse Rate of Temperature		50
	4.7	Equivalent Potential Temperature		50
	4.8	Variation of Saturation Vapour Pressure with Temperature		51
		4.8.1	The Clausius-Clapeyron Equation	51
		4.8.2	Melting Point of Ice – Variation with Pressure	53
	4.9	Co-existence of the Three Phases of Water – the Triple Point		53
	4.10	Stability of Moist Air		55
		4.10.1	Thermodynamic Diagrams	56
5	**Physics of Cloud and Precipitation**			**59**
	5.1	Introduction – Historical Perspective		59
	5.2	Cloud-Making in the Laboratory – Condensation Nuclei		60
	5.3	Atmospheric Nuclei – Cloud Formation in the Atmosphere		62
	5.4	Drop-Size Distribution in Clouds		64
	5.5	Rate of Fall of Cloud and Rain Drops		65
	5.6	Supercooled Clouds and Ice-Particles – Sublimation		66
	5.7	Clouds in the Sky: Types and Classification		68
	5.8	From Cloud to Rain		68
		5.8.1	Hydrodynamical Attraction	69
		5.8.2	Electrical Attraction	72
		5.8.3	Collision Due to Turbulence	72
		5.8.4	Differences in Size of Cloud Particles	73

5.8.5 Differences of Temperature Between Cloud Elements ... 73
5.8.6 The Ice-Crystal Effect ... 74
5.9 Meteorological Evidence – Rainfall from Cold and Warm Clouds . 75
5.9.1 Rainfall from Warm Clouds ... 76
5.10 Climatological Rainfall Distribution over the Globe ... 76

6 Physics of Radiation – Fundamental Laws ... 79
6.1 Introduction – the Nature of Thermal Radiation ... 79
6.2 Radiation and Absorption – Heat Exchanges ... 79
6.2.1 Conduction ... 79
6.2.2 Convection ... 79
6.2.3 Radiation ... 80
6.3 Properties of Radiation ... 81
6.4 Laws of Radiation – Emission and Absorption ... 82
6.4.1 Kirchhoff's Law ... 82
6.4.2 Laws of Black Body and Gray Radiation ... 83
6.4.3 Stefan-Boltzmann Law ... 84
6.4.4 Wien's Displacement Law ... 84
6.4.5 Planck's Law of Black Body Radiation ... 84
6.4.6 Derivation of Wien's Law and Stefan-Boltzmann Law from Planck's Law ... 85
6.5 Spectral Distribution of Radiant Energy ... 86
6.6 Some Practical Uses of Electromagnetic Radiation ... 88

7 The Sun and its Radiation ... 89
7.1 Introduction ... 89
7.2 Physical Characteristics of the Sun ... 90
7.3 Structure of the Sun – its Interior ... 90
7.3.1 The Core – Nuclear Reactions ... 91
7.3.2 The Radiative Layer ... 92
7.3.3 The Convective Layer ... 93
7.4 The Photosphere ... 94
7.4.1 Sunspots ... 95
7.5 The Solar Atmosphere ... 95
7.5.1 The Reversing Layer ... 96
7.5.2 The Chromosphere ... 96
7.5.3 The Corona ... 96
7.6 The Solar Wind ... 97
7.7 The Search for Neutrinos ... 98

8 The Incoming Solar Radiation – Interaction with the Earth's Atmosphere and Surface ... 99
8.1 Introduction – the Solar Spectrum ... 99
8.2 Interactions with the Upper Atmosphere (Above 80 km) ... 100

         8.2.1   Interaction with the Solar Wind: Polar Auroras
                 and Magnetic Storms 100
         8.2.2   Interaction with the Solar Ultraviolet Radiation 102
  8.3   The Mesosphere (50–80 km Layer) 102
  8.4   Interaction with Ozone: the Ozonosphere (20–50 km) 102
         8.4.1   Formation of Ozone 103
         8.4.2   Destruction of Ozone: the Ozone Hole 103
         8.4.3   Warming of the Stratosphere 104
         8.4.4   Latitudinal and Seasonal Variation of Ozone 104
         8.4.5   Ozone and Weather 105
  8.5   Scattering, Reflection and Absorption of Solar Radiation
         in the Atmosphere... 107
         8.5.1   Scattering and Reflection 107
         8.5.2   Atmospheric Absorption............................ 107
  8.6   Incoming Solar Radiation (Insolation) at the Earth's Surface 108
         8.6.1   The Solar Constant 108
         8.6.2   The Transparency of the Atmosphere – Effects of
                 Clouds and Aerosols 109
         8.6.3   Distribution of Solar Radiation with Latitude – The
                 Seasonal Cycle.................................... 109
         8.6.4   Seasonal and Latitudinal Variations of Surface
                 Temperature 110
         8.6.5   Diurnal Variation of Radiation with Clear and Cloudy
                 Skies 111
  8.7   Reflection of Solar Radiation at the Earth's Surface – The Albedo. 113

**9 Heat Balance of the Earth's Surface – Upward and Downward
  Transfer of Heat** 115
  9.1   Introduction: General Considerations 115
  9.2   Heat Balance on a Planet Without an Atmosphere 116
  9.3   Heat Balance on a Planet with an Atmosphere: The Greenhouse
         Effect 117
         9.3.1   The Greenhouse Effect 118
  9.4   Vertical Transfer of Radiative Heating – Diurnal Temperature
         Wave.. 119
  9.5   Sensible Heat Flux 120
         9.5.1   Vertical Transfer of Sensible Heat into the Atmosphere .. 120
  9.6   Evaporation and Evaporative Heat Flux from a Surface 123
         9.6.1   Bowen's Ratio 124
         9.6.2   Evaporative Cooling 124
  9.7   Exchange of Heat Between the Earth's Surface
         and the Underground Soil 125
         9.7.1   Amplitude and Range 126
         9.7.2   Time Lag 126
         9.7.3   Velocity 127

	9.7.4	Wavelength 127
	9.7.5	Diurnal Wave 127
	9.7.6	Annual Wave 128
9.8	Radiative Heat Flux into the Ocean 129
	9.8.1	General Properties of Ocean Water 129
	9.8.2	Optical Properties of Ocean Water – Reflection and Refraction 131
	9.8.3	Absorption and Downward Penetration of Solar Radiation in the Ocean 132
	9.8.4	Vertical Distribution of Temperature in the Ocean 133
9.9	The Thermohaline Circulation – Buoyancy Flux 134
9.10	Photosynthesis in the Ocean: Chemical and Biological Processes . 135

10 Heat Balance of the Earth-Atmosphere System – Heat Sources and Sinks 137

- 10.1 Introduction – definition of heat sources and sinks 137
- 10.2 Physical Processes Involved in Heat Balance 138
- 10.3 Simpson's Computation of Heat Budget 139
- 10.4 Heat Balance from Satellite Radiation Data 141
- 10.5 Heat Sources and Sinks from the Energy Balance Equation 145
- 10.6 Computation of Atmospheric Heating from Mass Continuity Equation 149

Part II Dynamics of the Earth's Atmosphere – The General Circulation

11 Winds on a Rotating Earth – The Dynamical Equations and the Conservation Laws 155

- 11.1 Introduction 155
- 11.2 Forces Acting on a Parcel of Air 156
  - 11.2.1 Pressure Gradient Force 156
  - 11.2.2 Gravity Force 157
  - 11.2.3 Force of Friction or Viscosity 157
- 11.3 Acceleration of Absolute Motion 158
- 11.4 Acceleration of Relative Motion 159
  - 11.4.1 Coriolis Force 161
- 11.5 The Equations of Motion in a Rectangular Co-ordinate System ... 162
- 11.6 A System of Generalized Vertical Co-ordinates 162
- 11.7 The Equations of Motion in Spherical Co-ordinate System 164
- 11.8 The Equation of Continuity 167
- 11.9 The Thermodynamic Energy Equation 168
- 11.10 Scale Analysis and Simplification of the Equations of Motion 169
  - 11.10.1 The Geostrophic Approximation and the Geostrophic Wind 170
  - 11.10.2 Scale Analysis of the Vertical Momentum Equation 171

12 Simplified Equations of Motion – Quasi-Balanced Winds 173
12.1   Introduction 173
12.2   The Basic Equations in Isobaric Co-ordinates 173
    12.2.1 Horizontal Momentum Equations 173
    12.2.2 The Continuity Equation 174
    12.2.3 The Thermodynamic Energy Equation 174
12.3   Balanced Flow in Natural Co-ordinates 175
    12.3.1 Velocity and Acceleration in Natural Co-ordinate System 175
    12.3.2 The Gradient Wind 176
    12.3.3 The Geostrophic Wind 177
    12.3.4 Relationship Between the Geostrophic Wind
                   and the Gradient Wind 178
    12.3.5 Inertial Motion 178
    12.3.6 Cyclostrophic Motion 179
12.4   Trajectories and Streamlines 180
12.5   Streamline-Isotach Analysis 182
12.6   Variation of Wind with Height – The Thermal Wind 183

13 Circulation, Vorticity and Divergence 187
13.1   Definitions and Concepts – Circulation and Vorticity 187
13.2   The Circulation Theorem 188
13.3   Absolute and Relative Vorticity 191
13.4   Vorticity and Divergence in Natural Co-ordinates 191
13.5   Potential Vorticity 193
13.6   The Vorticity Equation in Frictionless Adiabatic Flow 196
13.7   The Vorticity Equation from the Equations of Motion 196
    13.7.1 Vorticity Equation in Cartesian Co-ordinates (x, y, z) 196
    13.7.2 The Vorticity Equation in Isobaric Co-ordinates 198
13.8   Circulation and Vorticity in the Real Atmosphere
       (In Three Dimensions) 199
13.9   Vertical Motion in the Atmosphere 200
    13.9.1 The kinematic Method 200
    13.9.2 The Adiabatic Method 201
    13.9.3 The Vorticity Method 201
13.10 Differential Properties of a Wind Field 202
    13.10.1 Translation, (u_0, v_0) 203
    13.10.2 Divergence, Expansion (D) 203
    13.10.3 Deformation 204
    13.10.4 Rotation 205
13.11 Types of Wind Fields – Graphical Representation 205

14 The Boundary Layers of the Atmosphere and the Ocean 207
14.1   Introduction 207
14.2   The Equations of Turbulent Motion in the Atmosphere 208
14.3   The Mixing-Length Hypothesis – Exchange Co-efficients 211

Contents

- 14.4 The Vertical Structure of the Frictionally-Controlled Boundary Layer ... 212
  - 14.4.1 The Surface Layer ... 212
  - 14.4.2 The Ekman or Transition Layer ... 214
- 14.5 The Secondary Circulation – The Spin-Down Effect ... 217
  - 14.5.1 The Nocturnal Jet ... 220
  - 14.5.2 Turbulent Diffusion and Dispersion in the Atmosphere ... 221
- 14.6 The Boundary Layer of the Ocean – Ekman Drift and Mass Transport ... 222
- 14.7 Ekman Pumping and Coastal Upwelling in the Ocean ... 224

15 Waves and Oscillations in the Atmosphere and the Ocean ... 227
- 15.1 Introduction ... 227
- 15.2 The Simple Pendulum ... 228
- 15.3 Representation of Waves by Fourier Series ... 229
- 15.4 Dispersion of Waves and Group Velocity ... 230
- 15.5 The Perturbation Technique ... 231
- 15.6 Simple Wave Types ... 232
- 15.7 Internal Gravity (or Buoyancy) Waves in the Atmosphere ... 239
  - 15.7.1 Internal Gravity (Buoyancy) Waves – General Considerations ... 240
  - 15.7.2 Mountain Lee Waves ... 244
- 15.8 Dynamics of Shallow Water Gravity Waves ... 244
  - 15.8.1 The Adjustment Problem – Shallow Water Equations in a Rotating Frame ... 245
  - 15.8.2 The Steady-State Solution: Geostrophic Adjustment ... 246
  - 15.8.3 Energy Transformations ... 247
  - 15.8.4 Transient Oscillations – Poicaré Waves ... 250
  - 15.8.5 Importance of the Rossby Radius of Deformation ... 251

16 Equatorial Waves and Oscillations ... 253
- 16.1 Introduction ... 253
- 16.2 The Governing Equations in Log-Pressure Co-ordinate System ... 254
  - 16.2.1 The Horizontal Momentum Equations ... 254
  - 16.2.2 The Hydrostatic Equation ... 255
  - 16.2.3 The Continuity Equation ... 255
  - 16.2.4 The Thermodynamic Energy Equation ... 255
- 16.3 The Kelvin Wave ... 255
- 16.4 The Mixed Rossby-Gravity Wave ... 257
- 16.5 Observational Evidence ... 259
- 16.6 The Quasi-Biennial Oscillation (QBO) ... 260
- 16.7 The Madden-Julian Oscillation (MJO) ... 262
- 16.8 El Niño-Southern Oscillation (ENSO) ... 266
  - 16.8.1 Introduction ... 266
  - 16.8.2 El Niño/La Niña ... 266

	16.8.3 Southern Oscillation (SO)	268
	16.8.4 The Walker Circulation – ENSO	270
	16.8.5 Evidence of Walker Circulation in Global Data	270
	16.8.6 Mechanism of ENSO?	271

17 Dynamical Models and Numerical Weather Prediction (N.W.P.) 275
17.1 Introduction – Historical Background 275
17.2 The Filtering of Sound and Gravity Waves 276
17.3 Quasi-Geostrophic Models 278
17.4 Nondivergent Models 279
17.5 Hierarchy of Simplified Models 282
    17.5.1 One-Parameter Barotropic Model 282
    17.5.2 A Two-Parameter Baroclinic Model 282
17.6 Primitive Equation Models 284
    17.6.1 PE Model in Sigma Co-ordinates 285
    17.6.2 A Two-Level Primitive Equation Model 288
    17.6.3 Computational Procedure 289
17.7 Present Status of NWP 290

18 Dynamical Instability of Atmospheric Flows – Energetics and Energy Conversions 293
18.1 Introduction 293
18.2 Inertial Instability 294
18.3 Baroclinic Instability 295
    18.3.1 The Model 295
    18.3.2 Special Cases of Baroclinic Instability 297
    18.3.3 The Stability Criterion – Neutral Curve 298
18.4 Vertical Motion in Baroclinically Unstable Waves 300
18.5 Energetics and Energy Conversions in Baroclinic Instability 302
    18.5.1 Definitions 302
    18.5.2 Energy Equations for the Two-Level Quasi-Geostrophic Model 304
18.6 Barotropic Instability 306
18.7 Conditional Instability of the Second Kind (CISK) 308

19 The General Circulation of the Atmosphere 311
19.1 Introduction – Historical Background 311
19.2 Zonally-Averaged Mean Temperature and Wind Fields Over the Globe 313
    19.2.1 Longitudinally-Averaged Mean Temperature and Wind Fields in Vertical Sections 313
    19.2.2 Idealized Pressure and Wind Fields at Surface Over the Globe in the Three-Cell Model 316
19.3 Observed Distributions of Mean Winds and Circulations Over the Globe 317

	19.4	Maintenance of the Kinetic Energy and Angular Momentum 319
		19.4.1 The Kinetic Energy Balance of the Atmosphere 319
		19.4.2 The Angular Momentum Balance – Maintenance of the Zonal Circulation 320
	19.5	Eddy-Transports 324
		19.5.1 Eddy Flux of Sensible Heat 324
		19.5.2 Eddy-Flux of Angular Momentum 324
		19.5.3 Eddy-Flux of Water Vapour 326
		19.5.4 Vertical Eddy-Transports 327
	19.6	Laboratory Simulation of the General Circulation 327
	19.7	Numerical Experiment on the General Circulation 331

Appendices 333
Appendix-1(A) Vector Analysis-Some Important Vector Relations 333
  1.1 The Concept of a Vector.. 333
  1.2 Addition and Subtraction of Vectors: Multiplication of a Vector by a Scalar 334
  1.3 Multiplication of Vectors 335
  1.4 Differentiation of Vectors: Application to the Theory of Space Curves 337
  1.5 Space Derivative of a Scalar Quantity. The Concept of a Gradient Vector... 338
  1.6 Del Operator, ∇ 339
  1.7 Use of Del Operator in Different Co-ordinate Systems 340
    1.7.1 Cartesian Co-ordinates (x, y, z) 340
    1.7.2 Spherical Co-ordinates (λ, ϕ, r) 340
Appendix-1(B) Motion Under Earth's Gravitational Force 341
Appendix-2 Adiabatic Propagation of Sound Waves 342
Appendix-3 Some Selected Thermodynamic Diagrams 343
Appendix-4 Derivation of the Equation for Saturation Vapour Pressure Curve Taking into Account the Temperature Dependence of the Specific Heats (After Joos, 1967) 344
Appendix-5 Theoretical Derivation of Kelvin's Vapour Pressure Relation for e_r/e_s 346
Appendix-6 Values of Thermal Conductivity Constants for a Few Materials, Drawn from Sources, Including 'International Critical Tables'(1927), 'Smithsonian Physical Tables' (1934), 'Landholt-Bornstein' (1923–1936), 'McAdams' (1942) and others 348
Appendix-7 Physical Units and Dimensions 349
Appendix-8 Some Useful Physical Constants and Parameters........... 350

References 353

Author Index 359

Subject Index 363

About the Book and the Author

This comprehensive text on the physics and dynamics of the earth's atmosphere covers a wide range of topics of interest not only to professional meteorologists but also scientists of several allied disciplines,such as oceanography, geology, geophysics, environmental sciences, etc, desirous of securing a working knowledge of our atmosphere and how it works.and what it means to life on earth. It relates how solar radiation received at the earth's surface together with the heat exchanged between the surface and the overlying atmosphere produces heat sources and sinks in the earth-atmosphere system and how the atmosphere and the ocean are set in motion by the differential heating on different time and space scales to neutralize the imbalance so created and restore the balance of mass and energy. Wind systems and circulations set up in the process include not only the general circulation but also perturbations in the form of waves and oscillations, such as Kelvin waves, Inertio-gravity waves, ElNino - Southern Oscillation, Madden-Julian Oscillation, Quasi-biennial Oscillation, etc. Written by an expert in atmospheric science in an easy-to-understand style, each topic is developed from first principles and brought up-to-date to the extent practicable keeping in view the size of the book. It is bound to appeal to a large circle of readers from an average undergraduate student of science to seasoned professionals.

A Doctorate in Physics (meteorology) from the University of Calcutta (now kolkata) in India in 1956, Kshudiram Saha, born 1918, received his school, college and university education in India and graduated from the University of Allahabad in physics in 1940. He joined the Indian meteorological service in I942. During the war and till 1961, he served as a meteorological officer with the Indian Air Force. He returned to Indian Meteorological Service in 1962, and joined the WMO-sponsored newly-established Institute of Tropical Meteorology (which later became the Indian Institute of Tropical Meteorology) in Poona in 1964 where he served first as Head of the Forecasting Research Division and later as Director of the Institute. During his tenure in the Institute, he not only involved himself in carrying out original research work, but also acted as advisor to several dedicated young research colleagues who later made their mark in their respective fields of work. Singly and in collaboration with many of them, he published more than 60 original research papers on different aspects of tropical meteorology in peer-reviewed standard journals. He also visited several educational and research institutions in U.S.A. and acted as member of several advisory groups of the WMO during this period. He retired from his position as Director of the Institute in 1976.

Dr. Saha has continued his research work in meteorology even after retirement from office, as evident from his recent publications. While he worked as an Emeritus scientist with the Council of Scientific and Industrial Research in India after retirement, he was invited by MIT Cambridge as a Visiting Scientist, in 1978, to work on a problem of monsoon depressions with Professors Shukla and Sanders. In November 1980, he was awarded a Senior Research Associateship by the U.S. National Academy of Sciences National Research Council to work at the Naval Post-graduate School, Monterey, California, where he collaborated with Professor C.–P.Chang in further studies of monsoon problems. He has been a Life Fellow of the Royal Meteorological Society, London, since 1946; a Professional (now Emeritus) Member of the American Meteorological Society, since 1947; a Life Fellow of the Indian Meteorological Society, since 1976; and a Foundation Fellow of the Maharashtra Academy of Sciences, also since 1976. Now in his late eighties, he is still active in the field of meteorology. His educational and scientific background and knowledge and vast experience in the field of atmospheric science over a period of more than half a century make him eminently qualified to write this book.

Guide to Systems of Numbering Diagrams and Equations in Text and Appendices

Text

Figures are numbered serially chapterwise. For example, Fig. 5.12 is Figure 12 of Chapter 5.
Equations are numbered serially, sectionwise and chapterwise. For example, Equation (5.12.4) is Equation 4 in Section 12 of Chapter 5.

Appendices

Figures and equations are both numbered serially appendixwise and distinguished from their counterparts in the regular chapters by adding a prime to the number. For example, Fig. 5.6' is Figure 6 in Appendix 5. Similarly, Equation (4.5') is Equation 5 in Appendix 4.

Part I
Physics of the Earth's Atmosphere

Chapter 1
The Sun and the Earth – The Solar System and the Earth's Gravitation

1.1 Introduction

From time immemorial, humans have wondered about their place in the universe. Those in early ages believed that the earth was flat and at the center of the universe and that all celestial bodies which they could see above and around them revolved around the earth. This geocentric view prevailed for a long time in human history and was even supported by Aristotle in 320 B.C. and Ptolemy in the second century A.D. It was not until 1514 A.D. that a Polish priest by name Copernicus challenged the Aristotle-Ptolemic theory and shifted the earth from its proud central position to a position where it revolved around the sun like any other planet of the solar system. Copernicus knew that his heliocentric view would meet violent opposition from the then orthodox church which was wedded to the geocentric view of Ptolemy. So, he withheld publication of his book, 'De Revolutionibus Orbium Coelestium' concerning the revolutions of the Celestial Orbs, until the end of his life. His fears were well founded, for Copernicus's theory was condemned as heresy and his book remained locked up in papal custody until 1835.

Meanwhile, the heliocentric theory of Copernicus received strong support from the work of the astronomers, Johannes Kepler (1571–1630) in Germany and Galileo Galilei (1564–1642) in Italy. Kepler using the careful astronomical measurements of his predecessor, Tycho Brahe, enunciated in 1609 his celebrated three laws of planetary motion as follows:

(i) The planets revolve round the sun in elliptical orbits with the sun occupying one focus;
(ii) The orbital velocity of a planet sweeps out equal areas in equal times; and
(iii) The squares of the periods of revolution of the planets are proportional to the cubes of their orbital major axes.

But both Kepler and Galileo were afraid of coming out with the truth for fear of being persecuted by the then orthodox church. However, after Kepler enunciated his laws of planetary motion which were soon followed by Newton's law of universal gravitation in 1687, the truth ultimately triumphed and the scientific world accepted the heliocentric theory of our solar system.

According to the accepted view, the earth is one of the inner planets of the solar system (for detailed information on the sun's planetary system, the reader may consult a book on the solar system or astronomy) which revolve around the sun and are held in their orbits by the gravitational force of the sun. The earth orbits around the sun in an elliptical orbit under a centrally-directed gravitational force at an average distance of about 149.6 million km from the sun once in about 365 days. It also rotates about its own axis once in about 24 h. Its angular velocity is about 7.29×10^{-5} radians per second and is usually denoted by Ω. The earth's equatorial plane is inclined to its orbital plane by an angle of $23.45°$, which is called the earth's obliquity to the sun. While the rotation of the earth makes day and night, the obliquity gives us the seasons.

The shape of the earth departs slightly from being a sphere. Its polar radius happens to be about 21 km shorter than its equatorial radius, the average radius being about 6371 km. To understand the likely cause of the departure of the earth's surface from sphericity, we need to consider the earth's gravitational force and rotation from the time of its birth from the sun.

1.2 Earth's Gravitational Force – Gravity

According to Newton's law of universal gravitation, every body in the universe attracts every other body with a force proportional to the product of their masses and inversely proportional to the square of the distance between them. This means that if the earth were truly spherical, its gravitational attraction **F** on a body of mass m placed at a point P on its surface (see Fig. 1.1) would be given by the relation

$$\mathbf{F} = (G\,Mm/r^2)\,(\mathbf{r}/r), \qquad (1.2.1)$$

where **r** is the radius vector of P, G is the Gravitational constant, M is the mass of the earth deemed to be concentrated at its center, and r is the mean radius of the earth.

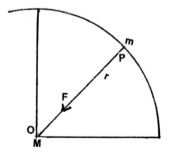

Fig. 1.1 Gravitational force of the earth

1.2 Earth's Gravitational Force – Gravity

[Vectors are indicated by bold letters; a summary of some commonly-used vector symbols and operations appears in Appendix-1A. Also, a list of some useful physical constants appears in Appendix-8]

The acceleration due to gravity, $\hat{\mathbf{g}}$ that corresponds to the gravitational force in (1.2.1), is given by

$$\hat{\mathbf{g}} = (GM/r^2)\,(\mathbf{r}/r) \tag{1.2.2}$$

We assume that the earth's surface was once spherical in shape when it was born and that the effect of its rotation moulded its shape only gradually after its birth. Two forces must have acted on a body of unit mass at its surface then, viz., a gravitational force pulling it towards the center of the earth and centrifugal force acting away from the axis of rotation of the earth (see Fig. 1.2).

The resultant of the two forces which we may call the earth's effective gravity $\mathbf{g}$, is given by the relation

$$\mathbf{g} = \hat{\mathbf{g}} + \Omega^2\,\mathbf{R} \tag{1.2.3}$$

where $\mathbf{R}$ is the radius vector of the body at a point P in a direction perpendicular to the axis of rotation and Ω is the angular velocity of the earth. Because of the centrifugal force, the resultant acceleration due to gravity $\mathbf{g}$ no longer passes through the center of the earth except at the poles and the equator. The reason simply is this: If the earth's surface were truly spherical, the effective gravity would have a component parallel to the earth's surface and directed towards the equator. This force is denoted by the vector $\mathbf{E}$ in Fig. 1.2. The earth's surface has adjusted to this equatorward component by taking up a spheroidal shape with a bulge at the equator and contraction at the poles so that the local vertical at all points on the earth's surface would be parallel to the resultant gravity. It is because of the equatorial bulge and the polar contraction that the polar radius is shorter than the equatorial radius by about 21 km. The transformation envisaged here must have had occurred long ago in earth's history when its surface layers were cooling off from a hot molten plasma state to a solid crust.

The acceleration due to gravity, g, varies with latitude along the earth's surface, with a maximum at the poles and minimum at the equator. It also varies with altitude. The value of g at latitude φ and height h meters above the earth's surface is empirically given by the approximate relation:

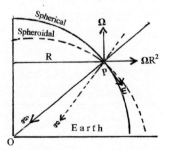

Fig. 1.2 Illustrating how the earth's rotation moulded the shape of its surface

$$g(h) = 9.80616\,(1 - 0.0026373\cos 2\varphi + 0.59 \times 10^{-5}\cos 2\varphi)(1 - 3.14 \times 10^{-7}h) \qquad (1.2.4)$$

where $9.80616\ \text{ms}^{-2}$ is the value of the acceleration due to gravity at mean sea level at latitude $45°$.

1.3 Geopotential Surfaces

If a body of unit mass is raised from the earth's surface to a height z in the atmosphere, the work that must be done against the earth's gravitational field is called its geopotential which is usually denoted by Φ and defined by the relation

$$\Phi(z) = \int_0^z g(z)\,\delta z \qquad (1.3.1)$$

where the geopotential $\Phi(0)$ at sea level is taken to be zero.

Due to the spheroidal shape of the earth and variation of the acceleration due to gravity, g, over the earth's surface, the surfaces of constant geopotential are not quite parallel to the surfaces of constant geometric height z above mean sea level which are called level surfaces. However, if we take a globally-averaged constant value of g at the earth's surface and call it g_0 ($= 9.81\ \text{m s}^{-2}$), it is possible to define a geopotential height Z given by

$$Z = \Phi(z)/g_0 = (1/g_0)\int_0^z g(z)\,\delta z \qquad (1.3.2)$$

The difference between z and Z, however, is small and sometimes ignored and the height of a geopotential surface in geopotential metres (gpm) is taken to be equal to its height in geometric metres.

1.4 Motion in the Earth's Gravitational Field – The Law of Central Forces

Kepler's discoveries of the laws of planetary motion around the sun, followed by Newton's law of universal gravitation, paved the way for a better understanding of several phenomena in the planetary atmospheres under the centrally-directed force of the planet's gravitation in the same way as for the motion of planets around the sun. Two phenomena immediately come to mind. The first is the present composition of the earth's atmosphere. Scientists believe that the primordial atmosphere billions of years ago had an abundance of many lighter gases, such as hydrogen and helium which exist in traces only in to-day's atmosphere (see chap. 2).

1.4 Motion in the Earth's Gravitational Field – The Law of Central Forces

What happened to all those lighter elements? Why and how did they escape or disappear? We shall have an occasion to look into this question in the next chapter where we discuss the composition of the present-day atmosphere. Secondly, there is the problem of space travel which we are by now all familiar with. Why must we use the booster rockets to go into space? A review of the general problem of motion under centrally-directed forces may throw light on some of these issues.

The discoveries of Kepler and Newton prescribed two conditions to be satisfied by a moving object in or above the earth's atmosphere. These are that: (a) the sum of its kinetic energy and potential energy has to remain invariant; and (b) the areal velocity of the object has to remain constant (Kepler's second law). These conditions in the case of a body of mass m moving with a velocity v in the earth's atmosphere, assuming there is no friction, are expressed by the following relations

$$(1/2)\, mv^2 - GM\, m/r = (1/2)\, m\, v_0^2 - GM\, m/r_0 \tag{1.4.1}$$

$$(1/2)\{r^2(d\theta/dt)\} = A \text{ (constant)} \tag{1.4.2}$$

where r is the radial distance of the object from the earth's center at time t, $d\theta/dt$ is the angular velocity of the object, and the variables with suffix 0 denote values at time $t = 0$ (see Fig. 1.3).

Using simple mathematics and eliminating t from the Eqs. (1.4.1) and (1.4.2), it can be shown (see Appendix-1B) that the path of the moving body will be a conic the polar equation of which is given by

$$r = k/\{1+ \in \cos(\theta + \alpha)\} \tag{1.4.3}$$

where $k = 4A^2/GM$, $\in = 2AB/GM$, with $B = [v_0^2 - 2GM/r_0 + (GM/2A)^2]^{1/2}$, which is called the eccentricity, and α = phase angle.

If θ be measured from the maximum value of r, $\alpha = \pi$. (1.4.3) then reduces to the familiar polar equation of a conic with the origin at one focus (Kepler's First law). The orbit is an ellipse, parabola or hyperbola, according as the numerical value of the eccentricity $\in$ is less than, equal to, or greater than, unity. This condition requires that in the expression for $\in$ above,

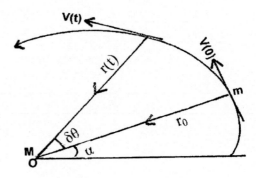

Fig. 1.3 Motion under a centrally-directed force

$$\begin{aligned} &< 2GM/r_0, \text{ for an ellipse} \\ v_0{}^2 &= 2GM/r_0, \text{ for a parabola} \\ &> 2GM/r_0, \text{ for a hyperbola} \end{aligned} \qquad (1.4.4)$$

The condition (1.4.4) implies that if the initial kinetic energy of the moving object is greater than its potential energy with sign changed, the object will move in an hyperbolic path and escape from the earth's gravitational pull. The minimum velocity required for such an escape, as calculated from the relation,

$$v_0 = (2GM/r)^{1/2}$$

is about $11.3 \, \text{km s}^{-1}$. This explains why a booster rocket is required to give an earth-orbiting satellite a velocity greater than the velocity of escape in order to propel it to space.

The relation (1.4.3) also enables us to derive an expression for Kepler's third law, in the case of an elliptical orbit.

The parameter of an ellipse is

$$k = 4A^2/GM = b^2/a = a(1- \epsilon^2) \qquad (1.4.5)$$

where a is semi-major axis, and b is semi-minor axis.

From (1.4.5), $(1- \epsilon^2) = 4 A^2/GMa$

If T be the orbital period,

$$AT = \pi ab = \pi a^2 (1- \epsilon^2)^{1/2} = 2\pi \, a^2 A/(GMa)^{1/2} \qquad (1.4.6)$$

Thus,

$$T^2 = (4 \pi^2/GM)a^3 \qquad (1.4.7)$$

This is Kepler's third law.

Chapter 2
The Earth's Atmosphere – Its Origin, Composition and Properties

2.1 Introduction: Origin of the Earth's Atmosphere

We know little about the origin of our atmosphere, just as we know little about the origin of our planet. In the absence of any reliable evidence, one can only speculate. According to cosmologists, our planet most likely originated from the sun about 4.6 billions of years ago in the wake of the latter's encounter with a passing star, following a cosmic event, popularly known as the Big Bang. After its separation from the sun, it started revolving around the sun under the effect of the sun's gravitational pull and rotating about its axis like a fiery ball, surrounded by an extremely hot gaseous envelope which may be called the primordial atmosphere.

However, it may be imagined that the primordial atmosphere at that stage must have been at a great upheaval as it hurled through space and its hot gaseous envelope rapidly cooled, condensed and solidified forming a solid crust in the surface layers after giving out considerable amount of volatile hot gases and vapours from the molten material at the surface. Part of the hot gases and vapours which cooled and condensed into water formed the world's oceans. The remainder formed a gaseous envelope around the planet or was stored in rocks. The atmosphere thus formed had a preponderance of hydrogen and little or no oxygen, so could not support life of the kind that we know on the earth to-day. However, as the earth cooled down further, complex chemical actions and reactions in the crust and the interactions between the crust and the atmosphere gradually led to the formation of an atmosphere which could support an early form of life such as single-celled microbes which required little oxygen for their survival. Such microbial forms of life, also known as blue-green algae, perhaps, first appeared in the oceans where they absorbed carbon dioxide and in the presence of water and sunlight released oxygen by a bio-chemical process known as the green-plant photosynthesis. The accumulation of oxygen in the atmosphere facilitated the evolution of more complex and multi-celled forms of life which we observe on our planet to-day. It is believed that all these developments occurred within the first one billion years of the earth's history and that, since then, our atmosphere has gradually stabilized to its present state.

2.2 Composition of the Atmosphere

The gaseous envelope that covers the earth's surface has no upper limit. It gradually merges into the interplanetary space. It exists today as a mixture of several gases, the composition of which within the first 25 km of the earth's surface is presented in Table 2.1.

As shown in the Table, nitrogen and oxygen are the two main constituents of the earth's atmosphere with their combined proportions approaching almost 99% by mass as well as by volume. Their compositions vary little with time, so they are treated as permanent gases. Other gases exist in small amounts only. The proportions of carbon dioxide and ozone are variable. Another constituent of the atmosphere which finds no place in the Table but is highly important for meteorology is water vapour which also occurs in small and variable proportion. But, as we shall see later, the three gases, viz., water vapour, carbon dioxide and ozone, though they occur in small proportions, play very important roles in atmospheric processes because of their radiative and thermodynamic properties.

The presence of water substance in the atmosphere is especially important, because it can exist in three phases, viz., vapour, liquid and solid. A change of phase involves either liberation or absorption of a large quantity of heat which affects atmospheric properties and behaviour. Evaporation of water from the world's oceans, condensation of water vapour into cloud and rain in the atmosphere, formation of polar ice caps on our planet, are all examples of change of phase of the water substance. Minor constituents at low levels of the atmosphere may include variable quantities of dust, smoke and toxic gases and vapours such as sulphur dioxide, methane, oxides of nitrogen, etc., some of which pollute the atmosphere and are highly injurious to health.

The above-mentioned composition of the earth's atmosphere undergoes changes from about 25 km upward, under the effect of the sun's ultra-violet radiation. The gases most affected by this process are oxygen and nitrogen. Their molecules gradually break up leading to formation of ozone in the middle atmosphere (20–50 km

Table 2.1 Composition of pure dry air

Constituent Gas	By Mass (%)	By Volume (%)	Molecular Wt
Nitrogen (N_2)	75.51	78.09	28.02
Oxygen (O_2)	23.14	20.95	32.00
Argon (Ar)	1.3	0.93	39.94
Carbon dioxide (CO_2)	0.05	0.03	44.01
Neon (Ne)	1.2×10^{-3}	1.8×10^{-3}	20.18
Helium (He)	8.0×10^{-4}	5.2×10^{-4}	4.00
Krypton (Kr)	2.9×10^{-4}	1.0×10^{-4}	83.7
Hydrogen (H_2)	0.35×10^{-5}	5.0×10^{-5}	2.02
Xenon(X)	3.6×10^{-5}	0.8×10^{-5}	131.3
Ozone (O_3)	0.17×10^{-5}	0.1×10^{-5}	48.0
Radon (Rn)	–	6.0×10^{-18}	222.0

2.2 Composition of the Atmosphere

approximately) and atoms and ions (charged particles) in the upper layers of the atmosphere (> 80 km). Further discussion about the changes in the composition of air in the upper layers of the atmosphere will be taken up in Chap. 8. Due to preponderance of diatomic molecules in the atmosphere, the mean gram-molecular mass of dry air is taken as 28.699. For water vapour, which is triatomic (H_2O), the mean gram-molecular mass is 18.016.

We may now, perhaps, address the question raised in Chap. 1 regarding the loss of lighter elements, especially hydrogen and helium, from the earth's atmosphere. According to the kinetic theory of gases, the mean square velocity of a molecule of a gas is directly proportional to the absolute temperature, as given by the relation 2.2.1 (see, e.g., Saha and Srivastava, 1931, Fifth corrected edition 1969, reprinted 2003),

$$c^2 = 3RT \quad (2.2.1)$$

where c is the molecular velocity, R is the gas constant and T is the absolute temperature of the gas. Table 2.2 gives the values of the mean molecular velocity of some of the gases in the earth's atmosphere at different temperatures. The values are taken from a paper entitled 'Is life possible in other planets?' by Saha and Saha (1939).

However, it may be assumed that our atmosphere had been quite different in the past from what it is to-day, especially at the time when the earth got separated from the sun and the temperature of the sun, in all probability, might have been much in excess of the present value of about 6,000 °C. At such a high temperature the mean velocity of the hydrogen atoms would be 12.8 km s^{-1}, and that for the hydrogen molecules would be about 9 km s^{-1}. So, a mass like the earth, just separated from the sun, engulfed in hot gases, would rapidly lose all hydrogen atoms and most of the hydrogen molecules.

But even if the temperature was lower, there would be steady loss of the lighter constituents, for according to Maxwell's law of distribution of velocities, all molecules in a gas do not move with the same velocity; there would be some whose velocities may even at the ordinary temperatures exceed the velocity of escape and such particles would escape. The rate of loss will increase with higher temperature and lower molecular weight. Jeans has in fact calculated the time required for loss

Table 2.2 Values of the mean molecular velocity c (km s^{-1}) at different temperatures (°C)

Gas	Temperatures (°C)		
	−100	0	300
Hydrogen	1.47	1.80	2.66
Helium	1.04	1.31	1.90
Water vapour	0.49	0.61	0.88
Nitrogen	0.39	0.49	0.71
Oxygen	0.37	0.46	0.67
Argon	0.33	0.41	0.59
Carbon dioxide	0.31	0.39	0.57

of planetary atmospheres from different planets and for different temperatures. He finds that if the mean molecular velocity of the gas is one-fourth the critical velocity of escape, the atmosphere would be lost in 50,000 years. But if the ratio is one-fifth, 25 million years would be needed for complete loss.

It is probable that the earth's atmosphere lost most of its primordial hydrogen, helium and other lighter gases quite early in the course of its geological history, while the heavier gases were retained.

2.3 Properties and Variables of the Atmosphere

In meteorology and thermodynamics, simplifying assumptions are made regarding the structure and behaviour of the gaseous molecules and the atmosphere is treated as an ideal gas. The main assumptions of an ideal gas concept are that the molecules do not occupy any finite space and hence have no volume and that there are no forces of attraction or repulsion between any two molecules. The properties of the air that find important applications in meteorology and thermodynamics under these assumptions are pressure, temperature, and volume or density.

As will be shown later in this chapter, in an ideal gas these variables are related to each other by an equation of state. The space-time distributions of these properties or variables, along with the distribution of water vapour to be introduced later, determine in large measure the stability and behaviour of the atmosphere. The following pages give an introduction to these variables and their climatological (1979–1996) mean distributions over the globe during January and July at the peak of winter and summer respectively in the northern hemisphere.

2.3.1 Pressure

2.3.1.1 Definition

Pressure (p) of the atmosphere at any level is defined as the weight of the overlying column of air per unit area of the surface at that level. It is found by integrating the weight of air in small segments of height δz upward from that level to the top of the atmosphere which is assumed to be at infinity and may be expressed by the integral

$$p(h) = \int_h^\infty \rho\, g\, \delta z \qquad (2.3.1)$$

where p(h) is the pressure at height h, ρ is the mean density of the air in the height interval δz and g is the acceleration due to gravity (the variation of which with height is neglected here) and the top of the atmosphere is assumed to be at infinity. According to the above definition, pressure is maximum at the earth's surface and decreases with height as more and more air is left below. In 1643,

2.3 Properties and Variables of the Atmosphere

an Italian scientist Torricelli was the first to measure atmospheric pressure at the earth's surface by balancing it against the weight of a column of mercury of height 76 cm in a vacuum tube. He used mercury mainly because of its high specific gravity, 13.6. If he had used water, the column would have been 13.6 times taller, i.e., about 10.3 m, which would have been very unmanageable. He called the instrument a barometer (weight-meter). Because of the convenient height and accuracy of the mercury barometer, it has been adopted as a standard pressure-measuring instrument throughout the world. A portable version of the barometer which can make continuous measurement of atmospheric pressure at any location has also been devised. This is called the 'aneroid' barometer, which means the barometer without fluid. It consists of a small partially-evacuated metallic box with a flat top which moves up or down with variations in atmospheric pressure. The movement actuates a pen arm which records pressure on a calibrated revolving drum called the barograph. Aneroid barometers are widely used in aviation and other areas of human activity where it may be inconvenient to carry a mercury barometer.

Several systems of units are in vogue for measurement of atmospheric pressure. In c.g.s system, pressure is measured in dynes per square centimetre. Thus, at mean sea level (msl), the atmospheric pressure at $0\,°C$ is $(76.0 \times 13.6 \times 981.0)$, or 1013.9×10^6 dynes cm^{-2}. The equivalence in other systems of units is

$$1 \text{ Atmosphere} = 1.013 \times 10^6 \text{ dynes cm}^{-2}$$
$$= 1013 \text{ millibars}$$
$$= 1.013 \times 10^5 \text{ Pascals}$$

2.3.1.2 Space-Time Variation of Pressure

The pressure of the atmosphere varies in time and space. At any particular location, especially over the tropics, there is a prominent diurnal and seasonal variation of pressure. Atmospheric pressure varies along the earth's surface as well as with height. The total variation, δp, following the movement of a parcel of air over an infinitesimally small time interval, δt, may be expressed, in the Cartesian coordinates(x, y, z), by the relation:

$$\delta p = (\partial p/\partial t)\, \delta t + \{(\partial p/\partial x)\, \delta x + (\partial p/\partial y)\, \delta y\} + \partial p/\partial z)\, \delta z \qquad (2.3.2)$$

where the operator $\partial/\partial t$ is used to denote partial derivative of p with respect to the independent variable t. In (2.3.2), on the right-hand side, the first term denotes the barometric tendency at a fixed location, while the terms within the second bracket represent the horizontal variations of pressure and the last term denotes the vertical variation of pressure in the atmosphere.

In the limit when $\delta t \rightarrow 0$, we may write (2.3.2) in the form

$$dp/dt = \partial p/\partial t + u\, \partial p/\partial x + v\, \partial p/\partial y + w\, \partial p/\partial z$$

where u,v,w are the components of the velocity vector **V** along the x,y,z co-ordinates respectively, d/dt denotes total derivative following motion and $\partial/\partial t$ is the partial derivative at a fixed location.

Alternatively, we may write (2.3.2) in the vector form,

$$dp/dt = \partial p/\partial t + (\mathbf{V_H} \cdot \nabla_H p) + w\, \partial p/\partial z \qquad (2.3.3)$$

where $\mathbf{V_H}$ denotes the horizontal velocity vector and ∇_H the horizontal Del operator. The horizontal variation of pressure at the earth's surface is important because it is primarily the horizontal pressure gradient that forces the air to move from a region of high pressure to that of low pressure with an acceleration given by the vector relation,

$$\text{Pressure gradient force} = -\nabla p/\rho$$

where ρ is mean air density. However, as will be shown in Chap. 11, this movement is considerably modified by the deflecting force of the earth's rotation and the retarding force of friction. In general, atmospheric pressure falls off with height. The fall of pressure with height causes an upward force, which, it can be shown (see Chap. 12), is approximately balanced by the downward force of gravity.

$$(\partial p/\partial z) = -\rho g \qquad (2.3.4)$$

The balance relation (2.3.4) is usually called the hydrostatic approximation.

2.3.1.3 Vertical Variation of Pressure

The pressure at any height in the atmosphere can be computed with the aid of (2.3.4) and the equation of state (see 2.4.12 which relates pressure, density and temperature), as shown below:

Let p be pressure and T temperature at a height z above the earth's surface. Then, using the equation of state, $p = \rho RT$, we get

$$\delta p = -\rho g\, \delta z = -(p/RT)\, g\, \delta z$$

or, $\qquad \delta p/p = -(g/RT)\, \delta z \qquad (2.3.5)$

Equation (2.3.5) can be integrated if the variation of T with height is known.

If T is taken as constant and equal to the value T_0 at the earth's surface, the integration yields

$$p(z) = p_0\, \exp(-gz/RT_0) = p_0\, \exp(-z/H) \qquad (2.3.6)$$

where $H = RT_0/g$ is called the scale height and p_0 is the pressure at the earth's surface. For a value of $T_0 = 273\,\text{K}$, $H = 8.0\,\text{km}$ approximately.

However, it is known from measurements that the temperature in the lower atmosphere normally decreases with height. If it is assumed that $T = T_0 - \beta z$, where T_0 is the temperature at the surface of the earth and β denotes a constant lapse rate of temperature with height, (2.3.5) may be written in the form

$$\delta p/p = -[g/\{R\,(T_0 - \beta z)\}]\, \delta z \qquad (2.3.7)$$

2.3 Properties and Variables of the Atmosphere

On integration, (2.3.7) gives

$$\ln p(z) = \ln p_0 + (g/\beta R) [\ln \{(T_0 - \beta z)/T_0\}] \qquad (2.3.8)$$

Equation (2.3.8) may be used to compute pressure at any height, provided the lapse rate of temperature in the intervening layer is specified.

2.3.1.4 Reduction of Surface Pressure to Mean Sea Level (msl)

The uneven topography of the earth's surface poses a problem for determination of msl pressure at high-level stations, if needed. The msl pressure at a high-level location is usually obtained by reducing the observed surface pressure to msl with the aid of the hydrostatic approximation (2.3.4). But the main problem here lies in finding the density distribution in the imaginary air column between the station level and the msl at the location. A mean value of density is usually assumed, based on the values of pressure and temperature at the station and the nearest msl or low-level station. The procedure applied may often lead to large errors at high mountain stations.

2.3.1.5 Msl Pressure Distribution Over the Globe

Fig. 2.1 shows the climatological msl pressure distributions over the globe during (a) January and (b) July.

Some noteworthy features of the pressure distribution are as follows:

In January (a), msl pressures in the northern hemisphere are generally high over the continents, with the highest pressure exceeding 1035 hPa located over central Asia, while they are generally low over neighboring oceans. Two intense low pressure areas appear over the northern oceans, one over the Icelandic area and the other over the Aleutian area. In the southern hemisphere, the distribution appears to be reverse, with relatively low pressures appearing over the continents of Africa(south of the equator), Australia and South America and relatively high pressures over the southern oceans.

In July (b), the pressure fields are more or less reverse of those in January. In the northern hemisphere, pressures are generally low over the continents with 'heat lows' appearing over most of the continents, while they are generally high over neighboring oceans. In the southern hemisphere, high pressures appear over the continents with relatively low pressures over adjoining oceans.

2.3.2 Temperature

2.3.2.1 Definition and Measurement

Temperature gives a measure of heat in a body. When a body is heated or cooled, its temperature changes. However, given the same quantity of heat, the rise in temperature is not quite the same for all bodies. It depends on a property of the body called

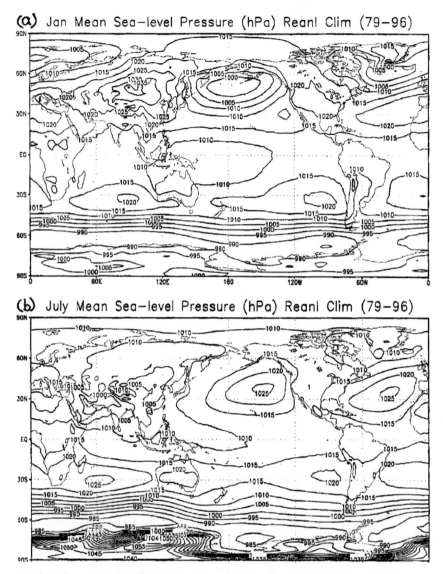

Fig. 2.1 Climatological (1976–1996) mean sea level (msl) pressure (hPa) over the globe during (**a**) January and (**b**) July (Courtesy: NCEP/NCAR Reanalysis)

its heat capacity which is given by the product of its mass and a quantity called specific heat. Specific heat of a body is defined as the quantity of heat required to raise the temperature of unit mass of the body through 1 °C. Thus, specific heat is related to a given quantity of heat by the relation:

Heat added (or subtracted) = mass × specific heat × rise (or fall) in temperature

2.3 Properties and Variables of the Atmosphere

The unit of heat is a calorie which is the quantity of heat required to raise the temperature of 1 g of water through 1 °C, usually from 15 to 16°C. Heat is measured by calorimeters.

Specific heat varies little with volume or pressure in the case of a liquid or solid but its variations in the case of a gas can be quite considerable, as will be shown later in Chap. 3.

Temperatures are measured by thermometers. There are different kinds of thermometers in use with different scales of measurements. Most thermometers are calibrated against some fixed points of temperature, usually the freezing point of water as the lower fixed point and the boiling point of water as the upper fixed point. Several scales of measurement are in use. In the Celsius or Centigrade (C) scale, the lower fixed point is marked as 0°C and the upper 100°C and the interval is divided into 100°. In the Fahrenheit (F) scale, the lower fixed point is marked 32°F, while the upper 212°F and the interval is divided into 180°. Thus, 1°C corresponds to 1.80°F. There is also the Reamur scale in which the interval between the fixed points is divided into 80°. Temperature of the air measured by any of these thermometers is called the dry-bulb temperature.

However, the scale of temperature that is widely used in meteorology is called the Absolute or Kelvin scale of temperature, which is related to the centigrade scale by the expression, $T = 273.16 + t$, where t is in degrees Centigrade and T in degrees Absolute or Kelvin. The absolute temperature is usually denoted by the capital letter T or K. From now on in this book we shall express temperature either in degrees of Centigrade (t), or Absolute (T or K), unless otherwise mentioned.

2.3.2.2 Temperature Distribution in the Atmosphere

Like barometric pressure, temperature of the air varies in time and space. The total change, δT, in the temperature of a parcel of air over time δt following motion in space may be expressed as

$$\delta T/\delta t = (\partial T/\partial t) + u\,(\partial T/\partial x) + v(\partial T/\partial y) + w\,(\partial T/\partial z) \qquad (2.3.9)$$

where u,v,w are the components of the velocity vector **V** along the co-ordinate axes x,y,z respectively. As in (2.3.3), (2.3.9) may be written in vector form

$$dT/dt = \partial T/\partial t + \mathbf{V}_H \cdot \nabla_H T + w\,\partial T/\partial z \qquad (2.3.10)$$

The first term on the right-hand side of (2.3.10) gives the temporal variation of temperature at a fixed point, while the others are the so-called advective terms. We are all familiar with the diurnal and seasonal variations of temperature at the place where we live. Inter-annual and secular fluctuations also occur. The second term denotes the horizontal variation of temperature along the earth's surface, for example, between continents and oceans or between the equator and the poles. The last term denotes the variation of temperature with height. Since the maximum temperature occurs at the earth's surface, there is a gradient in the vertical direction with temperature normally decreasing both upward in the atmosphere and downward below the surface. The upward gradient often leads to static instability and upward transfer of heat by convection currents, while the downward gradient makes the sub-surface layer thermally stable. In the latter case, heat can travel in the sub-surface layer mainly by conduction, though in the ocean other processes such as wind-driven turbulence, etc., may be at work to transport heat downward.

2.3.2.3 Temperature Distribution at the Earth's Surface

Like barometric pressure, there are several problems in getting a reliable set of air temperature values at the earth's surface and at different heights aloft, despite recent advances in measurement techniques. So far as the ocean surface temperatures are concerned, we have now a fairly reliable set of mean sea level temperatures over the global oceans. But the same cannot be said about the temperatures over the land where observations are complicated by orography and it is not easy to reduce the surface observations to mean sea level, if that were desired. If needed, the surface values are usually reduced to mean sea level by assuming a standard lapse rate of temperature for the height of a station but the procedure can lead to serious error for high-level stations. For this reason, Fig. 2.2 which shows the climatological surface temperature distribution over the globe during (a) January and (b) July gives the temperatures over land at 2m above the land surface, regardless of the altitude of the land station (Courtesy: NCEP/NCAR Reanalysis Project).
Salient features of Fig. 2.2(a, b) are as follows:

In January (Fig. 2.2a), with maximum solar heating in the southern hemisphere, the temperatures in the northern hemisphere are generally lower than those in the same latitudes in the southern hemisphere and the land surface (as indicated by the 2 m-level temperatures) is much colder than the adjoining oceans. In the southern hemisphere, however, the continents are much warmer than the oceans.

In July (Fig. 2.2b), with the maximum solar heating in the northern hemisphere, the temperature field is reversed not only between the hemispheres but also between continents and oceans. This means that in the northern hemisphere, temperatures over the continents are in general higher than those over the neighboring oceans, whereas in the southern hemisphere, continents are much cooler than the neighboring oceans.

In general, the temperature maxima over both continents and oceans follow the seasonal movement of the sun, though with a certain time lag which may vary from about a month over land to about two months over oceans. This difference is largely due to much lower thermal inertia (heat capacity) of land surface than that of ocean.

2.3.2.4 Vertical Temperature Distribution

Observations show that the vertical variations of pressure and temperature in the atmosphere follow entirely different pattern. This is evident from Fig. 2.3 which

2.3 Properties and Variables of the Atmosphere

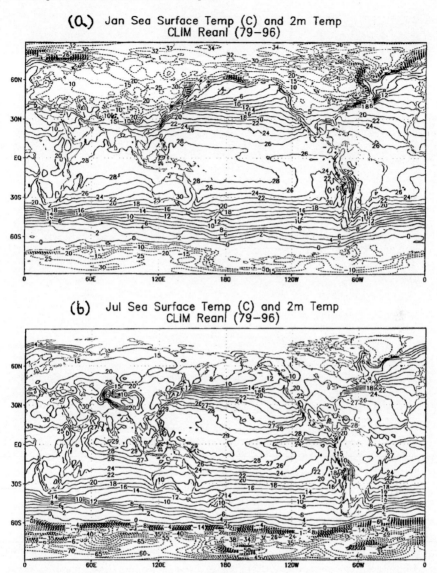

Fig. 2.2 Climatological (1979–1996) mean temperature (°C) at surface over the globe: (**a**) January, (**b**) July. Temperature over land is at height 2 m above land surface (Courtesy: NCEP/NCAR Reanalysis)

shows the vertical distribution of mean air temperature in the 1962 U.S. standard atmosphere. It shows that while pressure decreases almost exponentially with height, the temperature varies differently in different atmospheric layers.

The four principal layers of the atmosphere defined by Fig. 2.3 are: the troposphere, the stratosphere, the mesosphere and the thermosphere. The troposphere is

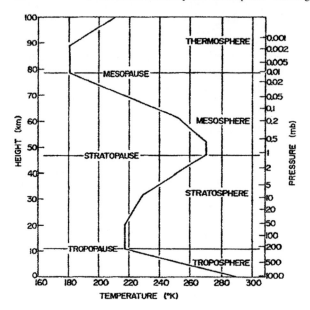

Fig. 2.3 Vertical distribution of temperature in the U.S. standard atmosphere

the lowest layer which extends to a height varying between about 9 km and 16 km depending upon the season and latitude. In this layer, the temperature drops from a value of about 15 °C at the earth's surface steadily with height at the rate of about 6.5 °C per km till a minimum value of about −55 °C to −60 °C is reached at a level called the tropopause. The temperature distribution in the troposphere is maintained by convective and turbulent transfer of heat due to absorption of solar radiation at the surface to the atmosphere.

Above the troposphere lies the stratosphere where the temperature increases gradually with height to reach a maximum of about 0 °C at a height of about 50 km. The increase of temperature in this layer is due to the presence of ozone which absorbs ultra-violet radiation from the sun and acts as a source of heat for the atmosphere. Since ozone controls the thermal structure of this layer, it is also sometimes called the ozonosphere. The top of the stratosphere is called the stratopause. Above the stratopause, the temperature drops again to reach a minimum of about −100 °C at about 80 km. The level of this minimum temperature is called the mesopause. The mesopause marks the level from where the temperature starts rising again, this time almost monotonously to large values at a great height of the atmosphere. This uppermost layer is called the thermosphere. The importance of this layer lies in the fact that it intercepts the highly-charged solar rays from space and the high-energy ultra-violet radiation from the sun which are both harmful to life at the earth's surface. The atoms and molecules of gases such as oxygen and nitrogen present in this layer absorb the high-energy short-wave radiation from the sun and get ionized. The ionized layer by reflecting radio waves helps in global telecommunication. For this reason, this layer is also sometimes called the ionosphere.

The temperature of the thermosphere varies greatly with solar activity, with a value of about 2000 K at the time of 'active sun' and 500 K at the time of 'quiet sun' at 500 km altitude (Banks and Kockarts (1973). Because of the large variation in the thermal structure of the thermosphere with active and quiet sun, this part of the thermosphere is often called the heterosphere. Above the heterosphere lies the exosphere.

2.3.3 Density

The density of the air is defined as its mass per unit volume and expressed in units of grams per cubic centimetre or kilograms per cubic metre. The inverse of density is specific volume which is volume per unit mass. The density of the air is not ordinarily directly measured by instruments but computed from the values of pressure and temperature to which it is related by the equation of state. Like pressure and temperature, the density of the air varies in time and space. The horizontal variation of density sets air in motion and causes the wind on the earth's surface. On account of the stronger gravitational pull of the earth, the heavier elements of the atmosphere have settled down in the lower layers, with the result that the density of the air falls off rapidly with height. In fact, it can be shown by using the hydrostatic approximation (2.3.4) and the equation of state (2.4.12) that the density decreases exponentially with height according to a relation similar to (2.3.6).

The vertical variation of density is important in connection with many human activities, such as mountain climbing, balloon and aircraft flights, and space travel. As the air gets rarefied with height, the oxygen content which is most vital for our existence diminishes and poses a problem for normal breathing which then requires additional supply of oxygen. It is a common experience in all balloon flights that the balloon bursts after reaching a certain height in the atmosphere. This is due to the continual expansion of the inside gas against external air pressure as the outside air density falls off with height.

It is a common experience to observe shooting stars in the sky on some nights. They are nothing but meteors which enter the earth's atmosphere from space and get evaporated in the intense heat generated by friction with the air molecules as they move into the denser layers of the atmosphere.

In the present space era, the density of the atmosphere plays a very important role during re-entry of space probes into the earth's atmosphere. As the satellite descends into the denser air of the lower layers at high speed, air resistance produces such intense heat as can almost burn out the space vehicle unless its body is made of materials which can withstand the burning heat. Recent mishaps with a few space vehicles, such as that with NASA space shuttle Columbia during its re-entry into the earth's atmosphere on 1 February 2003, when it got burnt out in mid-air with seven astronauts on board, have drawn attention to this potential danger of space travel for human beings.

It must, however, be mentioned that the density of the air, especially in the lower troposphere, is very much affected by the presence of water vapour which is usually lighter than dry air at the same temperature and pressure. We shall deal with the density of moist air in Chap. 4.

2.3.4 Other Variables of the Atmosphere

There are several other variables of the atmosphere which concern a meteorologist in his day-to-day work. Amongst these, perhaps, the most important are the wind and the clouds. Unlike the atmosphere in a closed laboratory which can be studied at will, the free atmosphere is constantly in motion. Also, in the open atmosphere, when air temperature decreases adiabatically with height, water vapour carried in the air undergoes change of phase resulting in condensation and formation of fog, clouds, rain or snow, etc.

2.3.5 Observing the Atmosphere

For understanding the atmosphere, especially its current state and future evolution, we need to observe and measure its different elements to the extent possible at surface and upper levels. However, there is a difficulty here. Unlike air in a closed laboratory, the free atmosphere is constantly in motion. So what is observed or measured at a station at a particular time is not representative of what comes over the station after an interval of time. To get over this difficulty and for practical requirements, every country maintains a network of meteorological observing stations where atmospheric variables are observed and measured at some standard hours as fixed by the World Meteorological Organization by international agreement. Most meteorological stations take surface observations but some take both surface and upper-air observations. Then there are a few stations which take upper-air observations only. Surface observations which are routinely taken and recorded at meteorological stations include pressure, temperature, wind (direction and speed), humidity, cloud, rainfall, weather and visibility. While instruments are used to measure most of these elements, some observations are made visually. Some selected stations which are earmarked as climatological stations observe and measure, besides routine elements, additional parameters, such as radiation (both incoming and outgoing), duration of sunshine, ozone content of the air, concentration of pollutants, etc. Upper-air observations are usually taken with the help of balloons carrying radiosonde and/or rawinsonde instruments which measure pressure, temperature, humidity and wind at different heights and which are tracked by ground-based instruments, such as theodolites and radars. Apart from balloons, other aerial platforms which have been used for upper-air sounding include high-flying kites, aircraft, rockets, etc.

2.4 Gas Laws – Equations of State

Table 2.3 List of principal meteorological elements, their methods of observation, and instruments used to observe or measure them

Elements observed	Method of observation	Instruments used
Pressure	Instrumental	Barometers
Temperature	Instrumental	Thermometer
		Radiometer
Humidity	Instrumental	Hygrometer
		Psychrometer
Visibility	Visual	–
	Instrumental	Light-meters
Wind	Instrumental	Anemometer and Wind-vane, Cloud motion
Clouds	Visual	–
	Instrumental	Nephoscope
		Satellites
Rainfall	Instrumental	Raingauge

After the introduction of polar-orbiting and geostationary satellites, remote sensing technique has been widely applied to observe and measure several atmospheric elements from space. The space-borne instruments intercept radiation upwelling from the earth's surface and atmosphere in the visible as well as infrared radiation channel at different wavelengths from which information is derived about clouds, temperature, humidity, winds, etc. at different levels of the atmosphere at frequent intervals of time every day. Prior to the advent of satellites, observations were confined to inhabited land areas only. Vast ocean areas were practically uncharted. With the help of satellites, it is now possible to observe the atmosphere over the whole globe including the oceans. However, so far as the data over the oceans are concerned, their accuracy still leaves much to be desired because of the presence of clouds and excessive water vapour in the boundary layers of the ocean. In Table 2.3, we give a list of the standard meteorological elements which are usually observed at a meteorological station, their methods of observation and the instruments used to measure them.

2.4 Gas Laws – Equations of State

In deriving these laws, we closely follow a treatment given by Joos and Freeman (1967):

2.4.1 The Equation of State – General

We treat dry air of the atmosphere as a thermodynamic system because its mass and chemical composition as a mixture of gases remain more or less invariant in the troposphere and the lower stratosphere and the only variables for a dry atmosphere are its pressure, temperature and volume. For an atmosphere which exists as a single-phase system, we consider an equation of state expressed by a function F of the variables:

$$F(p, V, t) = 0 \quad (2.4.1)$$

where p denotes pressure, V volume and t temperature in degrees C.

We can use Eq. (2.4.1) to derive some important properties of changes of state between any two of the variables by holding the third constant. This means that we can find the co-efficient of volume expansion with temperature by holding pressure constant, the co-efficient of pressure change with temperature by keeping volume constant and the co-efficient of compressibility by keeping temperature constant. If the above-mentioned co-efficients are denoted by α, β and χ respectively, they may be expressed as follows:

$$\alpha = (1/V_0)(\partial V/\partial t)_p; \quad \beta = (1/p_0)(\partial p/\partial t)_V; \quad \chi = -(1/V)(\partial V/\partial p)_t \quad (2.4.2)$$

Since in compressibility, volume decreases with increase of pressure, we put a negative sign before χ to make it positive. $1/\chi$ is called isothermal volume elasticity, E. It is easy to show that in the case of an ideal gas, $E = p$ (pressure). It may be noted that in the expression for χ, we have referred the change to the existing volume instead of that at $0\,^\circ\text{C}$. This is by convention and done for gases only. Since an equation of state exists, the three co-efficients of (2.4.2) are not independent. Their relationship may be found by solving the equation of state for any of the variables. For example, if we express V as a function of p and t, we may write

$$V = V(p, t)$$

and,

$$dV = (\partial V/\partial p)_t \, dp + (\partial V/\partial t)_p \, dt$$

For a constant V,

$$(\partial V/\partial p)_t (\partial p/\partial t)_V + (\partial V/\partial t)_p = 0 \quad (2.4.3)$$

Substituting the values of the co-efficients from (2.4.2) in (2.4.3), we have the relation,

$$\alpha\,\beta = (\chi p_0 V/V_0) \quad (2.4.4)$$

2.4.2 The Equation of State of an Ideal Gas

Two fundamental gas laws were experimentally determined for gases far removed from their point of liquefaction. These were: (a) the Boyle's law in 1660 and (b) the Charles' law in 1787 which was repeated by Gay-Lussac in 1802.

(a) Boyle's law

The Boyle's law states that at constant temperature, the pressure p of a gas varies inversely as its volume V, i.e.,

2.4 Gas Laws – Equations of State

$$p\,V(p, t) = p_0\,V(p_0, t) \qquad (2.4.5)$$

where t denotes the constant temperature (°C) and p_0 some standard pressure. In a (p, V) diagram, the isothermals have the shape of hyperbolas.

(b) Charles and Gay-Lussac's law

According to this law, the isobaric volume co-efficient α is the same for all gases independent of temperature and has a value of 1/273. Thus, at a constant pressure p_0, we may write

$$V(p_0, t) = V(p_0, 0)(1+\alpha t) \qquad (2.4.6)$$

Combining (2.4.5) and (2.4.6) we have

$$p\,V(p, t) = p_0 V(p_0, 0)(1+\alpha t) \qquad (2.4.7)$$

We now define a scale of temperature T which is related to the Centigrade scale t by the relation,

$$T = t + 273.16 = t + 1/\alpha \qquad (2.4.8)$$

Using (2.4.8), we can write (2.4.7) in the form

$$p\,V = p_0 V(p_0, 273.16)T/273.16 \qquad (2.4.9)$$

Now, since, according to Avogadro's law, the volume occupied by 1 mol (gramme-molecule) of a gas at a given temperature and pressure is the same for all gases, we may write (2.4.9) as

$$p\,v = (p_0 v_0/273.16)\,T = R^*T \qquad (2.4.10)$$

where v is taken as the volume of 1 mol of the gas and R^* stands for $(p_0\,v_0/273.16)$ which is constant and called the Universal Gas constant. The numerical value of R^* is found to be 8.314 Joules $mol^{-1}\,K^{-1}$.

If the volume V contains n mols of the gas,

$$p\,V = n\,R^*\,T \qquad (2.4.11)$$

It then follows that if we take R as the gas constant for a particular gas, it is related to the Universal Gas constant R^* by the relation, $R = R^*/M$, where M is the molecular weight of the gas.

Equation (2.4.11) may also be written as,

$$p = \rho\,RT, \qquad (2.4.12)$$

where ρ is the density of the gas.

2.4.3 The Equation of State of a Mixture of Gases

The equation of state of a mixture of gases may be found by making use of Dalton's law of partial pressures which states that in a given volume V, the total pressure exerted by a mixture of gases is equal to the sum of the partial pressures of the component gases, if each of the gases occupied the same volume.

Let $m_1, m_2, m_3 \ldots, m_k$ be the masses of the k gases of molecular weights $M_1, M_2, M_3 \ldots, M_k$ contained in a given volume V. Then, if p_1 is the partial pressure of the first component, $p_1 V = (m_1/M_1) RT = n_1 RT$ where n_1 is the number of mols of the first component.

Similarly, for the second and other components,

$$p_2 V = (m_2/M_2)RT = n_2 RT$$
$$p_3 V = (m_3/M_3)RT = n_3 RT$$

and, so on. Adding for all the components, we have from Dalton's law

$$p V = (m_1/M_1 + m_2/M_2 +)RT = nRT$$

and,
$$p_i = n_i p/(n_1 + n_2 + + n_k) \qquad (2.4.13)$$

Using these relationships, it can be shown that the average molecular weight M of the mixture is given by

$$M = (m_1 + m_2 + m_3 +)/(m_1/M_1 + m_2/M_2 + m_3/M_3 +) \qquad (2.4.14)$$

2.4.4 The Equation of State of a Real Gas

The concept of an ideal gas law which was stated earlier in Sect. 2.3 of this chapter was based on assumptions which do not hold in the real atmosphere in which molecules occupy some finite space, however small it may be, and there are also intramolecular forces. It was left to Van der Waals (1837–1923) who carefully reviewed these assumptions and suggested the following modified version of the equation of state for a real gas:

$$(p + a/v^2)(V - b) = R'T \qquad (2.4.15)$$

where a and b are constants and R' is a modified Gas constant, somewhat different from R^*, the universal gas constant. However, it can be shown that as long as the temperature T remains appreciably above the point of liquefaction of a gas, the difference between the ideal gas law and the Van der Waals' equation is negligible in meteorology.

Chapter 3
Heat and Thermodynamics of the Atmosphere

3.1 Introduction. The Nature of Heat and Kinetic Theory

According to the kinetic theory of gases, heat is a form of energy, i.e., the energy of motion of the molecules which constitute a gas and which are in constant agitation and colliding with one another and the walls in a completely random fashion in a closed system. In other words, it is the energy of the disorganized motion of the molecules inside the system that is called heat. This energy is to be distinguished from the kinetic energy of the system of molecules as a whole moving in a particular direction in which the kinetic energy is perceived as that of the centre of gravity of the molecules inside the system moving in that direction. The latter, therefore, is treated as the kinetic energy of organized motion and does not generate any heat.

The distribution of velocities of molecules in a perfect gas is given by the Maxwell-Boltzmann distribution law which states that the number of molecules of mass m having speeds between c and c + dc out of a total number of n molecules per cubic centimetre is given by (see, e.g., Saha and Srivastava, 1931, page 110).

$$dn = 4\pi n \, (m/2\pi\kappa T)^{3/2} \, c^2 \, \{\exp(-mc^2/2\kappa T)\} dc \qquad (3.1.1)$$

where κ is Boltzmann's constant and T is the absolute temperature.

[For values of κ and other important physical constants, see Appendix-8]

The theory also predicts that the mean square velocity of the agitating molecules is directly proportional to the absolute temperature of the gas, as stated in Eq. (2.2.1).

3.2 The First Law of Thermodynamics

The first law of thermodynamics states that in a thermodynamic system, the total energy consisting of internal heat energy as well as mechanical energy is conserved. The internal energy of a closed system, U, is a single-valued function of the independent thermodynamic variables, V and T, and, as such, its variation must be a perfect differential, because, otherwise, there will be a mechanism to create new energy. So, we may write

$$dU = (\partial U/\partial V)_T \, dV + (\partial U/\partial T)_V \, dT \qquad (3.2.1)$$

Now, if a quantity of heat δQ be supplied to the closed system, it will go to increase the existing internal energy by dU and do work by increasing the volume by dV against external pressure p on the system. Here it should be noted that δQ is not a perfect differential like dV, or dT, or dU. So, we may write,

$$\delta Q = dU + p \, dV \qquad (3.2.2)$$

The Eq. (3.2.2), therefore, is a mathematical statement of the First Law of thermodynamics.

3.3 Specific Heats of Gases

The First Law of thermodynamics does not state how the energy is to be supplied to a system. It simply states that the total energy remains conserved. But, in reality, the quantity of heat required to be supplied to the system to raise its temperature through 1 °C depends on whether the heat is added at constant volume or constant pressure. In the case of a liquid or solid, the specific heat does not vary appreciably with pressure or volume. So, the manner in which heat is added does not matter much. But it is not quite so with a gas where it can vary considerably depending upon whether heat is added at constant volume or constant pressure. It is obvious that when heat is added at constant pressure, some additional heat is required to do work against external pressure. So, prima facie, specific heat at constant pressure should be greater than that at constant volume. This can be shown as follows:

Let us take 1 mol of a gas and add a quantity of heat δQ to it. Also, let us denote the specific heat at constant volume by c_v and that at constant pressure by c_p. Then, using (3.2.2), we derive thermodynamic expressions for them as follows:

(a) Specific heat at constant volume, c_v

Since volume is kept constant, the second term in (3.2.2) vanishes and we get,

$$c_v = (\delta Q/\partial T)_v = (\partial U/\partial T)_v \qquad (3.3.1)$$

(b) Specific heat at constant pressure, c_p

In this case, the pressure is kept constant but the volume changes, so that the substitution for dU from (3.2.1) in (3.2.2) gives

3.3 Specific Heats of Gases

$$\delta Q = (\partial U/\partial T)_v \, dT + \{(\partial U/\partial V)_T + p\} \, dV$$
$$= c_v \, dT + \{(\partial U/\partial V)_T + p\} \, dV \qquad (3.3.2)$$

Expressing dV as a function of p and T, we obtain, for constant pressure,

$$dV = (\partial V/\partial T)_p \, dT \qquad (3.3.3)$$

Using this value of dV in (3.3.2) and dividing by dT, we get

$$c_p = (\delta Q/dT)_p = c_v + \{(\partial U/\partial V)_T + p\} \, (\partial V/\partial T)_p$$

Or,

$$c_p - c_v = \{(\partial U/\partial V)_T + p\} \, (\partial V/\partial T)_p \qquad (3.3.4)$$

In the case of an ideal gas, it was shown by the experiments of Gay-Lussac that the internal energy U does not vary with volume at constant temperature. So, in (3.3.4), we put $(\partial U/\partial V)_T = 0$ and get

$$c_p - c_v = p(\partial V/\partial T)_p \qquad (3.3.5)$$

On account of the equation of state for dry air, (3.3.5) simplifies to

$$c_p - c_v = p(\partial V/\partial T)_p = R \qquad (3.3.6)$$

This is known as Mayer's equation for an ideal gas.

Thus, the difference between the values of the specific heats of an ideal gas at constant pressure and constant volume is equal to value of the gas constant of that particular gas. Since dry air is composed largely of diatomic molecules, it can be shown that the values of its specific heats are:

$$c_p = 7R/2, \qquad c_v = 5R/2.$$

For dry air, $R = R^*/M_d$, where M_d is the molecular weight of dry air. Substitution of the value of $R (= 287 \, J \, Kg^{-1} \, K^{-1})$ gives

$$c_p = 1004.5 \, J \, Kg^{-1} \, K^{-1} \qquad c_v = 717.5 \, J \, Kg^{-1} \, K^{-1}$$

Integration of (3.3.1) gives the energy function U of an ideal gas as

$$U = \int_0^T c_v \, dT + U_0 \qquad (3.3.7)$$

where U_0 is the internal energy at $T = 0$.

3.4 Adiabatic Changes in the Atmosphere

There are several thermodynamical processes in the atmosphere in which heat is prevented from entering or leaving a system during the process. Changes taking place during such a process are called adiabatic changes. This condition can be met in two ways: either by thermally insulating the system, or by allowing the changes to occur so rapidly that there is no communication of the system with the outside, so far as heat transfer is concerned.

The following are some examples of adiabatic changes in the atmosphere:

3.4.1 Adiabatic Relationship Between Pressure, Temperature and Volume

Under adiabatic condition, $\delta Q = 0$, and (3.2.2) may be written

$$dU + p\, dV = 0 \qquad (3.4.1)$$

If U is expressed as a function of p and V, we can obtain a relationship between p and V by substituting for dU from (3.3.1) and making use of the equation of state (2.4.1). Thus:

$$dU = c_v\, dT \quad \text{and} \quad dT = (p\, dV + V\, dp)/R$$

But, by (3.3.6), $R = c_p - c_v$,
So,

$$(c_p/c_v)(dV/V) + dp/p = 0 \qquad (3.4.2)$$

If we denote the ratio of the specific heats, c_p/c_v, by γ, then by integrating (3.4.2) we get

$$pV^\gamma = p_o V_o^\gamma = \text{Const} \qquad (3.4.3)$$

Since γ has a value of 1.4 for dry air, the adiabatic curves are steeper than the isothermals in a pV-diagram.

The adiabatic relationships between T and V, or between p and T, can be found from (3.4.3) and the equation of state for dry air. These are:

$$T V^{\gamma-1} = T_o V_o^{\gamma-1} = \text{Const} \qquad (3.4.4)$$

$$T p^{(1-\gamma)/\gamma} = T_o p_o^{1-\gamma/\gamma} = \text{Const} \qquad (3.4.5)$$

3.4.2 Potential Temperature

When a parcel of air ascends or descends in the atmosphere, the changes that occur in its temperature are given by the adiabatic relation (3.4.5). Using this relation, we

3.4 Adiabatic Changes in the Atmosphere

may define a temperature, θ, which a parcel of air at temperature T and pressure p will assume if it is compressed (or expanded) adiabatically and brought to a standard pressure, p_o, which is usually taken as 1000 mb. It is called the potential temperature and defined by the relation

$$\theta = T(p_o/p)^\kappa \qquad (3.4.6)$$

where $\kappa = R/c_p$

The potential temperature θ is a conservative property of air in any adiabatic process and, as such, widely used for identification in air mass analysis. Lines of constant potential temperatures are called adiabats. So, when a parcel of dry air rises adiabatically in the atmosphere, it follows a dry adiabat. The concept of potential temperature is useful in studies of vertical stability of the atmosphere.

3.4.3 Dry Adiabatic Lapse Rate of Temperature with Height

The temperature of a parcel of dry or unsaturated air decreases when it ascends adiabatically in the atmosphere. If we take logarithm of (3.4.6) and differentiate it with respect to height and use the hydrostatic approximation and the ideal gas equation, we obtain the expression:

$$(T/\theta)\partial\theta/\partial z = \partial T/\partial z + g/c_p \qquad (3.4.7)$$

For an atmosphere in which the potential temperature does not vary with height,

$$-\partial T/\partial z = g/c_p = \Gamma_d \qquad (3.4.8)$$

where Γ_d is the dry adiabatic lapse rate of temperature with height. Since g and c_p vary little with height in the lower atmosphere, Γ_d remains approximately constant with height.

With $g = 9.81\,\text{m s}^{-2}$ and $c_p = 1004\,\text{J kg}^{-1}\,\text{K}^{-1}$, Γ_d is about $-10\,°\text{C}$ per km.

3.4.4 Static Stability of Dry Air – Buoyancy Oscillations

In view of (3.4.8), (3.4.7) may be written as

$$(1/\theta)\partial\theta/\partial z = \Gamma_d - \Gamma_e \qquad (3.4.9)$$

where we write Γ_e for $-\partial T/\partial z$ which is the actual lapse rate of temperature with height in the environment.

From (3.4.9), it follows that if the actual lapse rate Γ_e is less than the dry adiabatic lapse rate Γ_d, $\partial\theta/\partial z$ is positive, which means that the potential temperature increases with height. This makes the atmosphere statically stable. Thus, the atmosphere is vertically stable, neutral or unstable, according as Γ_e is less than, equal to or greater than Γ_d. In a stable atmosphere, any displacement (upward or downward) will restore the parcel to its original position through a series of vertical oscillations which are known as buoyancy oscillations. The frequency of such oscillations, which is known as Brunt-Väisälä frequency, may be derived as follows:

Let a parcel with pressure p, density ρ and potential temperature θ be displaced upward from its equilibrium level z through a small distance δz without disturbing the environment. If the corresponding variables in the environment be p_e, ρ_e and θ_e, $dp_e/dz = -\rho_e g$, and the acceleration of the parcel is given by the relation

$$d^2(\delta z)/dt^2 = -g - (1/\rho)(\partial p/\partial z) = -g(\theta_e - \theta)/\theta_e \quad (3.4.10)$$

where it has been assumed that since the environment is not disturbed,

$$p = p_e, \text{ and } \partial p/\partial z = \partial p_e/\partial z = -\rho_e g.$$

Now, since the parcel moves adiabatically, its potential temperature does not change with height. But, in the environment, the potential temperature varies with height, so

$$\theta_e(z+\delta z) = \theta_e(z) + (\partial \theta_e/\partial z)\,\delta z \quad (3.4.11)$$

Since, $\theta_e(z) = \theta(z)$, we have at height $z + \delta z$

$$\theta_e - \theta = (\partial \theta_e/\partial z)\,\delta z$$

Equation (3.4.10) may, therefore, be written as

$$d^2(\delta z)/dt^2 = -(g/\theta_e)(\partial \theta_e/\partial z)\,\delta z = -N^2\,\delta z \quad (3.4.12)$$

where $N = \{(g/\theta_e)\,(\partial \theta_e/\partial z)\}^{1/2}$ is called the Brunt-Väisälä frequency.

The general solution of (3.4.12) is of the form

$$\delta z = A\,\exp(i\,N\,t) \quad (3.4.13)$$

Therefore, if $N > 0$, the parcel will oscillate about its equilibrium level with a frequency N and period $2\pi/N$.

The value of N computed from vertical soundings of the mean tropical atmosphere is about $0.012\,s^{-1}$. This gives for the period of the oscillations about 8.8 min.

In the case where $N = 0$, there is no acceleration and the parcel will be in equilibrium with its environment at the end of the motion. However, if $N < 0$, the parcel will continue to rise, if displaced upward, and there is no equilibrium. Thus, the stability conditions are as follows:

$$\begin{aligned}\partial\theta/\partial z &> 0 \text{ Stable} \\ \partial\theta/\partial z &= 0 \text{ Neutral} \\ \partial\theta/\partial z &< 0 \text{ Unstable}\end{aligned} \quad (3.4.14)$$

The atmosphere in general remains in static equilibrium most of the time. However, if static instability develops in any part of the atmosphere at any time, rapid vertical motion occurs so as to restore stability over the region again The situation, however, changes in cases of moist convection due to release of latent heat of condensation and this will be discussed in the next chapter.

3.4.5 Adiabatic Propagation of Sound Waves

Another example of adiabatic change in the atmosphere is found in the propagation of sound waves. Sound waves are longitudinal waves which travel by compression and rarefaction of air parcels. Newton was the first to compute the velocity of sound in air by using the relation

$$v = (E/\rho)^{1/2} = (p/\rho)^{1/2} \qquad (3.4.15)$$

where E is isothermal volume elasticity which in the case of an ideal gas is equal to pressure, p.

However, the velocity of sound computed by (3.4.15) differed from observed values. The cause for the discrepancy was found by Laplace who pointed out that the process of movement of sound waves in the air by compression and rarefaction was not isothermal but adiabatic. The changes in pressure and temperature took place so rapidly that there was little time for heat to leave the system. Now, it is readily shown (see Appendix-2) that when the process is adiabatic, the adiabatic modulus of elasticity is γ times that of isothermal modulus of elasticity. The corrected velocity of sound is, therefore, given by the expression (3.4.16):

$$v = (\gamma p/\rho)^{1/2} \qquad (3.4.16)$$

A mean value of sound velocity in the atmosphere is about $330 \, \text{m s}^{-1}$.

3.5 The Concept of Entropy

It is a unique property of heat that it always flows in one direction, viz, from a body at higher temperature to one at lower temperature when they are in contact with each other either directly or through some intermediate conductor. To show this, let us heat a piece of metal, say iron, of mass m_1 and specific heat c_1 to a temperature T_1 and place it in water of mass m_2, specific heat c_2 and temperature T_2, with $T_1 > T_2$. In this case, heat will flow from the metal to the water and soon an equilibrium temperature T will be reached, which is given by the relation

$$T = (m_1 \, c_1 \, T_1 + m_2 \, c_2 \, T_2)/(m_1 \, c_1 + m_2 \, c_2) \qquad (3.5.1)$$

In the above process, an amount of heat Q was drawn from the metal at higher temperature T_1 and delivered to the water at lower temperature T_2 till they were both at the same equilibrium temperature T.

Max Planck divided thermo-dynamical processes, in general, into three distinct categories: natural, unnatural and reversible. The above-mentioned example of heat transfer from a warmer to a cooler body is a natural process. An unnatural process would be one in which heat would flow in the opposite direction, i.e., from a cooler to a warmer body. Experience tells us that the latter never happens. In most naturally-occurring processes, heat flows one-way only, i.e., down the temperature gradient till an equilibrium is reached. Unnatural processes move away with equilibrium and never occur. Sometimes, however, a process can be reversible and equilibrium can be achieved by only a slight change in external conditions. Consider, for example, evaporation from a liquid in contact with its vapour under an external pressure P. Let p be the saturation vapour pressure of the liquid. If $P < p$, some additional liquid will evaporate as a natural process, while if $P > p_s$ evaporation will be an unnatural process. In fact, under latter condition, some vapour will condense. At $P = p$, the process can go either way. A slight decrease (increase) in P will cause evaporation (condensation). Such a process is clearly reversible.

To provide a quantitative basis to such naturally-occurring processes as mentioned above, classical thermodynamics first introduced a function called entropy which may be defined as

$$S = \int \delta Q/T + \text{Const} \qquad (3.5.2)$$

where S is entropy and $\int \delta Q$ is a quantity of heat that is added to, or subtracted from, a working substance at temperature T.

Since the boundary value of S is not known and specified, S is usually expressed as a differential between two thermodynamic states. Its relationship with the other thermodynamic variables, derived from the First law of thermodynamics (3.2.1) and the equation of state (2.4.12), explains why dS should be an exact differential and not a total quantity like δQ to which it is related and which denotes only a change in heat content.

For,
$$\begin{aligned} dS &= (\delta Q/T) = (dU + p\, dV)/T \\ &= c_p\, d\, \ln\theta \\ &= c_v\, d\, \ln T + R\, d\, \ln V \\ &= c_p\, d\, \ln T - R\, d\, \ln p \end{aligned} \qquad (3.5.3)$$

Since the expressions on the right-hand side of (3.5.3) are all exact differentials, dS must be an exact differential. Thus, (1/T) in (3.5.2) is simply an integrating factor and $\delta Q = T\, dS$.

It follows from (3.5.3) that the change of entropy of a system may be expressed as functions of either T and V or T and p, as stated below:

3.5 The Concept of Entropy

$$\delta Q = TdS = c_v \, dT + p \, dV, \quad (3.5.4)$$

Or,

$$\delta Q = TdS = c_p \, dT - V \, dp = c_p \, dT - T \, (dV/dT)_p \, dp \quad (3.5.5)$$

If the constant in (3.5.2) is taken as S_0, where S_0 represents the zero energy state of the system at temperature T_0 of the thermodynamic scale, then S may be regarded as the absolute energy state at temperature T. However, since, in most cases, we are only concerned with the energy input or output between two thermodynamic states, we can get a measure of this energy simply by multiplying dS by T in a cyclic process, as in a T-S diagram shown in Fig. 3.1.

In Fig. 3.1, the entropy (S) is shown along the ordinate and the temperature T along the abscissa. If we visualize that the working substance is taken through a reversible cycle represented by the curve ABCD, then the energy drawn by it from a heat source in moving from P to P' is TdS (represented by the hatched area) and in going through the whole cycle is $\oint T \, dS$. The total amount of heat is treated as positive during the stage when entropy increases, i.e., in moving from A to B along the curve and negative when the entropy decreases, i.e., in moving from B to A, if B is taken at the top of the curve and A at the bottom T-S diagrams are routinely used in many meteorological services for estimation of available heat energy in the atmosphere (For further information on these diagrams, see Appendix-3).

It can be shown that in all naturally-occurring thermodynamical processes, entropy increases. Take, for example, the case of the heat transfer from the warmer to the cooler body, discussed earlier (3.5.1). Here, the warmer body loses an entropy, $-Q/T_1$, while the cooler body gains an entropy, $+Q/T_2$, so the net entropy of the system is $Q\{(1/T_2) - (1/T_1)\}$, which is clearly positive. Here, the transfer of heat takes place fast till a steady state is reached. This is an irreversible process. A reversible process is also a naturally-occurring process which passes through a continuous sequence of equilibrium states and takes place slowly so that the entropy remains constant. Thus, in all naturally-occurring processes,

$$dS \geq 0 \quad (3.5.6)$$

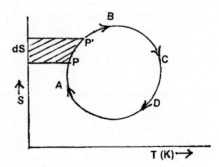

Fig. 3.1 Representation of energy in a temperature (T) – entropy(S) diagram

3.6 The Second Law of Thermodynamics

The Second law of thermodynamics makes an important statement about the heat balance in a thermodynamic system, such as the atmosphere. Like the First law, it also emphasizes that the total heat energy supplied to a system remains constant and goes partly to increase the internal energy of the system and partly to do work against external pressure. But there is a difference. The First law does not tell us how much of the given heat energy goes to increase the internal energy and how much to do the external work in any heat transfer process. The only stipulation is that the total of the two should remain constant. It was, however, known from early times that physical processes involving friction, turbulence, conduction, etc, dissipated a good part of the heat energy supplied to a thermodynamic system and only a partial conversion of heat into useful mechanical energy was possible by using even the best of heat engines. The manner in which heat was supplied to, or withdrawn from, a working substance made a lot of difference.

3.6.1 Carnot Engine

Sadi Carnot (1796–1832), a French physicist, devised an engine which working in a four-stroke cycle under given conditions could be said to be the best possible heat engine to secure maximum useful mechanical energy out of the heat energy supplied to the engine. His engine worked between two reservoirs of heat, one at temperature T_1 which we may call the source and the other at T_3 which we may call the sink, with $T_1 > T_3$. The working substance was 1 mol of a gas which was contained in a cylinder one end of which was closed with a conducting lid while through the other end a frictionless piston could move in and out. The cylinder could be placed in contact with the heat source and sink by turn at will and could also be insulated from them when not required. The four stages of the operations and the manner in which they were performed are shown by an indicator (p, v) diagram (Fig. 3.2).

The stages are as follows:

Stage 1. Place the cylinder in contact with the heat source T_1 and extract a quantity of heat Q_1 from it by a slow outward movement of the piston so that there is no change of temperature inside the cyclinder while the gas is being heated and its volume changing from v_1 to v_2 with the inside pressure being only a little higher than the outside pressure.

Stage 2. The cylinder is then removed from the heat source and placed in contact with the insulator and the gas is allowed to expand adiabatically under its own pressure from volume v_2 to v_3. The expansion leads to a fall of pressure as well as temperature from $T_2 (= T_1)$ to T_3.

Stage 3. The cylinder is then removed from the insulator and placed in contact with heat sink T_3. The piston is then pressed gently inward so that a quantity of heat Q_3 is delivered to the heat sink at temperature T_3, while the volume changes from v_3 to a volume v_4.

3.6 The Second Law of Thermodynamics

Fig. 3.2 The Carnot cycle: Indicator diagram

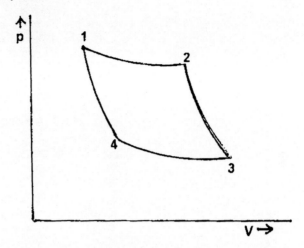

Stage 4. During this stage, an adiabatic compression will bring the volume v_4 back to the original volume v_1.

The work done by the gas during the different stages is as follows:

Stage 1 $\quad Q_1 = W_{12} = \int_{v_1}^{v_2} p\, dv = R\, T_1 \int_{v_1}^{v_2} dv/v = R\, T_1\, \ln(v_2/v_1)$ \hfill (3.6.1)

Stage 2 $\quad W_{23} = \int_{v_2}^{v_3} p\, dv = (p_2 v_2^\gamma) \int_{v_2}^{v_3} dv/v^\gamma = \{(p_2 v_2^\gamma)/(1-\gamma)\}\, [v_3^{1-\gamma} - v_2^{1-\gamma}]$

$\qquad = \{(RT_2/(1-\gamma)\}[(v_3/v_2) - 1] = \{R/(1-\gamma)\}\, [T_3 - T_2]$

$\qquad = \{R/(\gamma-1)\}(T_1 - T_3)$ \hfill (3.6.2)

Since, $(v_3/v_2)^{1-\gamma} = (T_3/T_2)$; and, $T_2 = T_1$

The cylinder is now placed in thermal contact with the heat sink T_3 and the gas compressed isothermally from volume v_3 to v_4, so that a quantity of heat Q_3 is delivered to the sink. The work done in this process is negative, since work is done on the gas and not by the gas. So, the work done in

Stage 3 $\quad Q_3 = W_{34} = \int_{v_3}^{v_4} p\, dv = R\, T_3 \int_{v_3}^{v_4} dv/v = RT_3\, \ln(v_4/v_3)$ \hfill (3.6.3)

The cylinder is now removed from the heat sink and placed on the insulator stand and compressed adiabatically from volume v_4 to v_1. The work done in

Stage 4 $\quad W_{41} = \int_{v_4}^{v_1} p\, dv = (p_4\, v_4^\gamma) \int_{v_4}^{v_1} dv/v^\gamma$

$\qquad = \{(p_4\, v_4^\gamma)/(1-\gamma)\}[v_1^{1-\gamma} - v_4^{1-\gamma}]$

$\qquad = \{RT_4/(1-\gamma)\}[(v_1/v_4)^{1-\gamma} - 1]$

$\qquad = \{R/(1-\gamma)\}(T_1 - T_3)$ \hfill (3.6.4)

Since, $(v_1/v_4) = (T_1/T_4)$; and, $T_4 = T_3$

The work done in stage 4 also is to be reckoned negative, since it is done on the gas.

Thus, the total work done by the gas, from (3.6.1–3.6.4), is

$$W = W_{12} + W_{23} + W_{34} + W_{41}$$
$$= R(T_1 - T_3) \ln(v_2/v_1) \tag{3.6.5}$$

since, the expressions in (3.6.2) and (3.6.4) cancel each other, and $(v_2/v_1) = (v_3/v_4)$.

In summary, the working substance in the Carnot engine drew a quantity of heat energy Q_1 from the source, performed the useful work W, and returned the balance of heat Q_3 to the sink before being adiabatically compressed to the original state.

In other words, $Q_1 - Q_3 = W$, and the efficiency η of the engine is given by

$$\eta = (W/Q_1) = (Q_1 - Q_3)/Q_1 = (T_1 - T_3)/T_1 = 1 - (T_3/T_1) \tag{3.6.6}$$

Also, it follows from (3.6.6) that

$$(Q_1/T_1) = (Q_3/T_3) \tag{3.6.7}$$

The Carnot engine being reversible is the most efficient of heat engines and it is not possible to invent an engine working reversibly between the temperatures T_1 and T_3 which can be more efficient than the Carnot engine. For, if it were possible to invent a super-Carnot machine, it would be possible to get work out of it continuously without transferring any heat to a sink. Such a possibility simply does not exist in nature. However, according to (3.6.6), the efficiency of the Carnot engine increases as the sink temperature T_3 is lowered, and attains unity when it is brought down to Absolute zero. Thus, the Absolute scale of temperature, which we introduced earlier (2.4.2), is a temperature scale which can be deduced from thermodynamical reasoning. For this reason, it is also sometimes called the thermodynamic scale of temperature.

3.6.2 Statement of the Second Law of Thermodynamics

We are now in a position to make a statement regarding the Second law of thermodynamics. The essence of a formulation of this law as given by Thomson and Planck is as follows:

It is impossible to construct a heat engine working between a source and a sink which draws heat from the source and converts the whole of it into useful work without producing any other effects. Such a formulation does not violate the First Law since the total energy remains conserved, but it does tell us heuristically that there can be no way by which heat can be converted fully into useful mechanical

3.6 The Second Law of Thermodynamics

energy without producing any other effects. Here 'any other effects' mean delivering any heat to a sink. In other words, what the Second law states is that there can be no machine which will convert the whole of the heat drawn from a source to useful work without delivering any of it to a sink. In the reversible Carnot cycle, the engine delivers to the sink as much heat as required to keep the entropy constant. Lessening the amount of heat to be delivered to the sink below this limit due to leakage by such physical processes as friction, turbulence, conduction, diffusion, etc., which is quite a common occurrence, makes the transfer process irreversible, resulting in increase of the entropy of the system. Such irreversibility and unavailability of useful work are exceedingly numerous in nature and in everyday experience. The following are some examples:

(i) A glass cylinder containing some water is rotated rapidly and then suddenly left to itself. The rotating water comes to a halt after a while due to friction, with a slight rise of temperature. The warming of the water is due to conversion of its kinetic energy into heat. However, the process is irreversible because water cannot be made to rotate again simply by extracting some heat from the water.

(ii) A high-speed bullet fired from a gun hits a target at a distance. The temperature rises at the target due to conversion of kinetic energy into heat. It is simply impossible to make the bullet to fly back simply by extracting some heat from the target. The process is clearly irreversible. The heat generated is unavailable for any useful work.

(iii) In a transparent cylinder, place a few layers of white marbles and then place a few layers of red marbles on top of the white ones. Shake the cylinder so that both types of marbles get mixed. It is impossible to get the red marbles back on top again simply by re-shaking the cylinder. In this case, order has given way to disorder. Experience tells us that a lot of external work will be needed to restore the original arrangement of the marbles. So the process is clearly irreversible.

(iv) A decorated glass filled with water is placed at the edge of a high table. It accidentally falls down and breaks into pieces with water splashed all over the floor. It is impossible to get the glassful of water, as it was originally, back on the table again. The process is clearly irreversible.

(v) A large ocean wave on striking the coastline breaks down into irregular whirls and foams. It is impossible to get back the energy of the wave by re-assembling the dissipated parts of the wave which have been lost to friction.

Numerous such examples may be cited about the way in which energy during transformations from one form to another becomes unavailable for useful work or for that matter order easily turns into disorder or organization into disorganization in physical processes, though the total of all forms of energy remains invariant. It is this unavailability of energy for work, tendency of nature to go from order to disorder or from organized state to a disorganized state that is highlighted by the Second Law of thermodynamics. In all such physical processes, entropy increases.

3.7 Thermodynamic Equilibrium of Systems: Thermodynamic Potentials

When a quantity of heat δQ is supplied to a thermodynamic system, its effect on the system depends on conditions of the variables of temperature, volume and pressure. It also depends on the conditions whether the system being considered is closed or open to the environment and if the system is a reversible or irreversible one. However, for all such cases, we may write from the laws of thermodynamics,

$$dS - (dU + \delta W)/T \geq 0 \qquad (3.7.1)$$

where W represents the external work done by the system and the sign of equality applies to the case where the system is reversible. In the reversible case, $\delta W = p\,dV$, where p is the equilibrium pressure on the system at any instant and dV is the change in volume. Thus, in the reversible case, we may define entropy in terms of the thermodynamic variables as

$$dS = (dU + p\,dV)/T \qquad (3.7.2)$$

3.7.1 Free Energy or Helmholtz Potential

Thermodynamic systems, however, are not closed in general and being open to the environment proceed at constant temperature. For T = constant and a reversible case, (3.7.2) may be written as

$$dW = -d(U - TS) = -dF \qquad (3.7.3)$$

where $F (= U - TS)$ is called the 'Free energy', a term first introduced by Helmholtz. For this reason, F is also sometimes called the Helmholtz potential. (3.7.3) simply states that at constant temperature, the decrease of free energy gives the maximum external work obtainable from a reversible system. In the general case,

$$\delta W \leq -dF \qquad (3.7.4)$$

where the sign of equality refers to the reversible case.

Now, if the volume does not change and the system is isothermal-isometric, $\delta W = 0$, and

$$0 \leq -\delta F$$

Or, $$\delta F \leq 0 \qquad (3.7.5)$$

For such a system, therefore, the free energy available for doing external work has its minimum value. This value is 0 at equilibrium.

3.7.2 Free Enthalpy, or Gibbs' Potential, or Gibbs Free Energy

If the pressure in an isothermal system is kept constant, (3.7.2) may be written in the form

$$d(U - TS + pV) = dG = 0 \tag{3.7.6}$$

where $G\ (= U - TS + pV)$ is called the 'Free enthalpy' or Gibbs potential, or Gibbs free energy.

As with free energy in an isothermal-isometric system, the Gibbs free energy in an isothermal-isobaric system tends to be at its minimum value at equilibrium, i.e.,

$$\delta G \leq 0 \tag{3.7.7}$$

where the sign of equality refers to a reversible system.

The Gibbs equilibrium condition (3.7.7) has been widely applied to study heat or energy transfer processes in many physical and chemical problems. It has been found to be particularly useful in studying the problems of evaporation and condensation in the atmosphere (see, e.g., Appendix-5).

3.8 The Third Law of Thermodynamics

The third law of thermodynamics states that the entropy of a thermodynamic system at a given temperature T vanishes at temperature 'Absolute zero'. This law was first stated by Nernst; hence it is also called the Nernst heat theorem. It was at first thought that the theorem applied to crystalline solids only but later studies showed that it was applicable to gases as well, or, for that matter, to any system.

The statistical view of this law is that every system may be regarded as a macrostate which corresponds to many microstates and that the thermodynamic probability of occurrence of that macrostate is simply given by the number of microstates favorable to the occurrence of that macrostate. The entropy of the system is, therefore, given in terms of this thermodynamic probability in accordance with the relation

$$S = k \log W \tag{3.8.1}$$

where S is the entropy, W is the thermodynamic probability, and k is Boltzmann constant which is equal to R^*/N, where R^* is universal gas constant and N is the Avogadro Number. At Absolute zero, the number of favorable microstates is reduced to one; hence the entropy becomes equal to zero.

Readers desirous of getting further details of the Nernst heat theorem are referred to an excellent treatment of the topic in Saha and Srivastava (1931).

3.9 The Atmosphere as a Heat Engine

It is open to question whether and, if so, to what extent the laws of thermodynamics, especially the principles of the Carnot engine and reversibility, are applicable to the earth's atmosphere in which the working substance which normally forms a portion of the atmosphere with boundaries exposed to the rest of the atmosphere is not really a closed system like the Carnot engine which draws a quantity of heat from a source, converts a part of it to work and delivers the balance to a sink under controlled conditions. The earth's surface is highly uneven and its different parts respond differently to the incoming net solar radiation, so that heat sources and sinks are created in the boundary layers of the atmosphere in an irregular fashion all the time. The atmosphere picks up the heat from the source, does work through a circulation and delivers the balance of heat to the sink, somewhat on the model of the Carnot cycle. However, there is a difference. The working substance in the atmosphere not being a perfectly closed system remains in direct communication with its environment and may share energy with it through conduction, convection, turbulence, etc., while transferring heat from the source to the sink. Thus, a loss of heat in transit will reduce the amount of heat to be delivered to the sink, thereby increasing the entropy of the system. Delivering less heat to the sink means that a part of the input energy becomes unavailable to the working substance for useful work. Further, the condition of reversibility of the cycle which is central to the working of the Carnot engine does not hold in the case of the atmosphere. The effects of buoyancy, friction, evaporation and condensation of water vapour are other complicating factors. All these add to make the atmospheric heat engine clearly irreversible and highly inefficient. Qualitatively, since the annual mean atmospheric circulation basically transfers heat from a low-level source in the equatorial region to a high-level sink in the polar region, a rough measure of the efficiency of the circulation which effects this transfer process may be obtained by an application of the Second Law of thermodynamics to the atmosphere. If we assume that a quantity of heat Q_1 is drawn from the equatorial heat source at temperature T_1 and a quantity of heat Q_3 is delivered to the sink over the polar region at temperature T_3, then, according to (3.6.6), the efficiency η of the heat engine is given by

$$\eta = (Q_1 - Q_3)/Q_1 = 1 - (T_3/T_1)$$

Substituting the approximate mean observed values of $T_1 = 300\,\text{K}$ and $T_3 = 230\,\text{K}$ in the earth's atmosphere, we obtain η to be about 23%, which turns out to be much less than that, for example, of a steam engine or an Otto engine which has an efficiency of about 30–40% or even higher (Saha and Srivastava, 1931). It is even doubtful whether the atmospheric heat engine attains the level of efficiency suggested by the above computation. We show later in Chap. 18 that of the total potential energy of the atmosphere, only a very small fraction is available for conversion to other forms and that out of the available potential energy, only a tiny fraction is used to generate useful kinetic energy.

Chapter 4
Water Vapour in the Atmosphere: Thermodynamics of Moist Air

4.1 Introduction

Water vapour exists in small and variable proportions in comparison with dry air in the atmosphere. However, its importance lies primarily in the fact that it can change its phase from a gaseous to a liquid or solid form, or vice versa, under certain conditions of temperature and pressure. A simple example of this change of phase is the well-known hydrologic cycle. A change of phase from liquid to vapour takes place when water evaporates from the wet ground, or a lake, or an ocean surface. The water vapour moistens the air and when the mixture of air and water vapour rises in the atmosphere due to convection and/or turbulence, it cools adiabatically resulting in the condensation of water vapour to form cloud and rain (liquid phase) and sometimes freeze into snow, ice and hail (solid phase) before returning to the surface and rising again in vapour form. In describing this life cycle of a 'cloud', Poet Shelley wrote:

> I am the daughter of Earth and water,
> And the nursling of the Sky;
> I pass through the pores of the ocean and shores;
> I change, but I cannot die.
> For after the rain when with never a stain
> The pavilion of Heaven is bare,
> And the winds and sunbeams with their convex gleams
> Build up the blue dome of air,
> I silently laugh at my own cenotaph,
> And out of the caverns of rain,
> Like a child from the womb, like a ghost from the tomb,
> I arise and unbuild it again.
>
> P.B. Shelly

The important point to note here is that at every change of phase, a large amount of heat called the latent heat is either withdrawn from or released to the atmosphere. A water substance withdraws heat from the environment when it changes from a solid to a liquid or vapour phase, thereby cooling the environment. Reversely, it releases heat to the environment when the vapour condenses into a liquid or sublimes into a solid phase, thereby warming the environment. It is this cooling or heating effect upon the environment that plays a dominant role in atmospheric thermodynamics. Further, water vapour plays important roles in radiative and heat balance processes of the earth-atmosphere system and several chemical and biological processes. Some of these will be discussed in later chapters of this book.

4.2 Humidity of the Air – Definitions

A measure of water vapour in the atmosphere is given either by its absolute density or the proportion of water vapour to dry air in a sample of moist air. The following are some of the terms which are in use.

Absolute humidity

This is simply the density or the mass of water vapour per unit volume in a sample of moist air, measured in unit of gm per c.c., or kg per cubic meter.

Humidity-mixing-ratio (h.m.r)

The amount of water vapour in grams mixed with 1 kg of dry air is called the humidity-mixing-ratio. Thus, if m_v grammes of water vapour are mixed with m_d kg of dry air, the h.m.r. is m_v/m_d

Specific humidity

The amount of water vapour in gms contained in 1 kg of moist air is called the specific humidity of air. If a sample of moist air of mass 1 kg contains m_v grams of water vapour and m_d grams of dry air, the specific humidity is $m_v/(m_d + m_v)$.

Relative humidity

This is the ratio of the partial pressure of water vapour actually present in the atmosphere at a certain temperature and pressure to the partial pressure of the maximum amount of water vapour that air can hold at the same temperature and pressure. It is usually expressed as a percentage. Thus, if the vapour pressure at a given temperature and pressure is e, the relative humidity is given by $(e/e_s) \times 100$, where e_s is saturation vapour pressure at the same temperature and pressure. It will be shown later in this chapter that the saturation vapour pressure of water is a function of temperature only by theory and increases rapidly with temperature, while the actual vapour pressure generally varies only slowly with temperature.

Dew-point

The dew-point is the temperature to which unsaturated air must be cooled at the existing pressure to produce saturation with the existing amount of water vapour. It is usually denoted by the symbol, T_d.

4.3 Density of Moist Air – Virtual Temperature

The difference between the actual dry-bulb temperature and the dew-point, which is called the dew-point depression, gives a measure of the humidity of the air. The depression varies inversely with the humidity of the air.

4.3 Density of Moist Air – Virtual Temperature

If x be the humidity-mixing-ratio in a given volume V of moist air at temperature T and total pressure p and if e be the partial pressure of water vapour, the density of moist air is the sum of the density of water vapour and the density of dry air and is found as follows:

The density of water vapour as given by the equation of state for water vapour may be written in the form,

$$\rho_v = x/V = e \in /R\,T \tag{4.3.1}$$

where ρ_v is the density and $\in$ the specific gravity of water vapour($= 0.622$) and R the gas constant for dry air($= R^*/M_d$) and the subscripts v and d refer to water vapour and dry air respectively.

Similarly, from the equation of state for dry air, the density of dry air ρ_d is given by the relation,

$$\rho_d = 1/V = (p-e)/RT \tag{4.3.2}$$

Adding (4.3.1) and (4.3.2), we get for the density of moist air ρ_m,

$$\begin{aligned}\rho_m &= (1+x)/V = p\{1 - e(1-\in)/p\}/RT \\ &= p(1 - 3e/8p)/RT \\ &= p/RT^* \end{aligned} \tag{4.3.3}$$

where T^* ($= T/(1-3e/8p)$) is called the virtual temperature of the moist air.

Thus, the virtual temperature is the temperature of the dry air which will have the same density as the moist air at the same pressure. The virtual temperature of moist air increases with increase of vapour pressure in the atmosphere.

The virtual temperature of moist air can also be expressed in terms of the humidity-mixing-ratio x, as shown below:

The equations of state of water vapour and dry air given in (4.3.1) and (4.3.2) may be written in the form

$$e\,V = xRT/\in$$
$$(p-e)V = RT$$

Adding the two relations above, we get

$$p\,V = (1+x/\in)\,RT \tag{4.3.4}$$

Substituting the value of V from (4.3.4) in the expression for the density of moist air, we get

$$\rho_m = (1+x)/V = (p/RT)(1+x)/(1+x/\epsilon) = (p/RT^*) \tag{4.3.5}$$

where,
$$T^* = T(1+x/\epsilon)/(1+x) \tag{4.3.6}$$

4.4 Measurement of Humidity – Hygrometers/Psychrometers

Many methods have been devised to measure humidity of the air. The instruments are generally called hygrometers or psychrometers. The most commonly used humidity-measuring instrument is that of wet and dry bulb thermometers, which consists of two similar mercury thermometers, one of which (dry-bulb) records the actual air temperature, while the other (wet-bulb) records the temperature of the air after it has been cooled by evaporation of water at the bulb which is kept wet by water supplied to it from a small reservoir by a muslin cloth wrapped round the bulb. The difference between the temperatures measured by the two thermometers is an index to the degree of wetness of the air in the atmosphere. Qualitatively, a small value of the depression of temperature at the wet bulb indicates that the air is already very moist, while a large value indicates extreme dryness of the air. The thermodynamics of this hygrometer is discussed below. The treatment given here is that due to Normand (1921).

Let x be the mixing-ratio of a sample of air at temperature T and pressure p, which flows over the wet-bulb thermometer and delivers a quantity of heat to the wet-bulb thermometer so as to evaporate some water and drop its temperature to T' before leaving the wet-bulb with a mixing-ratio of x' saturated at the wet-bulb temperature T' at the same total pressure p. The difference $(x'-x)$ is then the quantity of water that was evaporated at the wet-bulb and that brought down its temperature to T'. The heat exchange at the wet-bulb is then given by

$$(c_p + c_p'x)(T - T') = L'(x' - x) \tag{4.4.1}$$

where c_p, c_p' are the specific heat at constant pressure of dry air and water vapour respectively and L' the latent heat of vaporization of water at the wet-bulb temperature T'. Since $(c_p' x/c_p)$ is $\ll 1$, (4.4.1) may be written as

$$c_p(T - T') = L'(x' - x)$$

Or,
$$T + (L'x/c_p) = T' + (L'x'/c_p) \tag{4.4.2}$$

We now consider a temperature T'' of totally dry air $(x = 0)$ which has a wet-bulb temperature T', so that

$$T'' = T' + (L'x'/c_p) \tag{4.4.3}$$

4.4 Measurement of Humidity – Hygrometers/Psychrometers

The temperature T'' is known as the equivalent temperature. It is the temperature which a sample of moist air would assume if it were expanded saturated-adiabatically to have all its latent heat converted into sensible heat and then compressed dry-adiabatically to its original pressure.

If, in (4.3.1), we use the partial water vapour pressure e instead of the humidity-mixing-ratio x, and use the relationships, $x = \in e/(p-e)$, and $x' = \in e'/(p-e')$, where e' is the saturation vapour pressure of water at temperature T', we get

$$[\{\in (e'-e) p \, L'\}/\{(p-e)(p-e')\}] = [(T-T')\{c_p + \in c_p'e/(p-e)\}] \quad (4.4.4)$$

For small values of e compared to p, $\in c_p'/c_p$ has a value close to unity. The Eq. (4.4.4) may thus be simplified to

$$e' - e = Ap(T - T') \quad (4.4.5)$$

where,
$$A = (1 - e/p)(c_p/L' \in)$$

Since the value of e/p is only a very small fraction of unity, it may be neglected and the expression (4.4.5) may be used to measure the existing vapour pressure e of the air at temperature T, since e' which is the saturation vapour pressure at temperature T' can be found from standard meteorological tables. The relative humidity of the air may then be determined from the percentage ratio, $e \times 100/e_s$, where e_s is the saturation vapour pressure of water at temperature T, which can be found from the Tables.

Normand showed that the dry adiabat line through the dry-bulb temperature T, the saturated adiabat line through the wet-bulb temperature T' and the saturated-mixing-ratio line through the dew-point T_d of a sample of air all meet at a point P which came to be called the Normand point. The level of P is also called the lifting condensation level (LCL), as shown in Fig. 4.1.

In Fig. 4.1, the wet-bulb temperature T' lies in between the dry-bulb temperature T and the dew-point temperature T_d. This is due to the fact that saturation at the wet-bulb is attained partly by evaporative cooling of air and partly by addition of water vapour into it.

If the dry adiabat through the equivalent temperature T'' is extended to a standard pressure, say 1000 mb, the temperature reached at this pressure is called the equivalent potential temperature which is denoted by θ_e. Likewise, if the saturated adiabat

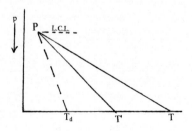

Fig. 4.1 The Normand diagram showing the Lifting Condensation Level (LCL)

through the wet-bulb temperature T′ is extended to a standard pressure 1000 mb, the temperature then attained is called the wet-bulb potential temperature, which is denoted by θ_w. The equivalent potential temperature θ_e and the wet-bulb potential temperature θ_w are both conserved during dry and moist adiabatic processes. Hence they are useful parameters in the identification of airmasses.

Psychrometers are instruments that make use of the principle of the dry and wet-bulb hygrometer as discussed above to measure humidity. Different types of psychrometers are in use. A portable variety called a sling psychrometer is very handy and in wide use. Other psychrometers include the Assmann psychrometer and the aspiration psychrometer.

The physical and chemical properties of many substances are affected by humidity of the air. For example, the physical dimension of a human hair changes with absorption of water vapour and this fact is made use of in making a class of hygrometers, such as a hair hygrometer, a torsion hygrometer or a goldbeater's skin hygrometer. A change in the physical and electrical properties of some substances due to absorption of water vapour forms the basis of another class of hygrometers, called absorption hygrometers, electrical hygrometers and carbon-film hygrometers. Besides these, we have the diffusion hygrometer the working of which depends upon the diffusion of water vapour through a porous membrane and the spectral hygrometer which depends upon measurements of the absorption spectra of water vapour.

4.5 Ascent of Moist Air in the Atmosphere – Pseudo-Adiabatic Process

When a stream of moist but unsaturated air rises in the atmosphere, it first cools by dry-adiabatic expansion till it reaches the lifting condensation level where it becomes saturated. Further ascent leads to condensation of water vapour on nuclei that may be present in large numbers in the atmosphere. Usually, hygroscopic particles act as effective nuclei. Experiments in the laboratory have shown that the equilibrium vapour pressure required for condensation depends upon not only the hygroscopic property of the nuclei but also their size and any electrical charge that may be carried by them. Condensation first starts on the hygroscopic nuclei such as salt particles at relative humidities even less than 100 percent. Further ascent to lower pressure and temperature leads to increased condensation and release of latent heat which diminishes the rate of cooling. In other words, the warming has the effect of making the saturated adiabatic lapse rate of temperature less than the dry adiabatic lapse rate. As the current rises further and condensation continues, droplets form and grow into cloud and rain. When it crosses the freezing level, drops are supercooled and start forming ice crystals. More and more ice crystals form as the temperature gets lower. It is observed that water can remain in supercooled state down to a temperature of about $-40\,°C$. But the fraction of water drops that turn into ice crystals increase with height. At temperatures below about $-20\,°C$, there

4.5 Ascent of Moist Air in the Atmosphere – Pseudo-Adiabatic Process

are more ice crystals than water drops. It, however, so happens that in the temperature range between 0°C and −20°C, the saturation vapour pressure over ice is less than that over water. So when they co-exist, the ice crystals grow at the expense of the water drops. It is widely believed that in the atmosphere large drops form by this mechanism of evaporation of water from supercooled cloud drops and its deposition on ice crystals in the upper layers of the atmosphere.

We, therefore, see that as moist air rises in the atmosphere, the water vapour contained in it changes into liquid and solid phases and that latent heat is liberated at every change of phase. The whole process would remain adiabatic and reversible as long as the products of the phase change are all carried along with the rising current and the heat liberated remains within the system. But the fact is that when the condensed particles grow and drop out as rain, snow or hail, they remove some heat out of the system and the process can no longer be treated as adiabatic and reversible.

In a real situation, however, not all condensed particles drop out and fall to the surface, as some remain suspended in the atmosphere as clouds. But, for practical purposes, the loss of heat due to fallout which is likely to be small is neglected and the process treated as pseudo-adiabatic. It is also assumed that the entropy of the whole system consisting of air, water vapour, rain and ice (though all the phases may not be present at the same time) always remains constant. Following Brunt (1944), we may derive an expression for the entropy of a mixture of dry air, water vapour, rain and ice, as follows:

Let x be the humidity-mixing-ratio of water vapour, y the amount in grammes of liquid water and z the amount in grammes of ice associated with 1 gramme of dry air. Then the total of water substances, $\xi = x + y + z$, remains constant throughout the thermodynamic process. If p and T be the pressure and temperature of the moist air and e the partial pressure of water vapour, then the entropy of 1 gramme of dry air S_1 is given by

$$S_1 = c_p \ln T - R \ln(p-e) \tag{4.5.1}$$

where c_p is the specific heat of dry air at constant pressure.

Let S_2 be the entropy of a mixture of x grammes of water vapour and y grammes of liquid water.
Then,

$$S_2 = (x+y)c_w \ln T + Lx/T \tag{4.5.2}$$

where c_w is the specific heat of liquid water and L the latent heat of vaporization of water.

Similarly, the entropy of ice, S_3, is the entropy of water minus the entropy involved in converting it into ice. Thus,

$$S_3 = zc_w \ln T - L_e z/T \tag{4.5.3}$$

where L_e is the latent heat of fusion of ice.

The total entropy S of the mixture is, therefore, given by

$$S = S_1 + S_2 + S_3 = (c_p + \xi c_w) \ln T + Lx/T - L_e z - R \ln(p-e) \tag{4.5.4}$$

4.6 Saturated Adiabatic Lapse Rate of Temperature

In a pseudo-adiabatic process, the lapse rate of temperature of a saturated parcel of air with height may be found from the entropy form of the First law of thermodynamics (3.2.1) as follows:

$$d(\ln T)/dz - \kappa\, d(\ln p)/dz = -(L/c_p T)dx_s/dz \qquad (4.6.1)$$

where $\kappa = R/c_p$, and x_s is the saturation mixing-ratio.

With the aid of the hydrostatic approximation and the equation of state, (4.6.1) may be written as

$$dT/dz + g/c_p = -(L/c_p)d\, x_s/dz = -(L/c_p)(dx_s/dT)(dT/dz) \qquad (4.6.2)$$

Since, by (3.4.8), the dry adiabatic lapse rate, $g/c_p = -dT/dz \equiv \Gamma_d$, we may write (4.6.2) as

$$\Gamma_s \equiv -dT/dz = \Gamma_d/[1 + (L/c_p)(dx_s/dT)] \qquad (4.6.3)$$

where Γ_s is the saturated adiabatic lapse rate.

Since dx_s/dT is always positive, $\Gamma_s < \Gamma_d$. In the tropics, Γ_s has a mean value of about 6°C per km as against about 10°C per km for Γ_d.

4.7 Equivalent Potential Temperature

When condensation occurs in a sample of moist air which is lifted, the heat liberated in the process amounts to $-L\, dx_s$, where L is the latent heat and x_s the saturation-mixing-ratio at the temperature at which the air becomes saturated. This heat is added to the air. The entropy equation, (3.5.3), may, therefore, be written as

$$-(L\, dx_s/T\, c_p) = d\ln\theta \qquad (4.7.1)$$

where T is the dry-bulb temperature at the level where air becomes saturated, c_p the specific heat of dry air at constant pressure, and θ the potential temperature of the dry air.

Now, the change of x_s following motion in a given time normally far exceeds that of L or T, so (4.7.1) may be written as

$$d\ln\theta \simeq -d(L\, x_s/Tc_p) \qquad (4.7.2)$$

Integrating (4.7.2) from the initial state (x_s, θ) to the state where $x_s = 0$, we get

$$\theta_e \simeq \theta \exp(Lx_s/T\, c_p) \qquad (4.7.3)$$

where θ_e is called the equivalent potential temperature.

The expression (4.7.3) may also be used for an unsaturated air provided the temperature T is the temperature at which the unsaturated air becomes saturated while ascending.

Thus, the equivalent potential temperature is conserved during both dry adiabatic and moist pseudo-adiabatic processes.

4.8 Variation of Saturation Vapour Pressure with Temperature

4.8.1 The Clausius-Clapeyron Equation

One of the most important and useful relationships in thermodynamics is what is known as the Clausius-Clapeyron equation which gives the variation of saturation vapour pressure of a liquid in equilibrium with its vapour with temperature. We may derive this equation by considering a reversible cycle of heat exchange between a liquid and its vapour, kept in a closed vessel, and applying the laws of thermodynamics to the system.

As we consider only two phases of a single substance which can exist in three phases, the system, according to Gibbs' phase rule, has only one degree of freedom. We take this to be the temperature. A change of volume at a fixed temperature has little or no effect on the saturation vapour pressure of water, as it would have if we were dealing with an ordinary gas. An increase or decrease of volume would only cause some evaporation or condensation. For the heat exchange, we consider a Carnot cycle, working between the temperatures $T + dT$ and T, as shown in a saturation vapour pressure e_s – volume v diagram (Fig. 4.2).

In Fig. 4.2, let the initial point in the cycle be A where a quantity of heat equal to the latent heat of vaporization of water $L(T + dT)$ is supplied to vaporize a mol of water at temperature $T + dT$. Let the saturation vapour pressure at this stage be $e_s + de_s$. The vapour is then allowed to expand at first isothermally upto the point B and then adiabatically to C where the temperature drops to T and vapour pressure to e_s. From C, the vapour is compressed isothermally to D from where an

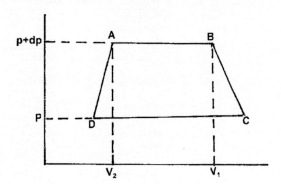

Fig. 4.2 The Carnot cycle for water vapour

adiabatic compression restores the system to its original state. During the process of isothermal compression (from C to D), a quantity of heat equal to the latent heat of condensation of vapour L (T) is released.

If all second-order quantities and the small differences of heat involved in passing from one temperature state to the other are neglected, the work done by the system during the cycle may be approximated to $de_s (v_1 - v_2)$, where v denotes specific volume and the subscript 2 refers to the liquid and 1 to the vapour. Now, according to the second law of thermodynamics, this work was done by a fraction of the heat supplied to the system, viz., by L (T) dT/T, since L (T+dT) ≈ L (T) and T+dT ≈ T.

Thus, equating the work done to the heat available, we arrive at the relation,

$$de_s/dT = L(T)/\{T(v_1 - v_2)\} \tag{4.8.1}$$

This is the celebrated Clausius-Clapeyron equation.

Before proceeding further with the above Eq. (4.8.1), we make the assumptions that v_2 is negligible compared to v_1 (since v_1 is 1674 times larger than v_2) and that the ideal gas law is applicable to the case of the saturated water vapour. The simplifications lead to the relation

$$de_s/dT = e_s L(T)/R_v T^2 \tag{4.8.2}$$

where R_v denotes the gas constant for saturated water vapour ($= R^*/M_v$).

The Eq. (4.8.2) can be integrated if we know the variation of L (T) with temperature. For this, we treat the specific heats c_p' of water vapour and c_w of liquid water as constant and consider the liquid to evaporate at temperature T = 0 and pressure e_s and the resulting vapour to warm up from zero to a temperature T. Alternatively, we may first warm up the water from temperature 0 to T and then evaporate it at T. Under these conditions, we have,

$$L(0) + c_p' T = c_w T + L(T)$$

Or,
$$L(T) = L(0) + (c_p' - c_w)T \tag{4.8.3}$$

where L(0) is the latent heat of vaporization at T = 0.

Substitution of this value of L (T) in (4.8.2) gives

$$de_s/dT = e_s\{L(0) + (c_p' - c_w)T\}/R_v T^2 \tag{4.8.4}$$

The Eq. (4.8.4) may now be integrated to yield

$$\ln e_s = -\{L(0)/R_v T\} + [\{(c_p' - c_w)\ln T\}/R_v] + A \tag{4.8.5}$$

where A is a constant.

Or,
$$e_s = A \exp\{-L(0)/R_v T\} T^{(c_p' - c_w)/R_v} \tag{4.8.6}$$

4.9 Co-existence of the Three Phases of Water – the Triple Point

Fig. 4.3 Saturation vapour pressure over a plain surface of water at different temperatures

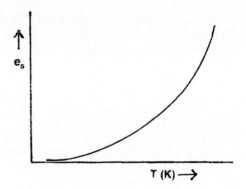

The influence of the second factor involving T being negligible, (4.8.6) states that the saturation vapour pressure of water increases rapidly, almost exponentially, with temperature.

In deriving (4.8.5), it was assumed for simplicity that the specific heats of the vapour or the liquid phase were constant. However, this assumption is not quite true. If this restriction is withdrawn, we arrive at a slightly different form of the saturated vapour pressure vs temperature relation. For derivation of this revised form, see Appendix 4.

The variation of saturation vapour pressure with temperature over a plain surface of water is shown in Fig. 4.3.

4.8.2 Melting Point of Ice – Variation with Pressure

The Clausius-Clapeyron equation (4.8.1) is exact and may be applied to the case of the variation of the melting-point of ice with pressure. For this, we invert the Eq. (4.8.1) and obtain

$$dT/de_s = T(v_2 - v_3)/L_e(T) \qquad (4.8.7)$$

where we write $L_e(T)$ for the latent heat of melting of ice (or fusion of water) and v_3 for the volume of 1 mol of ice. Since the specific volume of ice is greater than that of water, i.e., $v_3 > v_2$, the right-hand side of (4.8.7) is negative. This means that the melting-point of ice is lowered by increase of pressure.

4.9 Co-existence of the Three Phases of Water – the Triple Point

The Clausius-Clapeyron equation may be applied to study the variation of saturated vapour pressure with temperature between any two phases of water, for example, between vapour (V) and liquid water (W), or between liquid and solid ice (I), or even directly between vapour and solid. For this purpose, we integrate Eq. (4.8.2) from the initial temperature 273.16A where the saturation vapour pressure is experimentally known to be 6.11 mb, to temperature T, by noting that the latent heat of a water substance does not vary appreciably within the ranges of temperature normally encountered in the atmosphere (see Eq. 4.8.3) during a change from one phase to the other. The resulting approximate expressions for the various changes of phase are the following:

(i) **From vapour to liquid (condensation)**

$$\ln(e_s/6.11) = (L/R_v)\{(1/273.16) - (1/T)\} \tag{4.9.1}$$

where L is the latent heat of condensation of water vapour(or vaporization of water). Its value at 273.16 A is 2.496×10^6 J kg^{-1}.

(ii) **From liquid to solid (fusion)**

$$\ln(e_s/6.11) = (L_e/R_v)\{(1/273.16) - (1/T)\} \tag{4.9.2}$$

where L_e is the latent heat of fusion of water(or melting of ice). Its value at 273.16A is 3.33×10^5 J kg^{-1}.

(iii) **From vapour to solid (sublimation)**

$$\ln(e_s/6.11) = (L_z/R_v)\{(1/273.16) - (1/T)\} \tag{4.9.3}$$

where L_z is the latent heat of sublimation from vapour to solid (or solid to vapour). Its value at 273.16 A is 2.829×10^6 J kg^{-1}.

Figure 4.4 is a phase diagram which shows the saturation vapour pressure, e_s, plotted against absolute temperature T for all the above-mentioned changes of phase. It shows that the curves representing the three phases meet at a point P which we call the triple point. At P, all the three phases co-exist

The curves separating the three phases are as follows: The curve AP representing the change from the liquid to the vapour phase extends from the triple point to

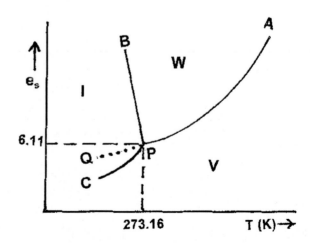

Fig. 4.4 Phase diagram in respect of water vapour – the Triple point

the critical point where the distinction between the liquid and vapour phases disappears. When the saturation vapour pressure equals the outside barometric pressure, water reaches its boiling point. The curve PB represents melting between water and ice. Since the specific volume of ice is greater than that of water, it follows from Eq. (4.8.1) that the melting line is negative and slightly curved. The curve PC which curves downward from P is the sublimation curve between the ice and the vapour phases. It may be noted that near P, the curvature of CP is greater than that of PA. This is due to the fact that the latent heat of sublimation of water vapour is greater than that of its condensation near the triple point. This follows from Eq. (4.8.1).

It is a fact of observation that water does not freeze immediately on cooling below 273.16A (or $0\,°C$) temperature. It can stay in the supercooled liquid phase down to a temperature of about 233 A. However, the degree of freezing increases with lowering of temperature. The saturation vapour pressure over supercooled water is not given by the sublimation curve PC but by the dotted curve PQ in Fig. 4.4, which may be regarded as an extension of the curve AP to lower temperatures. An interesting aspect of the curve PQ is that the saturation vapour pressure over supercooled water is greater than that over ice at temperatures below 273.16A. This means that when supercooled water drops and ice crystals co-exist in the upper layers of the atmosphere above the freezing level, water will evaporate from the supercooled drops and condense on the ice crystals, thereby leading to the growth of the ice crystals at the expense of the drops. This is the celebrated Bergeron-Findeisen mechanism for the formation of large raindrops in the atmosphere. More about this mechanism will be presented in Chap. 5.

4.10 Stability of Moist Air

In Chap. 3, Eq. (3.4.14), we derived the condition for the vertical stability of an atmosphere in which the presence of water vapour was ignored and the atmosphere was treated as totally dry. But the fact is that the terrestial atmosphere always holds some moisture which, when lifted, would produce condensation at some level which is called the lifting condensation level, whatever may be the mixing-ratio. A parcel of air rising in such a moist atmosphere would follow the dry adiabatic lapse rate only upto the level where condensation begins and the pseudoadiabatic lapse rate thereafter. More often than not, the actual atmosphere has a lapse rate which lies between the dry adiabatic and the saturated adiabatic lapse rates, rendering the atmosphere stable with respect to the dry adiabatic lapse rate but unstable with respect to the saturated adiabatic lapse rate. The atmosphere then is said to be conditionally unstable. Observations reveal that the tropical atmosphere tends to remain in conditionally unstable state most of the time. This is due to the fact that the boundary layer of the atmosphere exchanges heat and moisture with the surface of the earth, whereas the upper layers are relatively less affected by surface conditions. The degree of conditional instability, however, varies with location and season. For example, the lower troposphere over the continents is relatively cold and dry during the winter but warm and moist during the summer. In general, the seasonal difference makes the tropical atmosphere over the continents conditionally less unstable during the winter than during the summer.

4.10.1 Thermodynamic Diagrams

Several thermodynamic diagrams have been devised to study the static stability conditions of the atmosphere. Stability parameters in these diagrams vary but they all seem to have the same common objective: to find out by comparing the environment temperatures at different heights with the dry and moist adiabats at those heights whether stable and unstable conditions exist in any layer, and then, in some diagrams, if the atmosphere is conditionally unstable, to assess the amount of net instability energy that may be available for vertical development. Experience with these different diagrams shows that not all of them are suitable or convenient for practical use. In Appendix-3, we give brief particulars of a few thermodynamic diagrams which are in wide use. The T-S diagram (also known as the T-φ diagram) which is most widely used in meteorology is described in detail, while the main specifications only are given of others.

In the T-S diagram (see Fig. 3.1'), the abscissa is temperature T and the ordinate is entropy S (which is identified with $c_p \ln \theta$). The dry adiabats (Dry Adiabatic Lapse Rate of temperature, DALR) are horizontal, while the saturated adiabats (Saturated Adiabatic Lapse Rate of temperature, SALR) are curved lines sloping upward from right to left, the slopes varying with temperature, pressure and humidity-mixing ratio. The pressure (p) lines or isobars slope downward from right to left and space out with height, while the humidity-mixing-ratio (x) lines are indicated by broken lines. It is easy to obtain a rough estimate of the instability energy that can be realised from the atmosphere with a given case of radiosonde sounding from this diagram.

None of the thermodynamic diagrams in use, however, take into account the likely effect of water vapour on the vertical stability of the atmosphere. Some thermodynamic diagrams use the vertical profiles of equivalent potential temperature or wet-bulb potential temperature. The conditions for stability in such diagrams are found by following the standard procedure of comparing the potential temperature of the environment at any level with that of a parcel of air lifted pseudo-adiabatically from a lower level to that level, assuming that both the parcel and the environment possessed the same potential temperature at the lower level. In this procedure, let the potential temperature of the environment at level z_0 be θ_0. Then, at level $z_0 - \delta z'$, the potential temperature is $\theta_0 - (\partial \theta / \partial z) \delta z'$. Let us now lift a parcel of the environment air from this lower level to the level z_0. Then, if we denote the potential temperature of the parcel at level z_0 by θ_1,

$$\theta_1 = \theta_0 - (\partial \theta / \partial z) \delta z' + \delta \theta \qquad (4.10.1)$$

4.10 Stability of Moist Air

where $\delta\theta$ is the difference in the potential temperature of the parcel at the pseudo-adiabatic lapse rate between the two levels.

We can evaluate this difference from Eq. (4.7.2) written in the form

$$\delta\theta/\theta \simeq -\{\partial(L\,x_s/T\,c_p)/\partial z\}\,\delta z' \qquad (4.10.2)$$

Substitution for $\delta\theta$ from (4.10.2) in (4.10.1) gives the following expression which is proportional to the buoyancy of the parcel at level z_0,

$$(\theta_1 - \theta_0)/\theta_0 \simeq -[(1/\theta_0)(\partial\theta/\partial z) + (\theta/\theta_0)\{\partial(Lx_s/Tc_p)/\partial z\}]\,\delta z' \qquad (4.10.3)$$

If we now visualize a hypothetical fully-saturated atmosphere and write θ_e^* for θ_e in Eq. (4.7.3), we obtain

$$d\ln\theta_e^* = d\,\ln\theta + d(L\,x_s/Tc_p) \qquad (4.10.4)$$

If we assume that the parcel temperature is not too different from that of the environment when it arrives at level z_0, we may write (4.10.3) as

$$(\theta_1 - \theta_0)\theta_0 \simeq -[(1/\theta)(\partial\theta/\partial z) + \{\partial(Lx_s/Tc_p)/\partial z\}]\,\delta z' \qquad (4.10.5)$$

We use (4.10.4) to obtain the following buoyancy relationship in terms of the equivalent potential temperature of the saturated atmosphere,

$$(\theta_1 - \theta_0)/\theta_0 \simeq -(\partial\,\ln\theta_e^*/\partial z)\,\delta z' \qquad (4.10.6)$$

From (4.10.6), we observe that the parcel will be positively buoyant at z_0 if $\theta_1 > \theta_0$.

So we arrive at the following stability criteria:

$$(\partial\theta_e^*/\partial z) \quad \begin{array}{l}(<0 \text{ conditionally unstable})\\ (=0 \text{ Neutral})\\ (>0 \text{ absolutely stable})\end{array} \qquad (4.10.7)$$

Vertical profiles of potential temperature (θ) and equivalent potential temperature (θ_e) in the mean tropical atmosphere in the West Indies area (Jordan, 1958) are shown in Fig. 4.5, along with that of the equivalent potential temperature (θ_e^*) that would result if the same atmosphere were hypothetically saturated. While θ is found simply by reducing the observed temperature dry-adiabatically to 1000 mb, θ_e is found by lifting a parcel of air with its existing mixing-ratio x dry-adiabatically to saturation and then lifting it further pseudo-adiabatically to a level till all the water vapour condenses out and then reducing it dry-adiabatically from that level to 1000 mb. θ_e^* is calculated similarly as θ_e but with x replaced by x_s (the saturation value of the humidity-mixing-ratio at each level).

It is obvious from Fig. 4.5 that the tropical atmosphere is conditionally unstable in the lower and middle troposphere and stable above. However, it does not follow that this latent instability leads to convective overturning automatically. Since the humidity most often is less than 100%, low-level convergence is needed to lift the

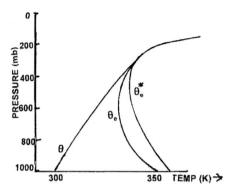

Fig. 4.5 Vertical profiles of potential temperature (θ) and equivalent potential temperature (θ_e) in the mean tropical atmosphere in the West Indies area (Jordan, 1958). The third profile gives the equivalent potential temperature (θ_e^*) of a hypothetically saturated atmosphere with the same p and T values at each level (Reproduced from Ooyama, 1969, with permission of American Meteorological Society)

unsaturated air to saturation. It is not surprising, therefore, that most of the tropical storms and cyclones originate and develop over the oceans. This is largely due to the fact that most of the moisture evaporated from the oceans remains stored in the boundary layer till a low pressure or depression arrives and produces the required low-level convergence to lift the moist air to higher levels. Experience shows that in the atmosphere, the low-level convergence of moisture must be supported by upper-level divergence for any deep convection to occur for development of severe local storms as well as synoptic-scale deep depressions and cyclones.

Chapter 5
Physics of Cloud and Precipitation

5.1 Introduction – Historical Perspective

It is generally believed that the formation of cloud and rain in the atmosphere is part of the simple hydrological cycle and that it can be easily explained and predicted by simple thermodynamical laws, as applied to a mixture of air and water vapour. But, in reality, it is not quite so. There is no guarantee that the water vapour that leaves the ocean surface and rises in the atmosphere is going to produce cloud and rain. In some situations, no clouds may form at all, and even if they do, they do not precipitate. More often than not, clouds may be seen floating in the sky for days together without shedding even a drop of rain and then suddenly there may be a cloudburst or they may completely disappear bringing back clear skies. These vagaries of Nature have puzzled humankind throughout the ages.

Early men who wondered about the above-mentioned vagaries of Nature especially in times of prolonged drought tried to influence nature in their favour by all possible methods. The earliest ones were those of magic. The African tribes, even till recently, had professional rain-makers, who at times when failure of clouds to shed rain became intolerable, performed magic rites accompanied by the blowing of horns and beating of drums to induce the spirit behind the clouds to shed the rains. If they were successful, the rain-maker would be greatly honoured and in some cases might be made the chief of the tribe, but, if unsuccessful, he would be as often sacrificed to appease the spirit behind the clouds. Similar rites were performed by other people in many other lands as well.

In the Rigvedas, the oldest extant literature of the Aryans, there are many beautiful verses describing the fight between the rain-god Indra who was also the king of Gods and the demon Vritra who was blamed for holding up rains from clouds for days together, until Indra, with his thunderbolt, would kill the demon, and the clouds would give beneficient and much-wanted rain. On the day of the summer solstice, the traditional day when monsoon rain was supposed to burst, a high flag would be raised in honour of Indra. When a region would suffer from continuous drought for successive years, it would be ascribed to some sin committed by the king or the people, and elaborate sacrifices would be undertaken to appease the wrath of Indra.

Scientific studies of the formation of cloud and rain may be said to have begun with laboratory experiments in the latter half of the nineteenth century and the early part of the twentieth century by such eminent investigators as John Aitken and C.T.R. Wilson in England, Coulier in France, Richarz in Germany, Kohler in Sweden, and several others. The laboratory studies clarified several important aspects of the process of condensation and formation of clouds. However, it was soon realized that conditions in the open atmosphere were not quite the same as in the laboratory. So, in recent years, attempts have been made to take measurements in the actual atmosphere, not only at or near the earth's surface but also at several levels of the atmosphere, using aircraft and satellites and remote-sensing techniques. The data so collected as well as the results of the laboratory experiments have led to an improved understanding of the physical processes involved in the formation of cloud and rain in the atmosphere. Also, in recent years, attempts have been made to induce clouds to shed rain, especially in drought-prone areas where rain is badly needed, by artificially seeding them with particles which are believed to aid the precipitation process in natural clouds. Both warm and cold clouds have been treated by releasing seeding materials from ground-based and airborne generators. Results of some of these experiments will be reviewed. In the latter part of the chapter, we give examples of a few common types of clouds which are observed in the sky and classified into four groups. These were first published in the 'International Cloud Atlas', by the International Meteorological Committee in 1932 which classified, catalogued and beautifully illustrated practically all types of clouds observed till then. The reader interested in more details about clouds may refer to this cloud atlas.

5.2 Cloud-Making in the Laboratory – Condensation Nuclei

Aitken (1923)'s method was to expand suddenly a closed volume of air standing over a water surface containing saturated water vapour into a larger volume. This is not exactly what happens in Nature, for when air saturated with water vapour goes up, say to a height of 2 km, the pressure falls and the mass of air expands gradually in volume and the expansion is not large. In the laboratory experiment, the expansion is sudden, and generally large.

However, in the atmosphere as well as in the laboratory process, the temperature falls and the air becomes super-saturated, i.e., it contains more water vapour than it can normally do and the excess amount of vapour is expected to be condensed in the form of small water droplets forming a fog-like cloud. This was actually found to be the case in the laboratory experiment; but what surprised Aitken and other physicists was the fact that if expansion was done with the same airmass in a limited volume,

5.2 Cloud-Making in the Laboratory – Condensation Nuclei

a number of times, then after a few expansions no fog would be formed even though these expansions were quite sufficient for producing supersaturation.

This was traced to the surprising fact that dust-particles which are normally present in the air are actually essential for the formation of a fog or cloud. In fact, if the expansion experiments are carried out from the very start with dust-free air, such as is obtained by sucking air through glass wool, no fog would be formed even at the first expansion. In the experiments of Aitken, ordinary air was taken which contains enough dust particles. After a few expansions, these are all precipitated. The air becomes dust-free and no further fogs are formed after expansion. The explanation of this surprising fact was speedily forthcoming.

Lord Kelvin (1870) had shown from thermodynamical considerations that the saturation vapour pressure of liquids over a curved surface, such as that of a small drop, of radius 'r', was considerably higher than that over a plane surface on account of the surface tension of liquids. The relation (a theoretical derivation of this relation is presented in Appendix-5) is

$$\ln(e_r/e_s) = (2\,\sigma/r)/(nkT) \qquad (5.2.1)$$

where e_r and e_s are the saturated vapour pressures over the curved surface and the plane surface respectively, σ is the surface tension of water, n is the number of molecules per unit volume, k is Boltzmann constant and T is the temperature in degrees Absolute.

For a water drop at $10°C$, the values of (e_r/e_s) calculated from the above formula for different dropsizes are given in Table 5.1.

Table 5.1 emphasizes the need of existence of nuclei of condensation. The H_2O-molecule has a radius of 2×10^{-8} cm. Ordinarily, in the process of cooling, nearly 100 molecules of water must come together if they were to form a tiny droplet of radius $\sim 10^{-7}$ cm. But this would not be stable, as the above figures show unless the degree of supersaturation exceeds 3.1, i.e., the cooled atmosphere contains three times more water vapour than is given by the saturation vapour pressure curve (Fig. 4.3). In the absence of such supersaturation, the drop evaporates as soon as it is formed. But if a dust particle having a radius of 10^{-6} cm is present, water molecules depositing on it form a droplet of radius of 10^{-6} cm and now the supersaturation needed is only 1.12 which is generally to be found in cooled air. So, droplets formed by deposition of water molecules on dust particles will continue to grow. The laboratory experiments prove that some kind of nuclei must exist if cooled water vapour is to form fogs or clouds under atmospheric conditions.

Table 5.1 Values of e_r/e_s for drops of water

r (cm)	2×10^{-8}	10^{-7}	10^{-6}	10^{-5}	10^{-4}
e_r/e_s	316.2	3.162	1.127	1.012	1.001

5.3 Atmospheric Nuclei – Cloud Formation in the Atmosphere

What are actually the nuclei on which clouds form in the atmosphere?

Dust is a very vague term which denotes minute particles of earth consisting mostly of sand or quartz carried upward by wind. In addition, there may be other types of particles which may act as nuclei for condensation. These are particles constituting smoke from industrial cities, particles of salt, like sodium chloride and magnesium chloride which are carried by wind over sea surfaces in the form of spray; these evaporating in the atmosphere, leave nuclei of minute particles of salt. There may be, in addition, particles composed of oxides of nitrogen or SO_3, which are formed by the action of sunlight on nitrogen and oxygen molecules, or on sulphur nuclei which are found to exist in industrial areas.

Some of these nuclei are hygroscopic, i.e., can easily draw water to them. The supersaturation needed for formation of drops on these hygroscopic nuclei is much smaller than on dust nuclei. In fact, laboratory experiments show that hygroscopic particles can gather moisture round them at relative humidities much below 100%. Owens in 1926 in a paper in the Proceedings of the Royal Society of London described occasions when nuclei began to draw moisture at a relative humidity of 74%. How do particles grow under such conditions? The earliest experiments carried out to answer this question were those of Köhler (1926) and the trend of his results is shown in Fig. 5.1.

The curve in Fig. 5.1 serves to show that condensation increases with rise of relative humidity slowly at first, but very rapidly when the relative humidity approaches 100%, or exceeds this value. It is natural to ask at this stage to what extent condensation can proceed on hygroscopic nuclei and what will be the order of the size of a drop formed in this way. From our previous considerations we can try an answer to this question. In the initial stages of absorption of water vapour the saturation vapour pressure on the surface of the drop decreases with increase of size, but at the same time the reduction of the salt concentration causes a rise in the saturation vapour pressure. Thus two opposing forces come into play at the surface of the growing nucleus and further condensation stops when they balance each other. The maximum size of a droplet formed in this way has been estimated to be of the order of 10^{-4} to 10^{-3} cm in radius which is usually the order of size of an atmospheric fog or cloud droplet.

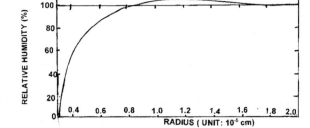

Fig. 5.1 Size of condensation nuclei at different relative humidities (After Köhler, 1926)

5.3 Atmospheric Nuclei – Cloud Formation in the Atmosphere

There is yet another factor which we have to consider while studying the physics of drop formation. It is the effect of electric charge on the condensation of vapour on a drop. J.J. Thomson showed that if a drop contains an electric charge 'E', the saturation vapour pressure on its surface is reduced. The relation is given by

$$\ln(e_r/e_s) = \{(2\sigma/r) - (E^2/8\pi r^4)\}/(nkT) \tag{5.3.1}$$

where the symbols have the same meanings as in (5.2.1). If we put the right-hand side of (5.3.1) equal to zero, it is easy to show that for every value of the radius there is what is called a critical charge which will make $e_r = e_s$, i.e., it will reduce the saturated vapour pressure over the drop to that over a plane surface. The value of the critical charge calculated for different drop sizes is given in Table 5.2.

C.T.R. Wilson in 1897 using his cloud chamber demonstrated the role of electrically charged particles in the formation of condensation nuclei in a supersaturated atmosphere. He sent a high-energy charged particle like α-ray from radium through a supersaturated vapour and by strongly illuminating the chamber photographed the track of the α-ray inside the chamber. This could be done because in passing through the gases inside the chamber the α-ray owing to its tremendous energy knocked off electrons from the gaseous molecules and it was these electrons which because of their electric charge rapidly gathered moisture round them and formed the minute droplets which constituted the path of the α-ray under strong illumination. Because of high electronic density, the saturation vapour pressure on the ions initially formed was considerably reduced and thus the supersaturation prevailing inside the chamber was sufficient to produce rapid condensation on them to form the visible drops.

What is the order of electronic charges that can aid rapid condensation on atmospheric nuclei? According to Table 5.2, the critical charge for a drop of radius 10^{-6} cm, which is the order of the size of average nuclei in the atmosphere, is as large as 130. Multiple electronic charges of this high order are seldom met with in the atmosphere. There is evidence of multiple electronic charges on fog droplets and rain drops but the charge on atmospheric nuclei, at least in the early stages of condensation, seldom exceeds one electronic charge. It is, therefore, rather unlikely that the effect of electric charge plays any important part in the formation of cloud drops in the atmosphere.

It is, therefore, recognized that some kind of 'nuclei' is necessary for the condensation of water vapour into droplets which composes a cloud, but which kind of nuclei – quartz particles (dust), Na Cl crystals obtained from sea-spray, or nitrous crystals play the predominant part is not yet decisively known. It is, however, found that only a small and variable fraction of the total number of particles that are measured in the continental and marine air act as cloud condensation nuclei. Observations made over the various parts of the globe do not suggest any systematic

Table 5.2 Critical electric charges for drops

r (cm)	4×10^{-8}	2×10^{-7}	4×10^{-6}	9×10^{-5}	2×10^{-3}	4×10^{-2}
E (charge)	1	10	10^3	10^5	10^7	10^9

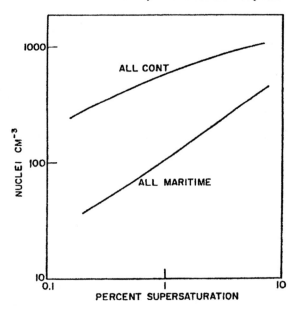

Fig. 5.2 Concentration of Cloud condensation nuclei (CCN) in continental and marine air near surface at different supersaturation (After Twomey and Wojciechowski, 1969, with permission of American Meteorological Society)

variations in the concentration of cloud condensation nuclei (CCN) with latitude or seasons. However, observations taken near the earth's surface show that the average concentration of CCN is much higher over continents than over oceans, as revealed by Fig. 5.2 reproduced from Twomey and Wojciechowski (1969).

However, the concentration of CCN over the land is found to fall off with height by a factor of about 5 between the surface and 5 km, whereas that over the ocean remains more or less constant with height. There also appears to be a diurnal variation in the concentration of CCN near the surface with a minimum in the morning and maximum in the evening.

5.4 Drop-Size Distribution in Clouds

Further elucidation of the problem depends upon our knowledge of the size of droplets forming clouds, and the actual size and nature of the nuclei on which these droplets are formed. 'Clouds' denote a wide variety of types, from low-lying fogs or stratus clouds which seldom give any precipitation to heavy rain-clouds enormous in extent and yielding large precipitation. Meteorologists have invented a system of classification which has been accepted by the International Meteorological Committee and is published in the International Cloud Atlas. The classification depends mostly on external physical appearance, levels of formation in the atmosphere (e.g., low, medium or high) and composition (whether they are made up of water droplets, or ice particles, or a mixture of both).

The point we have to discuss is why some clouds vanish without giving rain, while others give copious precipitation. This is answered by a close study of the distribution of droplet sizes in clouds.

5.5 Rate of Fall of Cloud and Rain Drops

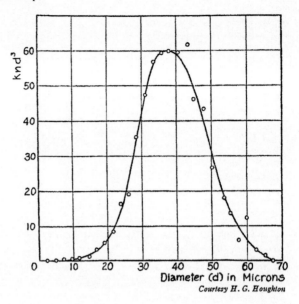

Fig. 5.3 Drop-size distribution in fog and low stratus clouds (Reproduced from G.F. Taylor, Aeronautical Meteorology, 1941, p 261)

Measurements have been made of the size of droplets in clouds by aeroplane ascents and other methods. They are found to range in radius from 10 to about 60 microns (1 micron = 10^{-4} cm, usually denoted by the symbol μ), with a mean value of about 20μ, in a fog or low-lying stratus cloud.

Figure 5.3, originally due to H.G. Houghton, shows the distribution of sizes of drops in a low-lying fog, as presented by Taylor (1941).

The droplets which are precipitated as rain have much larger radii. They range in value from 100μ to 1/4 of a centimetre. Drops larger than these generally get broken by air during the fall. Thus the difference between clouds that hover and disappear without giving any rain and those which give rain is, therefore, entirely one of the dimension of the drops of which they are composed, the hovering clouds consisting of droplets less than 100 microns in diameter, and the rain-giving clouds consisting of larger drops. We can understand the cause of this differential behaviour of the particles with the aid of the well-known Stokes' law which deals with the rate of fall of small particles through a viscous fluid, such as the air.

5.5 Rate of Fall of Cloud and Rain Drops

As soon as a drop is formed, it begins to fall, under gravity, but its fall is resisted by the viscous drag of the air. According to Stokes' law, the rate of fall of a very small particle of radius 'r' and density ρ through a fluid of density •, which is called the terminal velocity, is given by

$$v = 2g\, r^2 (\rho - \bullet)/9\,\mu \qquad (5.5.1)$$

Table 5.3 Particle sizes and their velocity and distance of fall

	Size of particles Diameter (μ)	Velocity of fall (cm s^{-1})	Time of fall through 1 km
Molecules	4×10^{-4}	–	–
Ions and nuclei	$(1–100) \times 10^{-3}$	–	–
Cloud particles	(4–20)	$(5 \times 10^{-2}–1.3 \times 10^0)$	21 hour
Drizzle	200	78	20 minutes (m)
Light rain	(450–600)	(200–260)	8 minutes
Heavy rain	$(1.5–2.0) \times 10^3$	(500–600)	3 minutes
Cloud burst	3.0×10^3	700	2.4 minutes
Largest raindrop	5.0×10^3	800	2 minutes

where v denotes the velocity, g is acceleration due to gravity and μ is molecular viscosity of the fluid.

For a cloud droplet of about 10μ in radius, the terminal velocity, as calculated from (5.5.1) is 1.3 cms^{-1} and it takes about 21 h to fall through a height of 1 km. It, therefore, evaporates before it reaches the ground. The smaller the particle, the more slowly it falls, and, therefore, more quickly it evaporates.

Table 5.3, adapted from a paper by Simpson (1941), makes these points clear. It shows that the largest raindrop that falls through the air at a velocity of 8 ms^{-1} has a diameter of 5 mm. It falls through a height of 1 km in about 2 min.

5.6 Supercooled Clouds and Ice-Particles – Sublimation

Not all clouds are, however, made up of water droplets. Observations show that high-level cirrus clouds, appearing at heights of 6 to 12 km, where the temperature is well below the freezing point (see chap. 2), are composed mostly of ice-particles. Between clouds consisting entirely of water droplets and those consisting entirely of ice particles, there are, however, many other types, composed partly of water droplets and partly of ice. The surprising fact is that clouds consisting even entirely of water droplets are found on high mountain tops and in aeroplane ascents even when the temperature is much below the freezing point) and are found to be of the same size as the fog droplets. These droplets are 'supercooled' and are, therefore, in unstable equilibrium. They generally transform themselves into ice-particles as soon as they strike against any hard surfaces or obstacles, like aeroplane sides.

How are these ice-particles formed? Do they require nuclei for condensation (we may call them 'sublimation nuclei'), just as water-droplets do at above-freezing temperatures? Why does not water vapour sublime directly to form ice particles when the temperatures are much below freezing point, but prefer, as observations seem to show, to condense to form supercooled drops which may persist indefinitely?

5.6 Supercooled Clouds and Ice-Particles – Sublimation

These questions also occurred to C.T.R. Wilson in 1897 when he repeated Aitken's cloud-chamber experiments. It is only in recent years that these questions have been satisfactorily tackled by means of laboratory experiments.

It is assumed that the cloud particles are actually liquid drops and not ice crystals, inspite of the fact that the condensation begins at temperatures much below the freezing point ($-15\,°C$) and the temperature when the particles are fully grown is, as we shall see, also slightly below the freezing point.

The laboratory experiments on the formation of ice crystals were taken up by Regener and Findeisen in Germany and by Cwilong (1947) a Polish refugee physicist working in the Clarendon Laboratory, Oxford, under Professor G.M.B. Dobson, during the World War II. The results of Cwilong given below are highly interesting for all theories of cloud formation:

The expansion chamber was placed in a cooling bath between $0\,°C$ and $-10\,°C$, with supercooled liquid water inside. The air was freed completely from dust. The following observations were recorded:

(1) As long as the expansion was less than 1.25, no fogs were formed, even when the temperature in the chamber fell to as low a value as $-41.2\,°C$. But when the temperature fell below $-41.2\,°C$, a shower of ice-particles was formed. (2) When the expansion ratio was between the Wilson limits, 1.25 and 1.38, only water droplets were formed as long as the chamber temperature is above $-41.2\,°C$. Apparently they were formed round negative ions. When T was less than $-41.2\,°C$, ice particles were also formed besides water droplets, and the proportion of ice- particles increased as the temperature was made lower. When the expansion ratio was more than 1.38, the C.T.R. Wilson limit for condensation on ions of all signs, and T was below $-41.2\,°C$, dense fogs consisting both of supercooled droplets and ice-particles were formed. With air containing dust, water droplets were formed at $-32\,°C$ at small expansion, but when temperature fell below $-32\,°C$, dense fogs containing both water and ice particles were found to be formed.

On first considerations, the results appear to be highly puzzling and unexpected, but they have been found to be in agreement with cloud observations. These observations have been confirmed by Schaefer (1946) working in the U.S.A. However, the questions that naturally arise here are:

(i) Why in the absence of sublimation nuclei, water vapour does not form into ice crystals until as low a temperature as $-41.2\,°C$ is reached?
(ii) How does the introduction of some kind of sublimation nuclei help in the formation of ice-particles at temperatures higher than $-41.2\,°C$?
(iii) What kind of nuclei is most effective as 'sublimation centres'?

All these questions have not yet been answered satisfactorily. An attempt is made here to answer them. In regard to (i), it is commonly said that water freezes into ice below the freezing point, but Langmuir in an article published in Fortune (Feb, 1947) thinks that this is not a correct description of facts. Very pure water has been shown to be capable of being supercooled to a temperature of $-50\,°C$. A more correct description would, therefore, be that once formed, ice cannot exist above $0\,°C$. According to Langmuir, the spontaneous condensation of water droplets into ice particles below $-41.2\,°C$ can be explained as follows: When we have a drop of water of radius r, the pressure inside the drop due to surface tension is equal to $2\,\sigma/r$; if we consider a droplet formed of 12 molecules of water, $2\sigma/r$ amounts to a pressure of 2000 atmospheres. Now Bridgmann has shown that at about $-36\,°C$, and 2000 atmosphere pressure, a variety of ice crystals called Ice-II appears. This is probably what happens at $-41.2\,°C$. If 12 molecules of water vapour come together at these temperatures, they form a unit crystal of Ice-II, and once such a sublimation nucleus is formed, water molecules deposit on it and the crystal grows. When it has grown sufficiently large, the Ice II transforms spontaneously into Ice-I.

As regards (ii) and (iii), probably the presence of a sublimation nucleus, on which a crystal of ice grows, helps in the prevention of evaporation just as in the case of water droplets. However, experience has shown that all kinds of nuclei are not equally effective, for injection of particles of quartz, salt, and many other substances were found to have no effect on production of ice particles in supercooled spaces. Cwilong found that supercooled drops do not transform into ice crystals when deposited on sufaces of mica, even at temperature of $-100\,°C$ but readily do so on zinc surfaces. Apparently, the nature of the surface and the crystal structure of the sublimation nuclei play a great role in this business.

Vonnegut, working in the M.I.T. laboratories, found in 1947 that crystals of silver Iodide (AgI) are particularly effective in promoting the growth of ice crystals. In fact, AgI crystals are reported to have been used with some success in artificial rain-making. He was led to the choice of AgI from his knowledge of principles of crystal growth that for the growth of crystals of some particular substance, say water, substances possessing crystal lattices of similar structure and nearly identical dimensions are helpful, and minute dusts of such crystals can act as very efficient sublimation centers for crystal growth.

5.7 Clouds in the Sky: Types and Classification

The 'International cloud atlas' of 1932 divides clouds into various families, genera, and species, but here we give only a few examples of cloud types which are more commonly observed. Broadly, clouds are classified into four families depending upon the height and the layer of their formation. These are high clouds (Fig. 5.4(a–c)), medium clouds (Fig. 5.4(d–e)), low clouds (Fig. 5.4(f–h)) and clouds with great vertical development (Fig. 5.4(i)). Table 5.4 gives a list of cloud types commonly observed in the sky.

5.8 From Cloud to Rain

But how do the cloud drops, whether water or ice, grow into rain drops? In Nature, the phenomenon of condensation cannot help us much, as it gives us particles which are so small that they hover and evaporate before reaching the ground. There must

5.8 From Cloud to Rain

be some process by means of which the cloud droplets must grow into rain drops. Taking the average radius of a cloud drop at 20 microns, we find that more than 125 of them are required to form the smallest rain drop, 100 microns in radius. How does this take place in Nature; by coalescence or by growth?

These matters have been hotly debated amongst meteorologists and the following processes have been proposed for the growth of cloud droplets into rain drops:

5.8.1 Hydrodynamical Attraction

Defant (1905), from measurement of raindrop sizes concluded that a large majority of raindrops he measured grouped themselves chiefly in the mass ratios 1:2:4:8. Coalescence of droplets of the same size was given as the explanation of this observation. W.Schmidt (1908) attributed the grouping to hydrodynamical forces and showed that if two droplets of the same size fall side by side a reduction of pressure occurs between them according to the Bernoulli principle and the droplets collide after descent through a certain height. Stickley (1940) has used Schmidt's original equation to compute the time required for collision using different values of concentration of particles inside clouds. These computations showed that with an average cloud drop concentration, it will require more than 7 days for drops of radii

Fig. 5.4 Pictures of clouds: (**a**) Cirrus cloud (NOAA Photo Library); (**b**) Cirrocumulus clouds http://www.windows.ucar.edu/tour/link=/earth/Atmosphere/clouds/cirrocumulus.html; (**c**) Cirrostratus clouds (Note halo around the sun); http://www.eo.ucar.edu/webweather/cirrus.html; (**d**) Altocumulus (NOAA Photo Library); (**e**) Altostratus (http://www.eo.ucar.edu/webweather/alto.html); (**f**) Stratus (NOAA Photo Library); (**g**) Fair weather cumulus (NOAA Photo Library); (**h**) Stratocumulus (NOAA Photo Library); (**i**) Cumulonimbus cloud (Note anvil-shaped top with mantle) (NOAA Photo Library)

Fig. 5.4 (Continued)

5.8 From Cloud to Rain

Fig. 5.4 (Continued)

10^2 microns to form and 75 days for drops of radii 10^3 microns. Even in the case of heavy cumulus in which the particle density is normally very high, it was found that more than three hours would be required for the formation of drops of radii 10^2 microns to form, and over 32 h for drops of radii 10^3 microns. It, therefore, appears that hydrodynamical attraction cannot be an effective enough factor to cause rain of any appreciable intensity in the atmosphere.

Table 5.4 Cloud types

Family	Genus	Height (h)	Form*
High clouds	Cirrus	h > 6 km	b
	Cirrostratus	-do-	b
	Cirrocumulus	-do-	c
Intermediate clouds	Altocumulus	2 < h < 6 km	a or b
	Altostratus	-do-	c
Low clouds	Stratus	h < 2 km	a or b
	Cumulus	-do-	a
	Stratocumulus	-do-	c
Clouds with vertical development	Cumulonimbus	0.5 < h < 6 km	a

Significance of small letters under Form* in Table 5.4
a : Isolated heap clouds with vertical development during their formation, and a spreading out when they are dissolving;
b : Sheet clouds which are divided into filaments, or rounded masses, and which are often stable or in process of disintegration;
c : More or less continuous cloud sheets, often in process of formation or growth.

Further, it was realized quite early in the history of rain formation that drops of about the same size in clouds can remain in some kind of colloidal stability which prevents their coalescence to form larger drops.

5.8.2 Electrical Attraction

It has been shown by Schmidt and Wigand from observations on fogs that electrically charged drops of the same sign do not coalesce and thus the uniformity of electrical charge is a strong stabilizing factor. Drops with charges of opposite signs attract each other but V. Bjerknes and his collaborators (1933) have shown that even with abnormally high charges and with the droplets almost touching each other the forces of attraction are rather small. These workers have also shown that the electrical attraction due to induced charges under strong electrical field as in thunderstorms is also negligibly small.

5.8.3 Collision Due to Turbulence

Arenberg (1939) has shown that collision effects resulting from sudden change of velocity of eddy turbulence in the atmosphere may be an important factor in the growth of cloud elements into rain drops. When an eddy carrying cloud particles of various sizes suddenly stops, the particles continue to move forward though the

5.8 From Cloud to Rain

eddy has stopped. The motion occurs in such a way that the loss of kinetic energy of the particle must equal the work done by the viscous drag in the medium.

The forward velocity ahead of the stopping eddy is, however, different for particles of different sizes. Arenberg calculated the distances that would be traversed by particles of different sizes and concluded that due to the path differences there would be numerous collisions to produce large drops.

In an average cloud, particles of all sizes exist and, therefore, collision would occur by eddy turbulence in the manner visualized by Arenberg. It appears highly likely that collision through eddy turbulence is an important factor in the formation of rain, especially in the tropics where large drops of rain are seen to fall from warm clouds in which convective turbulence is a dominant feature. But, in the absence of sufficient observational data, the matter is still one for debate.

5.8.4 Differences in Size of Cloud Particles

Non-uniformity in sizes of cloud elements can be an important factor in the growth of larger drops at the expense of smaller ones. Equation (5.2.1) gives the equilibrium vapour pressure over drops of different sizes. Therefore, when two drops of unequal sizes exist side by side, the differences in saturation vapour pressure between them would cause evaporation of water from the smaller drop and condensation on the larger. The larger drop would thus grow at the expense of the smaller till the resultant equilibrium vapour pressure between them is adjusted to a mean value between the two individual saturation vapour pressures.

In the atmosphere, droplets of different sizes are known to exist and it is, therefore, reasonable to hold that the effect of non-uniformity of sizes of cloud elements does play a part in the formation of rain. Squires (1958) who analyzed observations of cloud droplet spectra in different kinds of warm clouds found marked and systematic differences in their microstructure. He found that between different cloud types, the droplet concentration in a cloud can be used as an index of the type of droplet spectrum, low concentrations being associated with broad spectra and large maximum and average droplet sizes. Further, it was visualized that a spectrum characterized by relatively large average and maximum droplet sizes led to colloidal instability and operation of the coalescence process. However, Bergeron has pointed out that the contribution from this effect, by itself, is not likely to produce precipitation of any greater intensity than a light drizzle.

5.8.5 Differences of Temperature Between Cloud Elements

Saturation vapour pressure over water, as normally defined, varies with the temperature (Fig. 4.3). Hence if droplets of different temperatures co-exist, a difference of vapour pressure exists between them which would cause evaporation from the warmer particle and condensation on the cooler ones till the equilibrium

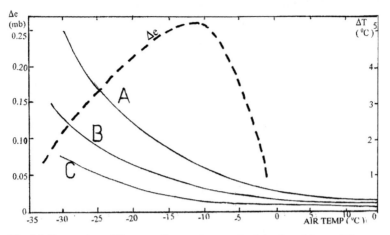

Fig. 5.5 Temperature-difference effect on the growth of cloud particles at different temperatures

vapour pressure between them is adjusted to a value intermediate between the original vapour pressures. The difference in saturation vapour pressure for the same temperature difference between cloud elements varies, however, with air temperatures. Fig. 5.5 shows the effect in the form of three curves.

Curve A shows the temperature differences that would produce a saturation vapour pressure difference of 0.27 mb at different air temperatures; Curve B shows the same effect for a saturation vapour pressure difference of 0.13 mb; and Curve C shows the temperature differences required for a saturation vapour pressure difference of 0.054 mb at different air temperatures. It will be seen from Fig. 5.5 that at air temperatures above freezing, a smaller difference in temperature between cloud elements is required to produce a given vapour pressure difference than at sub-freezing temperatures.

In the actual atmosphere, temperature difference between cloud elements occurs when different portions of the cloud are brought together by vertical mixing, or when drops from the upper part of the cloud descend through the cold lower part. Osborne Reynolds showed that the difference may result from the radiative cooling of the top surface of the cloud or the differential heating by the Sun of the different portions of the cloud. This so-called Reynolds effect is probably an important factor in the formation of rain in the tropics where a temperature difference of the order of 2 °C or 3 °C between neighboring cloud elements and intense vertical mixing through great heights occur frequently in large cumulus, stratocumulus, or cumulonimbus clouds.

5.8.6 The Ice-Crystal Effect

When drops of supercooled water and ice particles exist side by side, as in many clouds, it can be shown that the ice crystals will grow at the expense of the water

Table 5.5 Saturation vapour pressure (mb) over water and ice at different temperatures

Temp (°C)	0	−10	−20	−30	−40	−50	−60
Over water	6.11	2.87	1.26	.527	.189	.064	.020
Over ice	6.11	2.62	1.04	.284	.129	.039	.011

droplets owing to a difference in vapour pressure between the two cloud elements at certain temperatures. Values of saturated vapour pressure for the two phases at temperatures down to −60 °C are presented in Table 5.5.

Suppose we have supercooled water droplets and ice crystals in close proximity at −30 °C. The saturation vapour pressure (S.V.P.) at this temperature is 0.527 mb over water, and 0.284 mb over ice. If the vapour pressure in the space between them is 0.400 mb, the space is undersaturated with respect to water but supersaturated with respect to ice. So, water drops will evaporate and ice particles will grow at their expense.

Bergeron (1933) was the first to postulate that in such a co-existence of water droplets and ice particles lies the answer to the growth of ice particles to a size, when they can fall and after melting reach the ground as rain drops. According to him, rain is nothing but 'melted snow'. The Curve (Δe) in Fig. 5.5 shows graphically the difference in saturation vapour pressure over water and ice at different temperatures. It will be seen that the maximum difference, 0.27 mb, occurs at temperatures about −12 °C, below which the difference falls off rather slowly. Thus over a considerable range of temperatures in which supercooled water drops and ice crystals are found to co-exist in the upper regions of the atmosphere, the ice-crystal effect is probably the most potent factor to produce drops of rain large enough to reach the ground. Bergeron's ice-crystal theory received strong support from the work of Findeisen three years later.

5.9 Meteorological Evidence – Rainfall from Cold and Warm Clouds

Since 1933, there has been a large body of evidence in favour of Bergeron's ice-crystal theory. Bergeron showed from considerations of colloidal instability in clouds that effects other than the ice-crystal effect could not possibly account for precipitation of any greater intensity than light drizzle. Surely those could not explain the sudden release of heavy precipitation, sometimes accompanied by large hailstones, from towering cumulonimbus clouds, which is so common in all latitudes particularly in the tropics.

Bergeron showed that such releases could be easily explained by his theory in that the upper portion of these clouds is almost always glaciated before heavy precipitation occurs. He also points out the example of winter rain in high latitudes from low and thin clouds and absence of rain from clouds of greater vertical depth in the same latitudes in summer or from winter cumulus in low latitudes. The difference is due to the lower ice-nuclei level in high latitudes in winter. He maintains that any cloud from which appreciable rain is observed to fall must have contained ice crystals in its upper parts. This means that in addition to the cumuliform clouds, clouds like nimbostratus or altostratus, which also give rain, must contain ice crystals in those portions which are vertically developed in relation to the general top surface and extend into the layer of sub-freezing temperatures. Ice crystals falling from cirriform clouds into layers of medium clouds, as shown by Mare's tails or Virga, may also cause precipitation in accordance with Bergeron's theory.

But by far the strongest evidence in support of Bergeron's theory has been assembled by Stickley (1940) in America who showed from an analysis of 360 cases of reported rain observed by aircraft and on the ground that of the 324 effective observations practically all could be explained by the ice-crystal theory. Only 10 cases did not appear to support the theory, but, upon detailed investigations, even these 10 cases could be explained away by the Bergeron theory.

5.9.1 Rainfall from Warm Clouds

Notwithstanding very strong observational evidence in favour of the ice-crystal theory in cases of rainfall from clouds the top of which reaches heights well above the freezing level, there have been reports of appreciable rainfall from clouds the top of which never crossed the freezing level, especially in the tropics where the freezing level in summer usually lies near 5 km asl.

Substantial rain has been observed to fall from large but unglaciated cumulus clouds over tropical oceans and continents where factors other than the ice-crystal effect might have contributed to formation of rain. It is believed that growth of cloud drops in such clouds occurs by a process of collision-coalescence during violent upward and downward motion inside the clouds.

5.10 Climatological Rainfall Distribution over the Globe

There is no limit to the varieties of forms in which precipitation can occur at a place on the earth's surface. It can occur in the form of rain, drizzle, shower, snow, sleet and so many diverse forms. The total amount of rain registered at a given place may consist of contributions from one or more of these various forms and may result from diverse atmospheric situations. There are also extensive areas on the earth's surface where little or no precipitation occurs at any time of the year, even if some clouds may appear at times. These latter areas are mostly the desert areas. In the tropics, rainfall varies greatly with season and between continents and oceans.

In January, climatological mean rainfall maxima are generally to be found in the southern hemisphere, usually centered over the continents of South America, Southeastern Africa, and the northeastern part of Australia, with little rainfall over the

5.10 Climatological Rainfall Distribution over the Globe

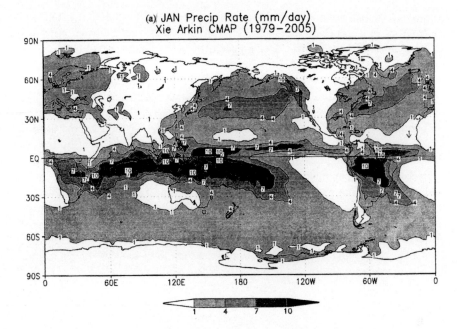

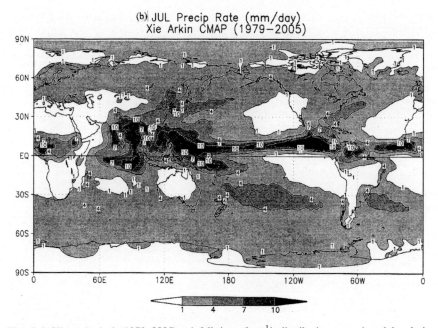

Fig. 5.6 Climatological (1979–2005) rainfall (mm day^{-1}) distribution over the globe during (**a**) January, and (**b**) July (Xie and Arkin, 1996) [Courtesy:NCEP/NWS]

oceans except the extreme southwestern parts of the major oceans, while the fields are reversed in the northern hemisphere where the continents are usually dry and, perhaps, more rain occurs over oceans than over land. Whatever rain falls over the continent and oceans is usually brought about by the passage of baroclinic waves.

In July, there is greater rainfall over the continents than oceans in the northern hemisphere, while the fields are reversed in the southern hemisphere. Over the globe as a whole, the belt of rainfall maxima appears to follow the seasonal movement of the Intertropical Convergence Zone (ITCZ). These aspects of the seasonal distribution of mean rainfall over different parts of the globe are brought out by maps of climatological rainfall (Fig. 5.6 a, b).

However, a proper interpretation of the observed distribution of rainfall at the earth's surface is not easy, as it requires consideration of several physical and dynamical factors, characteristic of the region. Qualitatively, evaporation from the underlying surface, net moisture convergence into the overhead vertical column and its lifting by the atmospheric circulation and orography are amongst the more important parameters to determine the amount of precipitation likely to occur at a place. These factors are related by the well-known water balance equation

$$P = E - \int_0^{p_0} \nabla \cdot (q\mathbf{V}) \delta p \qquad (5.10.1)$$

where P denotes precipitation, E is evaporation, q is specific humidity of the air, $\mathbf{V}$ is three-dimensional wind vector and p is pressure with p_0 at the surface. The integration is from surface to top of the atmosphere.

Chapter 6
Physics of Radiation – Fundamental Laws

6.1 Introduction – the Nature of Thermal Radiation

In the present book, we deal with radiation which is emitted by a body because of its heat or temperature only and not for any other reason. For example, a body may emit radiation when an electrical discharge passes through it, or is subjected to other sources of light such as phosphorescence or fluorescence, or when it goes into flames where chemical action takes part. Radiation from all such other sources is outside the scope of the book.

6.2 Radiation and Absorption – Heat Exchanges

It is known from experience that heat energy can travel from one place to another in any one of the three ways: conduction, convection or radiation.

6.2.1 Conduction

Conduction usually takes place in solids in which the tightly-held molecules pass on the heat by vibration or molecular agitation. Take, for example, the case of a long iron bar one end of which is heated. On receiving the heat energy, the molecules at the heated end which are now more agitated than those next to them on the cooler side pass on the energy through increased vibration to lower-energy molecules down the bar till the heat reaches the cooler end. Thus, conduction requires a temperature gradient and a material medium for the heat to pass through.

6.2.2 Convection

Convection, on the other hand, usually occurs in liquids and gases where the intra-molecular forces are relatively less strong with the result that the molecules have greater freedom to move away from one another and carry heat from one part of the medium to another by actual physical movement. Bulk movements carrying liquid or gas from the hotter to the cooler part of the medium can occur. Thus convection also requires a temperature gradient and a material medium for the heat transfer.

6.2.3 Radiation

By contrast, radiation is an entirely different process of heat transfer. In it, the energy is transmitted by electromagnetic waves emitted by the atoms and molecules inside the hot body. According to the quantum theory, energy from a hot body is radiated in quanta either by electronic transitions inside an atom, or by vibration of atoms about their mean positions in a molecule or by rotation of molecules about their center of mass.

In electronic transitions, the energy of the emitted radiation corresponds to the difference in energy between two quantized energy states. This means that when an electron jumps down from an orbit with higher energy to one with lower energy, the difference in energy is given out in the form of an electromagnetic radiation which travels with a velocity which in vacuum equals the velocity of light. The emitted energy travels in waves the frequency of which is given by the Planck relation, $E = h \nu$, where E is energy of the emitted radiation, ν is wave frequency and h is Planck's constant the experimental value of which is 6.62×10^{-34} J s. Thus, if we designate the velocity of light by c and the wave-length of the emitted radiation by λ,

$$E = c\, h/\lambda \tag{6.2.1}$$

where $c = 2.998 \times 10^8$ m s^{-1}.

Thus, in a spectrum of the radiation emitted, the energy due to any particular electronic transition appears as a line corresponding to its frequency or wave-length. In vibrational transition, the atoms in a polyatomic molecule which are bound together by electrical forces vibrate about their mean position with energy states which are quantized. A transition from a higher to a lower state of energy radiates a quantum of energy corresponding to the difference in energy between the two quantum states.

Rotational transition occurs when molecules rotating about their center of mass with different energies which are quantized jump down from a higher to a lower level of energy thereby giving out a quantum of radiation whose energy equals the difference in energy between the two levels.

In vibrational and rotational transitions, the radiated energies extend over a wide range of wave-lengths. So, in a spectrum, they do not appear as clear-cut lines as in electronic transitions but in broad bands spread over a range of wave-lengths.

Spectral analysis shows that electronic transitions account for the maximum amount of energy radiated by the atoms and molecules of a hot body, while vibrational transitions account for intermediate amounts and rotational transitions for minor amounts only.

6.2.3.1 Absorption of Radiation

When a body absorbs radiation, transitions between energy states inside an atom or molecule occur in the opposite direction, i.e., from a lower to a higher level of energy. But, here also, the condition is that only such radiation will be absorbed whose energy corresponds to the difference in energy between any two quantum states. If not, the incident radiation may simply pass through the body without being absorbed. However, if the energy of the incident radiation exceeds that required to knock off an electron from an atom, the energy may be absorbed, thereby leaving the atom in an ionized state. It is conceivable that such thermal ionizations may be taking place continuously in the atmospheres of the sun and other bright stars where the temperatures are extremely high (see Saha and Srivastava, 1931, fifth edition 1969, reprinted 2003, p658).

6.2.3.2 Prevost's Theory of Heat Exchanges

A question that plagued the minds of many in the early days of thermodynamics was whether a material medium or surrounding was necessary for transmission of heat energy by radiation from one place to another, as was required in conduction and convection. For a long time, people seemed to believe so, for to think otherwise would have meant going against their apparent experience that heat transmitted to a cooler body appeared to be more intense than that sent to a warmer body. But it was the work of Prevost in 1792 which cleared up the issue. According to Prevost, the rate of radiation from a body is independent of the surroundings and depends only on its temperature. He stated that every body radiates energy independently of any other body and that the apparent dependence on the environment was only due to the fact that the warmer body gives out more radiation than what it receives from the cooler body, or reversely, the cooler body receives more energy than what it radiates to the warmer body. Prevost's discovery marked an important advance in our understanding of the true nature of thermal radiation and the mode of exchange of thermal energy between two bodies at different temperatures.

6.3 Properties of Radiation

We are all familiar with the properties of light. It can be easily demonstrated with the aid of simple apparatus that radiant energy and light obey identical physical laws. This is suggested by the following:

(a) Like light, radiation can travel in straight lines and through vacuum with the speed of light and cast shadow of objects which obstruct its passage.

(b) The laws of reflection and refraction of light hold good in the case of radiation as well. In fact, a spectrum of radiation can be obtained by passing it through a glass prism in the same way as light.
(c) When radiation falls on matter, part of its energy may be reflected from the surface, part absorbed by the matter and part transmitted through it, in the same way as light.
(d) Like light, the intensity of radiation falls off with distance according to the inverse square law.
(e) Radiation can be polarised like light by passing it through a tourmaline crystal or nicol.

One may conclude from the above examples that radiation and light are identical in nature and properties. In fact, the spectral distribution of energy of an extremely hot body like the sun demonstrates that what we call light, or visible light to be exact, forms only a small part of the whole radiation spectrum, which has the capacity to affect the retina of the human eye and thus produce the sensation of vision and colour. Thus, radiation is the more general or expressive term for heat energy and visible light forms only a special part of it, both obeying the same or identical physical laws, as stated above and in the following section.

6.4 Laws of Radiation – Emission and Absorption

6.4.1 Kirchhoff's Law

It was Kirchhoff (1824–1887) who stated that the ratio of the emissive to the absorptive power of a body for any radiation is a function of the wave-length of the radiation and the temperature of the body and is the same for all bodies. That is

$$E_\lambda / A_\lambda = F(\lambda T) \qquad (6.4.1)$$

where the absorptive power A_λ of the body is defined as the rate at which a unit surface of the body at temperature T absorbs radiation of wave-length λ falling upon it and the emissive power E_λ as the rate at which the body emits radiation of the same wave-length per unit area of its surface.

The relation (6.4.1) does not preclude the possibility of both A_λ and E_λ of a body at a particular temperature being zero for any particular wave-length, meaning thereby that the body may be totally transparent to radiation of that wave-length and also emit no radiation of that wave-length. The same body, however, may absorb radiation of other wave-lengths and emit radiation of those wave-lengths. This type of selective absorption and radiation is quite common with several atmospheric gases. Water vapour, for example, is almost transparent to short-wave solar radiation but absorbs heavily in the long-wave radiation emitted by the earth's surface and atmosphere. Other examples include carbon dioxide and ozone.

6.4.2 Laws of Black Body and Gray Radiation

Experiments show that there exists an upper limit to the amount of radiation that a body can emit at a given temperature and wave-length. This can be shown to be directly proportional to the absorptive power of that body for that wave-length at that temperature. If the body absorbs 100% of that radiation falling upon it, i.e., if $A_\lambda = 1$, then E_λ will be maximum for all wavelengths at that temperature. Such a body is called a black body.

However, it is important to understand here the difference between the absorptive power of a body which takes into account the effective thickness or volume of the absorbing body and its absorption co-efficient which assumes absorption by a surface of unit thickness. For example, a gas which ordinarily has a low absorption co-efficient for radiation of any wave-length can absorb all the radiation of that wave-length falling on it when it is contained in a large volume such as a big sphere. Despite low absorption co-efficient, such a gaseous body has an absorptive power of unity and, therefore, constitutes a perfect black body. From this viewpoint, our sun may be treated as a black body because of its large volume. Further, the surface of a blackbody is not necessarily black in colour, so far as radiation is concerned. There are black substances which absorb radiation completely in some wave-lengths but not in others. For example, platinum black absorbs waves in the visible part of the spectrum completely but not in the longer waves.

A close approximation to a black body can be constructed by taking a large enclosure and making a hole on its side so small compared to the size of the enclosure that the amount of radiation that can escape through it is negligible. A beam of radiation of any wave-length entering such an enclosure will be almost wholly absorbed. Once inside, the radiation will undergo reflections at the walls of the enclosure and at every reflection a certain amount of energy will be absorbed at the walls. After several reflections, the enclosure will completely trap all the energy of the beam. The radiation emitted by such an enclosure and going out through the hole will then be at the maximum intensity possible at the given wave-length and temperature.

It follows from Kirchhoff's law that a good absorber is also a good radiator. However, one required to know the exact functional form of $F(\lambda, T)$ to determine the intensity of radiation at any given temperature and wave-length. A body which at a given temperature emitted radiation the intensity of which at every wave-length was in a fixed proportion of that emitted by a black body was called a gray body and the radiation as 'gray radiation'.

Following Kirchhoff's work, several attempts were made to find the functional form, $F(\lambda, T)$, of radiation. Though these attempts were not successful in their objective, they led to some important laws of radiation. These were: the Stefan-Boltzmann law (1879–1894) and the Wien's displacement law (1893). A brief statement of these laws follows.

6.4.3 Stefan-Boltzmann Law

Stefan in 1879 found experimentally that if he integrated all the energy radiated by a body at a given temperature, the total energy was proportional to the fourth power of the absolute temperature of the body. In 1894, Boltzmann, from purely theoretical considerations, proved the correctness of Stefan's experimental result. This means that at a given temperature T, the total energy E radiated by a body is given by

$$E = \sigma T^4 \qquad (6.4.2)$$

where σ is a constant, called the Stefan-Boltzmann constant. Its value is $5.6687 \times 10^{-8}\,\text{Wm}^{-2}\text{K}^{-4}$.

6.4.4 Wien's Displacement Law

Wien found that at a given temperature T, there is always a particular wave-length λ_m at which the radiated energy is at its maximum. He also found that the location of this maximum changes with a change of temperature, moving to the side of the longer (shorter) wave-lengths with a decrease (increase) of temperature. The displacement law, as stated by him, is

$$\lambda_m T = 2896 \text{ (approx.)} \qquad (6.4.3)$$

where λ_m is in microns (1 micron = 10^{-4} cm = 10^{-6} m, usually denoted by μ), and T is in degrees Kelvin.

Amongst the other notable attempts made to determine the functional form of radiation during the period, 1890–1900, was one called the Wien's radiation law, and another called the Rayleigh-Jeans radiation law. These laws were only partially successful, in so far as they could explain the distribution of energy over some sections of the wave-lengths but not the whole range of wave-lengths.

6.4.5 Planck's Law of Black Body Radiation

Finally, in 1900, Planck, using the results of careful experiments and on the basis of quantum theory came out with his celebrated law of black body radiation which states as follows:

6.4 Laws of Radiation – Emission and Absorption

$$E_\lambda d\lambda = (2hc^2/\lambda^5) \left[1/\{\exp(hc/\kappa\lambda T) - 1\}\right] d\lambda \quad (6.4.4)$$

where E_λ is the intensity of radiation from a black body at temperature T at wavelength λ, and h is Planck' constant, c the velocity of light and κ the Boltzmann's constant.

Planck's law (6.4.4) satisfactorily explains the distribution of energy amongst the whole range of wave-lengths radiated by a black body at a given temperature. It includes within its ambit the Stefan-Boltzmann law (6.4.2) as well as the Wien's displacement law (6.4.3), which may be readily deduced from (6.4.4), as follows:

6.4.6 Derivation of Wien's Law and Stefan-Boltzmann Law from Planck's Law

Wien's displacement law is found by differentiating the right-hand side of (6.4.4) with respect to λ and setting the result to zero

$$-10hc^2/[\lambda^6\{\exp(hc/\kappa\lambda T) - 1\}] + 2hc^2[\{\exp(hc/\kappa\lambda T)\}/\lambda^5\{\exp(hc/\kappa\lambda T) - 1\}^2] (ch/\kappa\lambda^2 T) = 0 \quad (6.4.5)$$

Simplifying (6.4.5) further and substituting x for $hc/\kappa\lambda T$, we get the transcendental equation,

$$x\, e^x/(e^x - 1) = 5$$

which, it is easy to see, has a value of x close to 5.

An application of the method of approximations gives $x = 4.965$ as the exact value. By drawing the curve, it can be seen that this is the only real root. Since $x = hc/\kappa\lambda T$, we have

$$\lambda_m T = (hc/4.965\,\kappa) = 2896 \text{ micron. deg K}$$

which is Wien's law (6.4.3) which states that in the spectrum of radiation from a black body, the wave-length λ_m of the maximum emission shifts towards the side of longer wave-lengths with decrease of temperature.

The Stefan-Boltzmann radiation law can be obtained by integrating (6.4.4) for black body radiation at a given temperature T over the whole range of wave-lengths from 0 to infinity. Thus, if E is the total radiation,

$$E = \int_0^\infty E_\lambda d\lambda = c_1 T^4 \int_0^\infty [(\lambda T)^{-5}/\{\exp(c_2/\lambda T) - 1\}] \, d(\lambda T)$$

where we have put c_1 for $2hc^2$ and c_2 for hc/κ, both of which are constants.

If we put x for λT, the above expression is reduced to

$$E = c_1 T^4 \int_0^\infty [x^{-5}/\{\exp(c_2/x) - 1\}] \, dx$$
$$= \sigma T^4$$

which is Stefan-Boltzmann radiation law (6.4.2) which states that the total unpolarized radiation emitted by a black body is proportional to the fourth power of its absolute temperature. The Stefan-Boltzmann constant σ may be shown to have a value

$$= 2\pi^5 \kappa^4/(15 c^2 h^3) = 5.6687 \times 10^{-8} \, \text{Wm}^{-2}\text{K}^{-4}$$

6.5 Spectral Distribution of Radiant Energy

Planck's law enables us to determine the complete spectral distribution of energy emitted by a black body at different temperatures at various wave-lengths. Prevost's finding that in the physical world, every body, regardless of its surroundings and temperature, emits its own radiation makes it possible for us to examine the characteristic spectrum of radiation from any body of interest to us. In fact, the spectral analysis has been used widely as a powerful tool to examine the physical conditions of radiating bodies in almost all branches of science. Since, in the present book, we are mainly concerned with the radiation emitted by the sun and the earth-atmosphere system, let us examine Fig. 6.1, reproduced from Brunt (1944).

Figure 6.1 which shows the spectral distribution of energy radiated at the temperature of the sun (T = 6,000 K) as well as those of the earth's troposphere (T = 300 K) and stratosphere (T = 200 K), as determined from Planck's law, assuming that they all radiate energy as a black body at their respective temperatures. From it, it is evident that although the horizontal scale has been designed to ensure a uniform scale of λT, the radiant energy from the three sources lie in entirely different ranges of wave-lengths. The curve shows less than even 1% of the total energy at λT-values below about 1,000 and above 24,000. From low values of λT, the energy rises steeply to a maximum near $\lambda T = 3,000$ and then falls gently to the higher wave-length side. At the temperature 6,000 K, most of the energy lies in the short-wave part of the spectrum between about 0.17μ and 4.0μ, with the maximum in the visible part near 0.5μ. At 300 K, the energy lies in the long-wave part of the spectrum between about 3μ and 80μ with a maximum near 10μ. At 200 K, the energy shifts further to the long-wave side, with the maximum near 15μ. Thus, with decrease of temperature, not only does the wavelength of the maximum emission shift towards the longwave side in accordance with Wien's displacement law, but also the intensity of the short-wave emission decreases.

We, therefore, conclude that radiant energy can be of any wavelength from 0 to infinity depending upon the temperature of the radiating body. Light energy forms only a very small part of it, even less than 3/4th of an octave in the continuous

6.5 Spectral Distribution of Radiant Energy

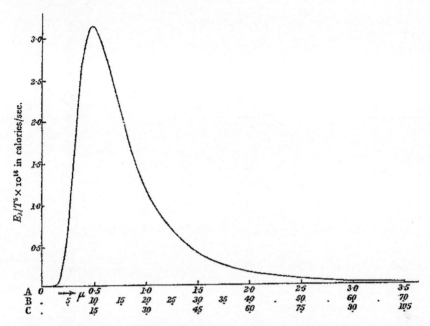

Fig. 6.1 Theoretical curve showing the distribution of black-body radiation at scales of wavelengths, μ in microns, corresponding to different temperatures: Scale- A corresponds to T = 6,000 K, scale- B to T = 300 K and scale- C to T = 200 K (Reproduced from Brunt, 1944, © Cambridge University Press, with permission)

spectrum. Saha and Srivastava (1931) in their book on 'A treatise on heat', Fifth edition, reprinted 2003, gives a list of electromagnetic waves discovered so far in the electromagnetic spectrum The range of waves appears to have no limit. The shortest waves discovered so far are the cosmic rays, while on the long-wave side the longest waves are the Hertzian or broadcasting waves. In Table 6.1, we present an extract from this list, showing the broad ranges of electromagnetic waves in the spectrum of radiation.

Table 6.1 Electromagnetic waves in the spectrum of radiation

Name of the wave	Wavelength (μ) Range	Frequency (Hertz) Range
Cosmic ray	$< 10^{-6}$	$> 3 \times 10^{20}$
Gamma ray	$10^{-6} - 10^{-4}$	$3 \times (10^{20} - 10^{18})$
X-ray	$10^{-5} - 10^{-1}$	$3 \times (10^{19} - 10^{15})$
Ultraviolet	$10^{-2} - 4 \times 10^{-1}$	$3(1 - 1/40) \times 10^{16}$
Visible light	$(4-8) \times 10^{-1}$	$3(1/4 - 1/8) \times 10^{15}$
Infrared	$8 \times 10^{-1} - 4 \times 10^{2}$	$(3/8) \times 10^{15} - (3/4) \times 10^{12}$
Microwaves	$10^{2} - 10^{7}$	$3 \times (10^{12} - 10^{7})$
Radio-waves	$> 10^{7}$	$< 3 \times 10^{7}$

Table 6.2 Ranges of wavelengths of the electromagnetic spectrum used for different remote sensing techniques

Technique	Range of wavelengths (μ)
1. Visual photography	(0.4–0.7)
2. Multispectral imagery	(0.47–1.11)
3. Infrared imagery and spectroscopy	(0.7–1400)
4. Radar imagery, scatterometry and altimetry	$(0.03–10) \times 10^4$
5. Passive microwave, radiometry and imagery	$(10–100) \times 10^6$

6.6 Some Practical Uses of Electromagnetic Radiation

Electromagnetic radiation has been applied using remote sensing techniques to several fields of human activity. These include:

   (i) Agriculture and forestry,
  (ii) Land surface mapping and analysis, and cartography;
 (iii) Wildlife ecology;
  (iv) Hydrology and water resources;
   (v) Meteorology and oceanography;
  (vi) Geology and mineral exploration;
 (vii) Snow and ice monitoring;
(viii) Coastal resources management;
  (ix) Monitoring biological activity in the ocean;
   (x) Military surveillance, etc.

However, owing to near-total absorption of ultra high frequency waves in the atmosphere, only a limited part of the electromagnetic spectrum is used for remote sensing purposes and this lies largely in the visible, infrared, microwave and radiowave parts of the spectrum. Selection of the specific part of the spectrum to be used in any particular case depends upon the photon energy, frequency and atmospheric transmission characteristics of the spectrum. Table 6.2 gives the ranges of wavelengths normally used for different remote sensing techniques.

Chapter 7
The Sun and its Radiation

7.1 Introduction

Our sun, a G-2 class dwarf star glowing with a surface temperature of about 6000 K and located at the center of the solar system, radiates energy in all directions. Only a tiny fraction ($\simeq 4.54/10^{10}$) of this radiation is intercepted by the earth at an average distance of 150×10^6 km. Solar radiation is received at the outer boundary of the earth's atmosphere at the rate of about $1368\,\mathrm{Wm}^{-2}$. Since the long-term variation of this rate is small, it is called the solar constant.

We know from experience that this tiny energy input, however small it may be by cosmic measure, is all-important for our living planet and without it everything will be dead and dark. In this chapter, therefore, we look at the sun and its radiation with the objective of finding out what goes on inside and outside its surface and at the surface itself. Since, at the high temperature of the sun, no matter can exist in liquid or solid state, the sun must be an entirely gaseous star with its constituent gases existing in either atomic form or a highly ionized state called plasma.

Using the observed properties of the star as a whole, such as its mass, density, volume and luminosity and the known properties of gases, it is possible to calculate the structure of the sun's interior, which will be consistent with the observations. A standard model based on such calculations suggests a layered structure and extremely high pressure, high density and high temperature inside the sun near the center, conditions in which thermonuclear reactions can take place. We also look into the structure of the solar atmosphere and its different layers through which solar radiation travels to the earth and some of the well-known solar activities, such as sunspots, solar magnetism, flares, prominences, solar wind, etc. We start by reviewing some of the physical characteristics of the sun.

Table 7.1 Physical characteristics of the sun

Characteristic (unit)	Value	Times of the Earth value
Mass (kg)	1.991×10^{30}	333,050
Volume (km^3)	1.41×10^{18}	1,300,000
Radius (km)	6.955×10^5	109.2
Density (kg m^{-3})		
(Center)	151.3×10^3	27.5
(Mean)	1.409×10^3	0.256
Pressure (bars)		
(Center)	2.334×10^{11}	2.30×10^{11}
(Surface)	1.0×10^{-4}	9.87×10^{-5}
Temperature (°A)		
(Center)	15.55×10^6	
(Photosphere)	5,780	
(Corona)	2.50×10^6	
Surface gravity (m s^{-2})	273.7	27.9
Total energy output (W) (Luminosity)	3.86×10^{26}	
Energy Flux at surface (Wm^{-2})	6.34×10^7	
Mean Period of rotation ~ 27 days (25 days at equator, 36 days at poles)		27
Chemical composition (% of total number of atoms)		
Hydrogen	92.1	
Helium	7.8	
All others	0.1	

7.2 Physical Characteristics of the Sun

Some of the vital statistics of the sun are given in Table 7.1 (After Lang, 1999, in 'The New Solar System' edited by Beatty, Petersen and Chaikin, 4th edn., with permission of Cambridge University Press). The final column gives a comparison of the value of the characteristic with that on the earth, wherever relevant.

7.3 Structure of the Sun – its Interior

The sun has a layered structure which is suggested by the results of helioseismic soundings, measurement of neutrino flux, continuous monitoring of X-rays and gamma rays emanating from the sun's interior and inferred from the standard theoretical model. The layered structure is shown in Fig. 7.1.

Some details about the layers in the interior of the sun are as follows:

7.3 Structure of the Sun – its Interior

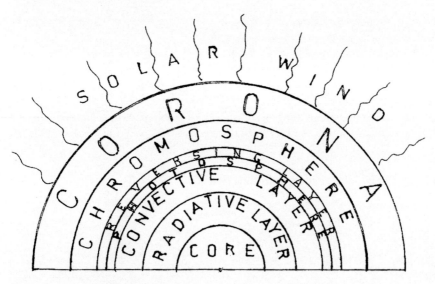

Fig. 7.1 The layered structure of the sun

7.3.1 The Core – Nuclear Reactions

The Core is the central region of the sun. Its boundary extends from the center to a distance of about 25% of the sun's radius. Its volume is only 1.6% of that of the sun, but, because of its high density, its mass is almost 50% of that of the sun. The constituents of the core region are predominantly hydrogen (92%) and helium (7.8%).

Under the intense pressure and temperature prevailing in the core, the atoms of these elements are completely ionized with their nuclei and electrons existing in a state of plasma. In such a state, hydrogen (H^1) gets converted into helium (He^4) (where the superscripts 1 and 4 denote respective atomic mass number) by thermonuclear fusion and a resulting mass defect which, according to Einstein's special theory of relativity ($E = mc^2$, where E is the energy, m the mass defect and c the velocity of light), leads to release of an enormous amount of energy from each conversion by the following process (see McGraw-Hill Encyclopaedia of Science and Technology, vol. 17, 9th edition, 2002):

7.3.1.1 Proton – Proton Chain Reaction

$$H^1 + H^1 = D^2 + e^+ + \nu + 1.44\,\text{Mev}$$
$$D^2 + H^1 = He^3 + \gamma + 5.49\,\text{Mev}$$
$$He^3 + He^3 = He^4 + H^1 + H^1 + 12.85\,\text{Mev}$$

where e^+ denotes a positron, ν a neutrino, γ a Gamma ray, D^2 a deuteron and the superscript denotes the atomic mass number.

The above proton–proton chain reaction which shows how hydrogen ions which are in great abundance in the interior of the sun get converted into helium accounts for nearly 98.5% of the total solar energy. The remainder is made up by another chain reaction known as the carbon–nitrogen chain reaction in which hydrogen ions get converted into helium through a chain reaction with carbon and nitrogen ions to form helium, as shown below(details are taken from Saha and Srivastava (1931):

7.3.1.2 Carbon–Nitrogen Chain Reaction

	Energy evolved	Half-life
$C^{12} + H^1 = N^{13} + \gamma$	2.0 Mev	
$N^{13} = C^{13} + e^+ + \nu$	0.5 Mev	10.5 min.
$C^{13} + H^1 = N^{14} + \gamma$	8.2 Mev	
$N^{14} + H^1 = O^{15} + \gamma$	7.5 Mev	
$O^{15} = N^{15} + e^+ + \nu$	0.7 Mev	2.1 min.
$N^{15} + H^1 = C^{12} + He$	5.2 Mev	
Total	24.1 Mev	

$$(1\,\text{Mev} = 1.6 \times 10^{-13}\,\text{J} = 4.45 \times 10^{-20}\,\text{KWH})$$

Thus, the core is virtually a nuclear furnace and serves as the source of all of sun's energy. Of the source of energy of stars in general (remember that the sun is also a star), Prof. H.N.Russell writes: "Hydrogen forms the fuel and helium forms the ashes of the process of combustion which keeps the stars shining through ages". The combustion also produces a large flux of neutrinos, particles with no mass or charge, at several points in the above chain reactions. About the neutrinos, Saha and Srivastava (1931) write: "It is assumed that 7% of the gross amount of energy evolved is carried away by the neutrinos, but even this may also ultimately be converted to energy by some process not yet known to us". Since the theory of nuclear fusion in the Sun was first formulated, the search for neutrinos on earth has continued (see further in sect. 7.7).

7.3.2 The Radiative Layer

The energy generated at the core which is mostly in the form of high-frequency X-rays and Gamma rays flows out into the next spherical layer which is called the radiative layer. It is quite a broad layer extending from the core boundary to almost 71% of the sun's radius. Moving with the velocity of light, the outflowing energy is expected to cross the radiative layer fast but on account of multiple deflections, absorption, re-radiation, etc., in the radiative layer, the actual movement is very slow

7.3 Structure of the Sun – its Interior

and it may take millions of years in some cases for radiation to work its way up to the top surface. The temperature drops from the core outward as the heat energy expands into an ever-increasing volume from a value of about 15.5 million K at the core to about 2.5 million K at the top of the radiative layer.

7.3.3 The Convective Layer

By the time, the outgoing radiation reaches the top of the radiative layer,the solar atmosphere becomes somewhat opaque to the outgoing radiation with the consequence that the heat energy piles up in a narrow transition zone at the top of the radiative layer causing the material below to be extremely hot as compared to that above. The pent-up energy of the transition zone then bursts into violent convection which rises to great heights, delivers the energy to the solar surface and then sinks. Across the convective layer, the temperature drops enormously from a value of about 2 million K at the bottom to about 6,000 K at the top.

7.3.3.1 Helioseismology

Helioseismology which may be described as a new branch of solar physics, has thrown considerable light on what all goes on in the sun's convective layer. In fact, it has brought out that the root causes of several observed features of the sun and its atmosphere, such as sunspots, flares, prominences, coronal discharges, etc., may be traced to the happenings in this layer, particularly the periodic upheavals and constant churnings and turbulent motions of the plasma in this layer which produce intense and far-reaching electric and magnetic fields. While vertical oscillations generate pressure waves, differential rotation between the equator and the high-latitude belts produce complex circulations in this layer. Unlike seismic waves inside the earth, the pressure waves in the sun's interior do not travel in straight lines. They form a kind of standing waves between the surface of the sun and the lower surfaces of the convective layer, but how deep they penetrate or how far they travel depend on the wave-length of the waves.

Helioseismological investigations further inform us that the deeper part of the sun's interior from the core center to the upper boundary of the radiative layer may be in solid- body rotation, but the convective layer and the photosphere above have a strong latitudinal shear between the equator and the higher latitudes, which, through increase of vorticity, may give rise to a meridional component of the horizontal velocity. This may mean that combined with vertical motions, the horizontal motion may actually cause a mean meridional circulation of the kind found on the earth, with rising motion near the solar equator, poleward motion above, sinking motion in high latitudes and a possible return equatorward flow at the bottom of the layer. A sketch of this likely circulation is shown in Fig. 7.2. However, the formation of such a circulation is only an inference at present. Further studies will be required to prove or disprove its existence.

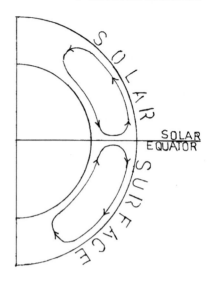

Fig. 7.2 Schematic showing likely mean meridional-vertical circulation in the Sun's interior (After SOHO/SOI/MDI Consortium) (Reproduced from Lang, 1999, in The New Solar System, 4th edn, edited by Beatty, Petersen and Chaikin, with permission of Cambridge University Press)

7.4 The Photosphere

At the top of the convective layer lies the 500 km-thick surface layer of the sun, called the photosphere (photos – the Greek word for light), where the temperature is about 5760 K. It is from here that electromagnetic radiation in the form of heat and light as we know them on earth along with the very high-frequency radiation that wells up from the interior move out into space through the solar atmosphere. The photosphere constitutes the visible surface of the sun and the base of the solar atmosphere. Normally, its dazzling brightness prevents any direct eye observations. However, its surface can be seen, though for a few moments only, at the time of solar eclipse or using a coronagragh at any time.

Ever since the time of Galileo, scientists have studied the sun using several direct and indirect methods of observation, such as powerful telescopes, spectrometers, photometers, polarimeters, etc. A lot of details about the conditions of the solar surface and its interior have been gathered using space-borne probes designed specifically for solar studies, such as the multinational Solar and Heliospheric Observatory(SOHO), soft X-ray telescopes carried onboard the Japanese satellite 'Yohkoh'(sunbeam) or the NASA satellite Ulysses in the 1990s.

Prominent amongst the findings by these probes are: the great unsteadiness of the sun with periodic rises and falls of its surface, regular occurrences of sunspots with an 11-year cycle in regions of strong magnetic fields and frequent solar flares and prominences which often rise to great heights in the sun's atmosphere. Helioseismic soundings suggest that the vertical oscillations of the solar surface may be due to the standing sound waves in the convection layer below in which the upward-moving wave is reflected downward by the lower surface of the photosphere and the downward moving wave is refracted upward by the steep rise of temperature at the base of the convection layer. The observed period of these waves is found to be between 3 and 6 min and their wave lengths range from a few thousand kilometers to almost the full circumference of the sun.

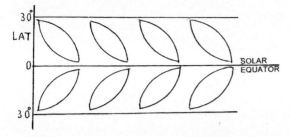

Fig. 7.3 The 11-Year sunspot cycle showing migration of the sunspots (After David Hathaway, NASA/Marshall) (Reproduced from Lang, 1999, in The New Solar System, 4th edn, edited by Beatty, Petersen and Chaikin, with kind permission of Cambridge University Press)

7.4.1 Sunspots

These are relatively dark areas on the visible surface of the sun, as against the general brightness of the photosphere. The intense magnetic field within a sunspot acts as some kind of a filter or control valve, which does not let any heat or energy from the interior to flow outward. This keeps the sunspots a few thousand degrees cooler than their surroundings.

At the center of a sunspot the strength of the magnetic field may be a few thousand Gauss. Sunspots occur in pairs of opposite magnetic polarity, which are joined by loops of magnetic field which often rise high into the solar atmosphere. These loops appear very bright in X-ray images of the sun and often appear as flares on the visible surface of the sun. They constitute a kind of hot spots on the solar surface in which the temperature can shoot up to 1,000,000 K or more, as against about 3,000–4,000 K in the sunspots.

The number of sunspots varies from a maximum to a minimum in an approximately 11-year cycle (Fig. 7.3). They first appear in high latitudes around 35°–40° and then migrate equatorward to about 5° before a new cycle begins in high latitudes.

It is observed that the brightness or luminosity of the sun varies with the sunspot cycle, being maximum at the sunspot maximum and minimum at the sunspot minimum. This appears to be somewhat paradoxical, since sunspots are relatively cooler areas of the solar surface and one would expect that the intensity of solar emission would be minimum at the sunspot maximum. But what really happens is that the coolness of the sunspot regions is offset by the extreme heat released by the large number of flares that accompany the sunspots over the remaining surface of the sun. The net result, therefore, is an increase (decrease) in luminosity at the sunspot maximum (minimum).

7.5 The Solar Atmosphere

The sun has an extensive atmosphere with several layers the boundaries of which are not clearly defined in every case, especially in the case of the corona which is the outermost layer and which extends into interplanetary space.

7.5.1 The Reversing Layer

Immediately above the photosphere lies a thin layer of the solar atmosphere, approximately 600 km thick, which is comparatively cooler than the photosphere and contains gases which can selectively absorb some of the radiation coming out of the photosphere. This is called the reversing layer, because the absorption gives rise to appearance of dark lines in the emission spectrum of the sun as received by the earth, at wavelengths characteristic of the gas present in the reversing layer. These lines were first observed by Fraunhofer, after whom they are called Fraunhofer lines. By identifying the Fraunhofer lines with emission lines in the spectrum of a terrestrial gas, one can get to know about the presence or absence of the particular gas in the solar atmosphere.

7.5.2 The Chromosphere

The reversing layer gradually merges into the chromosphere (from chromos, the Greek word for colour) which is about 1000 km-thick and becomes visible a few seconds before and after a total solar eclipse as a narrow crimson or ruby coloured band at the extreme limb of the sun. The temperature of the sun rises to about 10,000 K in the chromosphere. The gases are highly rarefied in this layer. Spectroscopic examination has shown that the colour of the chromosphere is caused by strong hydrogen emission, thereby demonstrating the presence of hot hydrogen in the layer.

7.5.3 The Corona

Above the chromosphere lies the corona (Latin word for crown). Like the photosphere, the corona is also self-luminous, but the intensity of its light is so feeble that it cannot be seen with the naked eye except at the time of total solar eclipse when the dazzling light of the sun's face is blocked by the moon. However, it can be routinely observed by coronagraphs carried on board space satellites. The hot plasma of the corona emits most of its energy at ultraviolet, extreme ultraviolet and X-ray wavelengths. Ultraviolet radiation also comes out of the chromosphere and the transition region at its base. Radiation at these wavelengths is almost totally absorbed by the earth's atmosphere, so it must be observed through telescopes in space. Observations reveal that there is a sharp rise in temperature in a transition layer at the base of the corona where it jumps from about 10,000 K to 1,000,000 K or even more through a height of less than 100 km. The density of the gaseous plasma decreases sharply as the temperature shoots up. This is evident from Fig. 7.4, which shows the radial distribution of temperature and density in the solar atmosphere.

What causes this phenomenal rise in temperature in the solar corona and how heat can travel from a cooler to a hotter layer, without violating the second law of ther-

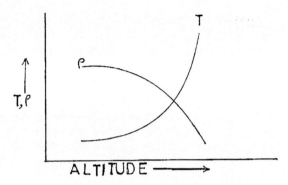

Fig. 7.4 Schematic showing radial distribution of temperature (T) and density (ρ) in the solar atmosphere

modynamics, remains a mystery. However, scientists feel that one of the plausible heating mechanisms may be the release of tremendous amount of heat energy in the gaseous plasma when oppositely-directed magnetic fields associated with magnetic loops interact by coming together in the great turmoil of the solar surface. The interaction would produce powerful electrical and magnetic 'short circuits' to heat up the coronal plasma. The gigantic flashes produced during these short circuits may also explain the origin of 'flares' on the solar surface.

The corona extends deep into space. During a total solar eclipse, it appears as a ghost-like white halo round the darkened face of the sun. However, the appearance of the corona changes with solar activity in an 11-year cycle, very much like the sunspots. During peak activity, it looks spiky and jagged with streaks of light flying off in all possible directions, while in quiet periods, it looks more even and rounded. On closer scrutiny, the space satellite Ulysses found that the observed jaggedness at maximum solar activity was due to luminous gases erupting out of the chromosphere in what are known as 'prominences'. The space satellite SOHO on 15 January 1996 observed magnetically-energized violent eruptions from the sun for a period of 8 h. Such Coronal Mass Ejections (CME) expand at high speed as they propagate outward from the sun. Like sunspots, the intensities of CMEs, flares and prominences also vary with the sun's 11-year cycle of magnetic activity.

The extreme ultraviolet and X-ray images of the sun taken by Yohkoh and SOHO have revealed the presence of large dark areas near its poles where there appears to be little hot material. These areas are known as 'coronal holes'. Their presence has been confirmed by the space satellite Ulysses which flew over both the poles of the sun in 1994–5.

7.6 The Solar Wind

The corona is for ever expanding into space and pouring out high-energy electrically-charged particles, such as electrons and protons as well as heavier ions and atomic nuclei in streams varying in speed from 300 to 1000km s^{-1}. These streams constitute what is known as the solar wind. The faster particles are generated during particularly energetic bursts of solar activity and ejected from solar latitudes higher than about 20°N and S.

The solar wind being under the control of the magnetic field of the sun which rotates with a mean period of about 27 days is ejected from the sun in a direction somewhat similar to that of water being ejected from a rotary garden sprinkler. Being constrained to move along the magnetic lines of force, the solar wind has no fixed direction. It rushes through space in a most complex and unpredictable manner filling up the whole interplanetary space with the solar radiation. As radiant energy gushes out from the outer layers of the sun's atmosphere as the solar wind, new radiation wells up from the interior of the sun to take its place. Ulysses in the mid-1990s found that the fast component of the solar wind pours out from the polar coronal holes where there are no magnetic field loops to prevent the escape of the charged particles from the sun's strong gravitational and magnetic pulls. The slow component emerged from equatorial corona.

Recent theoretical studies have suggested that although the solar wind is composed of charged particles of both positive and negative signs, it must travel through interplanetary space as an electrically neutral beam consisting of equal number of positive- and negatively charged particles. It is only on entering the earth's magnetic field that the charges get separated and follow the magnetic lines of force to enter the earth's atmosphere around the magnetic poles. In their downward passage, they pass through several layers of trapped radiation, such as the well-known van Allen radiation belts at altitude of about 5 earth radii, and then further down through the different ionization layers of the earth's upper atmosphere.

7.7 The Search for Neutrinos

In recent years, a good deal of research work on solar physics has been devoted to finding evidence for the presence of neutrinos in the solar beam arriving on earth, because it was believed that a positive finding would constitute a direct proof of the theory of nuclear fusion in the sun as the source of its energy. Attempts have been made by astrophysicists in several countries in this direction. After decades of search, the first positive evidence was reported from the United States of America where Raymond Davies, Jr., of the Brookhaven National Laboratory who later became a research professor in the University of Pennsylvania, succeeded in capturing solar neutrinos in a tank deep underground in a South Dakota mine. He started his research work in 1961 by placing a large tank of perchloroethylene – commonly used as a drycleaning fluid – in a deep mine 2300 ft underground in the State of Ohio. With promising results from this work, he later installed a 100.000-gallon tank filled with perchloroethylene 4850 ft underground in the Homestake Gold Mine in Lead, South Dakota. After over 30 years of sustained work, he successfully observed as many as 2000 neutrino events which conclusively demonstrated the occurrence of nuclear fusion in the sun. Dr. Davies, Jr., was awarded a Nobel Prize in Physics in 2002 for his epoch-making discovery. He died in 2006 (The Washington Post, June 4, 2006).

Chapter 8
The Incoming Solar Radiation – Interaction with the Earth's Atmosphere and Surface

8.1 Introduction – the Solar Spectrum

An examination of the spectrum of the solar radiation received at the earth's surface shows that it departs significantly from the theoretical curve, shown in Fig. 6.2, in respect of radiation from a black body at the temperature of the sun that is at about 6,000 K. The difference becomes evident from Fig. 8.1, reproduced from Brunt (1944), which presents the intensity of solar radiation at different wavelengths received at surface at Washington with the sun at different solar altitudes (curves II, III, IV and V at solar altitudes 90°, 30°, 19.3° and 11.3° respectively) and also at the outer boundary of the atmosphere in the latitude of Washington, after allowing for estimated absorption and scattering of the atmosphere (Curve I). Curve VI gives the relative brightness of the different parts of the spectrum. Curve VII gives the intensity distribution of the skylight at different wavelengths.

Two aspects of the departure from Fig. 6.2 stand out in Fig. 8.1. These are:

(a) A sharp cut-off of energy in the short-wave part of the spectrum below about 0.3μ; and
(b) Appreciable loss of energy in passing through the terrestrial atmosphere at all wave-lengths. The loss becomes more and more pronounced with decrease of solar altitude.

The intensity of the net solar radiation that we receive at the earth's surface is just about half of what is intercepted by the earth at the outer limit of its atmosphere. The short wave part of the solar radiation does not reach the earth's surface, since it is totally absorbed by the outer layers of the earth's atmosphere which may be called the upper atmosphere. So, it is the radiation from the sun's surface at wavelengths above 0.3μ that comes down to the lower layers in the form of heat and light. But, before solar radiation can reach the earth's surface, it has to pass through the different layers of the atmosphere. Several important effects are produced by the physical and chemical processes that occur in these layers as the radiation passes through them to the earth's surface. Some of these are the following:

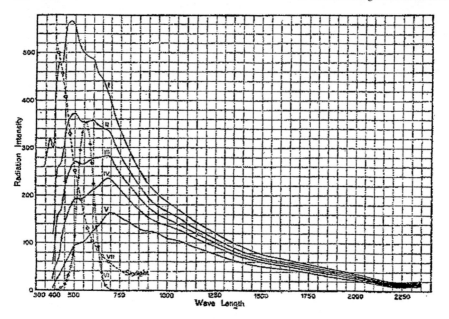

Fig. 8.1 Spectrum of the sun at Washington (Reproduced from Brunt, 1944, with kind permission of Cambridge University Press)

(a) Absorption, heating and ionization in the upper atmosphere (above about 80 km) at wavelengths $< 0.2\mu$);
(b) Absorption and heating by ozone in the stratosphere, between about 10 and 50 km, at wavelengths between about 0.2μ and 0.3μ; and
(c) Scattering, reflection and absorption by air molecules, particulate matter suspended in the atmosphere and by clouds;

The above-mentioned processes are briefly described in the sections that follow.

8.2 Interactions with the Upper Atmosphere (Above 80 km)

8.2.1 Interaction with the Solar Wind: Polar Auroras and Magnetic Storms

The solar wind which travels in interplanetary space at a supersonic speed (400–770 km per second) as a fully ionized and magnetized gas under the magnetic field of the sun, on approaching the earth, interacts with the earth's magnetic field while at a distance of several earth-radii, producing a comet-shaped cavity called a magnetosphere around the earth, while the lines of force of both the magnetic fields interconnect. The charged particles of the solar wind moving across the interconnected magnetic lines of force generate electric power of as much as 10^9 kW

8.2 Interactions with the Upper Atmosphere (Above 80 km)

near the boundary of the magnetosphere. A part of the electric current, so generated, flows down the interconnected magnetic lines of force towards the polar regions of the earth with sufficient energy to bombard the atoms and molecules of the gases present there till it is stopped by the atmosphere at an altitude of about 100 km. The gases affected by the process are mostly oxygen and nitrogen. The atoms and molecules of these gases absorb the radiation in the extreme ultraviolet and X-ray wavelengths and in the process undergo chemical dissociation or excitation and ionization. The excited and ionized atoms and molecules then give out their own characteristic radiation in the form of brilliant colourful lights known as polar auroras.

Auroral displays take varieties of forms, such as arcs, rays, curtains, crowns (corona), draperies, bands or even diffuse luminous surfaces. They may appear over latitudes poleward of about 45°, but the frequency of their occurrence is maximum around the magnetic poles. People have observed auroras over the Polar Regions, one in each hemisphere, for ages. They are known as 'aurora borealis' in the northern hemisphere and 'aurora australis' in the southern hemisphere. Recently, the NASA's Dynamic Explorer 1 satellite, flying over the earth's polar-regions, found that the auroras formed around the magnetic poles as glowing ovals or rings, about 500 km wide, 4500 km in diameter, and centered on the magnetic poles. Most auroral displays occur in the layer between 100 and 250 km above the earth's surface and are dominated by the green light emitted by the oxygen atoms at wave-length 5577 Å (1 Angstrom = 10^{-8} cm) and ruby-red light emitted by the oxygen ions at wave-lengths 6300 and 6364 Å. Atomic and ionized molecular nitrogen also add emissions in several bands of wavelength. The presence of these spectral lines provides unmistakable evidence of the presence of oxygen and nitrogen in both molecular and atomic form in the upper atmosphere. In the case of oxygen, atomic oxygen is produced by the strong absorption of ultraviolet radiation in the Runge-Schumann band (wave-lengths 1751–1200Å) by the oxygen molecules:

$$O_2 + h\nu \rightarrow O \text{ (excited)} + O \text{ (excited)}$$

It is likely that similar absorption of ultraviolet radiation by the nitrogen molecules leads to their dissociation into nitrogen atoms. However, the exact process in the case of atomic nitrogen is not yet clearly established.

The auroral activity undergoes sudden changes at times, for example during solar flares, when the solar wind becomes gusty and the orientation of the solar wind magnetic field with respect to the earth's magnetic field changes. The disturbed magnetic fields then cause what are known as geomagnetic storms. During such storms, sometimes cracks occur in the earth's magnetic shield and the dangerous solar wind leaks through them. Some recent studies with space probes such as NASA's IMAGE craft and four European Cluster satellites flying at different altitudes to obtain simultaneous measurements of one of the cracks in the shield found evidence of large volumes of the solar wind flowing through it towards the earth. The data indicated that the opening was twice the size of the earth at an altitude of about 38,000 miles, narrowing to about the size of California at the height of the earth's upper atmosphere.

Such storms have been found to interfere with the earth's radio communication.

8.2.2 Interaction with the Solar Ultraviolet Radiation

When the electromagnetic radiation from the sun's photosphere enters the earth's atmosphere, the high-frequency part of the radiation lying in the extreme ultraviolet and X-ray wavelengths ($< 0.2\mu$) impinges on the gases present in the upper atmosphere almost in the same manner as the solar wind, producing similar effects of dissociation, excitation and ionization of atoms and molecules, though there is a difference in the manner of their entry into the earth's atmosphere. While the solar wind is guided by the magnetic fields to move towards the magnetic poles only, there is no such embargo on the solar radiation coming from the sun's photosphere. But for this difference, they both interact with the upper atmosphere almost the same way.

There is enough observational evidence to hold that both the radiations contribute importantly to the creation of an extensive ionized layer around the earth known as the ionosphere which plays a fundamental role in the propagation of radio waves around the earth. They are also responsible for heating of the upper atmosphere above 80 km which has been designated as the thermosphere. In the thermosphere, the temperature may rise with altitude to a peak that varies from about 500 K to 2000 K depending upon solar activity. The thermosphere is stable to vertical mixing.

8.3 The Mesosphere (50–80 km Layer)

Below 80 km, there is a layer about 30 km deep, called the mesosphere, in which the temperature decreases with altitude, though at a slower rate than in the troposphere, from a value of about 400 K at 50 km to about 130 K at 80 km. The top of the layer is called the mesopause. The appearance of thin, noctilucent clouds at these heights, that reflect twilight to the earth long after sunset or long before sunrise, provides some evidence of the presence of water vapour in this layer.

The mesosphere may be described as the earth's intermediate or middle atmosphere, which separates the thermosphere above from the ozonosphere (or stratosphere) below, both of which are responsible for absorption of most of the ultraviolet part of the solar radiation and preventing it from reaching the earth's surface.

8.4 Interaction with Ozone: the Ozonosphere (20–50 km)

Below the mesosphere lies the stratosphere, or the ozonosphere as it is sometimes called, which extends from about 50 km down to about 20 km with maximum concentration of ozone between about 25 and 30 km. The total content of ozone in the layer amounts to hardly 1–2 mm when reduced to standard pressure and temperature, but the small quantity is sufficient to absorb all the ultraviolet radiation between 0.2μ and 0.3μ that enters the layer. Ozone has a strong absorption band, known as the Hartley band, between 0.21μ and 0.32μ. It also absorbs in several other bands, such as, Huggins bands between 0.32μ and 0.36μ, the Chappuis band between 0.45μ and 0.65μ and three absorption bands in the red and infra-red part of the spectrum with maximum absorption co-efficients at 4.7μ, 9.6μ and 14μ. But, compared to Hartley bands, the co-efficients of absorption in these other bands are quite small.

The content of ozone in the stratosphere is maintained by an approximate balance between its formation and destruction by the physical processes stated below.

8.4.1 Formation of Ozone

Ozone forms by the photodissociation of oxygen molecules into atoms by the ultra-violet radiation. The oxygen atoms formed thereby then combine with the oxygen molecules to form ozone. Thus,

$$O_2 + h\nu \to O + O \qquad (8.4.1)$$
$$O + O_2 + M \to O_3 + M$$

where ν is the frequency of the radiation and M is a third body which absorbs the extra energy and momentum released during the dissociation. The third body here simply acts as a catalyst.

The collision of an excited oxygen molecule with a neutral oxygen molecule may also produce ozone, as shown below.

$$O_2(\text{excited}) + O_2 \to (O_2 + O) + O \to O_3 + O \qquad (8.4.2)$$
$$O + O_2 + M \to O_3 + M$$

The oxygen atom released by the dissociation process in the first line of (8.4.2) combines with the oxygen molecule in the presence of a third body M to form ozone.

There are reasons to believe (Mitra, 1952) that although the concentration of ozone is found to be maximum in the 25–30 km layer, the formation of ozone takes place more efficiently at a much higher level in the stratosphere and that the process represented by (8.4.1) is more effective than (8.4.2).

8.4.2 Destruction of Ozone: the Ozone Hole

Ozone is destroyed by the process of photolysis (dissociation by collision with light waves) as well as by collision with atomic oxygen.

In photolysis, a light wave of frequency ν interacting with an ozone molecule produces an oxygen molecule and excited oxygen

$$O_3 + h\nu \to O_2 + O \text{ (excited)} \qquad (8.4.3)$$

However, in the ozonosphere, collision with atomic oxygen produced by the ultraviolet radiation at wavelengths of the Hartley band appears to be a more effective mechanism in destroying ozone. Thus,

$$O_3 + O \rightarrow O_2 (\text{excited}) \tag{8.4.4}$$

Other gases (or gas radicals) that may get involved in the destruction of ozone in the stratosphere include NO, NO_2, H, OH, and Cl. Their role is mostly that of a catalyst. For example, in the case of Cl, the destruction mechanism may be stated as follows:

$$Cl + O_3 \rightarrow ClO + O_2 \tag{8.4.5}$$
$$ClO + O \rightarrow Cl + O_2$$

Since stratospheric ozone protects us from the harmful effects of ultraviolet radiation, any reduction in the ozone content by such destructive processes as (8.4.5) poses a great danger to humankind and, perhaps, also to other forms of life on earth. The safe limit for humans has been set at 220 Dobson units and a reduction in overhead ozone content below this level is described as an ozone hole. Studies have shown that in recent years there has been an increase in the content of Cl in the atmosphere by human activities on earth and that stratospheric ozone has been depleted over the Antarctic region sufficiently to cause alarm. An international treaty (The Montreal protocol, 1987) has called for a ban on the use of products such as chlorofluorocarbons (CFCs) which release Cl into the atmosphere.

8.4.3 Warming of the Stratosphere

The ozonosphere is virtually a heat reservoir in the stratosphere, corresponding to the thermosphere above 80 km in the upper atmosphere. Both the layers absorb solar energy in the short wave part of the spectrum, the thermosphere at wavelengths below 0.2μ and the ozonosphere between 0.2μ and 0.3μ and together must be held responsible for the observed sharp cut-off of energy below 0.3μ in the solar spectrum (Fig. 8.1).

In the stratosphere, the energy balance shows that the dissociative heating due to absorption of ultraviolet radiation more than balances the energy lost during destruction and radiation. Hence there is a net warming of the atmosphere in the upper layers of the stratosphere, though ozone is concentrated in the lower layers.

8.4.4 Latitudinal and Seasonal Variation of Ozone

Figure 8.2 shows the latitudinal and seasonal variations of ozone content in the atmosphere (after Dobson, 1931).

8.4 Interaction with Ozone: the Ozonosphere (20–50 km)

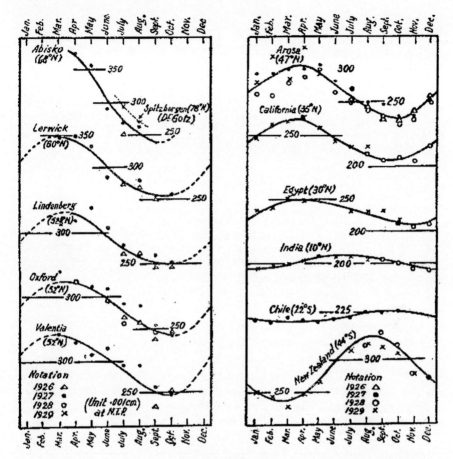

Fig. 8.2 Seasonal variation of atmospheric ozone content in different latitudes Unit: 0.001 cm at S.T.P. (Reproduced from Mitra, 1952, published by the Asiatic Society of Bengal, India)

In general the ozone content in the stratosphere is much greater in high latitudes than that in low latitudes. Also, in higher latitudes, there is a large annual variation with very large values in spring as compared to those in autumn. In the equatorial stratosphere, not only is the ozone content small but it has also a small annual variation.

8.4.5 Ozone and Weather

(a) Association of ozone with high-latitude cyclonic disturbances

In middle and high latitudes, the ozone content in the stratosphere appears to vary significantly with the movement of fronts associated with tropospheric disturbances.

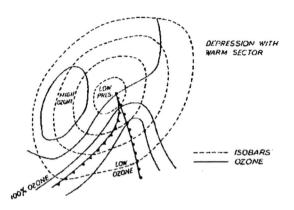

Fig. 8.3 Illustrating the distribution of stratospheric ozone around a typical young depression with a very marked warm sector. Note that the ozone content is low over the warm sector and high above the advancing cold airmass behind (After Dobson, 1931; Dobson et al., 1946). (Reproduced from Mitra, 1952, published by the Asiatic Society of Bengal, India)

For example, Dobson and his coworkers found that in the case of an eastward-propagating cyclone, if we take a vertical section at right angles to the fronts, there is a definite fall in ozone content ahead of an advancing warm front and a definite rise behind an advancing cold front, as shown in Fig. 8.3, reproduced from their paper (Dobson et al., 1946).

The observed variations of stratospheric ozone in the field of a midlatitude cyclone are not difficult to understand if we consider the latitudinal distribution of ozone in relation to the circulation. The southerly winds in the warm sector of the cyclone advect air from the south which has relatively less ozone content, whereas the northerly winds in the cold sector usher in airmass from high latitudes which is rich in ozone content. The effects are, however, observed only when the fronts extend well into the stratosphere. No such correlation has been found so far over the tropics or low latitudes where fronts are rare.

(b) Correlation of ozone with other meteorological parameters

Dobson (1931) found that in temperate latitudes, the ozone content varies inversely with the mean atmospheric pressure at 9–16 km level and with the mean temperature of the troposphere.

Meetham (1936) has shown that ozone has a high correlation with the potential temperature and density in the stratosphere.

(c) Variation of ozone with solar activity and terrestrial magnetic storms

Studies conducted so far seem to indicate that the average variation of the ozone content does not have any firm relationship with the eleven-year solar cycle.

A definite association has been found between the ozone content in the middle atmosphere and terrestrial magnetic storms. The ozone value always increases when there is a magnetic storm on the earth.

8.5 Scattering, Reflection and Absorption of Solar Radiation in the Atmosphere

8.5.1 Scattering and Reflection

The intensity of the incoming solar radiation is depleted by the processes of scattering, reflection and absorption by air molecules or other particulate matter that may be suspended in the atmosphere and by clouds. In scattering, a straight parallel beam of radiation changes direction either sideways or backwards. However, the degree of scattering depends upon the size of the molecules or particles compared to the wavelength of the incident beam. If the size is equal to or comparable to the wavelengths of the visible light, a good amount of light will be scattered, depending upon the scattering co-efficient for that wavelength. The well-known Rayleigh-Cabannes law of molecular scattering states that the light scattered is inversely proportional to the fourth power of the wavelength of the incident beam. Thus, if $I_{0\lambda}$ be the intensity of the light radiation of wavelength λ falling on a scattering particle of about the same size, the amount of light scattered I_λ is given by the following relation:

$$I_\lambda = k_\lambda \, I_{0\lambda}/\lambda^4 \qquad (8.5.1)$$

where k_λ is the scattering co-efficient.

Since, according to this law (8.5.1), the blue light is scattered more than the yellow or red, it accounts for the observed blue colour of the sky on a clear day. At sunrise or sunset, the sky turns yellow or red because most of the blue is scattered away and lost from view, yielding place to yellow or red. Also, at these times, sunlight has to pass through a much thicker and denser layer of the atmosphere than at noon or other times of the day.

If the scattering particles are not small compared to the wavelength, the Rayleigh-Cabannes law does not hold any more. In such cases, scattering is said to be neutral to wavelengths and no particular colour dominates. For example, scattering by large water droplets being neutral, clouds look white. For large-size droplets or particles, there is more of diffuse reflection or absorption of radiation than true scattering. In fact, a cloud layer is capable of reflecting nearly 80 percent of the incident radiation and plays a most decisive role in controlling the net inflow of solar radiation to the earth's surface.

Scattering and reflection by particulate matter thrown up into the atmosphere by major volcanic eruptions, which remains suspended in the air for a long time, can markedly reduce the intensity of the incoming solar radiation.

8.5.2 Atmospheric Absorption

Let a beam of radiation of intensity I at wavelength λ pass through a layer of the atmosphere and let a fraction dI be absorbed by the mass dm of the layer. Then,

$$dI/I = -a_\lambda\, dm \qquad (8.5.2)$$

where a_λ is the co-efficient of mass absorption.

Let us consider the case when the absorbing gas is at such a low temperature that its radiation may be neglected. Then, on integrating (8.5.2.), we get

$$I/I_0 = \exp\left(-\int a_\lambda\, dm\right) \qquad (8.5.3)$$

where I_0 is the intensity of the incident radiation.

The relations (8.5.2) and (8.5.3) are the differential and integral forms respectively of an absorption law known as the Beer's law. (8.5.3) may also be written in the form

$$I/I_0 = \exp\left(-\int d\tau\right) \qquad (8.5.4)$$

where $d\tau$ is called the optical thickness of the absorbing layer and stands for $a_\lambda dm$..

If the incident beam is inclined to the vertical at an angle θ, then (8.5.4) takes the form

$$I/I_0 = \exp\left(-\int \sec\theta\, d\tau\right). \qquad (8.5.5)$$

We have seen in the preceding sections that almost all the energy in the ultraviolet and extreme ultraviolet part of the solar spectrum is absorbed by atmospheric gases, such as, oxygen, nitrogen and ozone, in the upper atmosphere. Besides this absorption, there is very little absorption of the incoming solar radiation by other gases in the atmosphere. However, water vapour, carbon dioxide and ozone absorb some energy in the longwave part of the spectrum at lower levels of the atmosphere. It is estimated that with a cloudless sky and the sun at the zenith, the total loss of intensity of solar radiation due to absorption in the atmosphere may not exceed 6–7% of the incident radiation.

8.6 Incoming Solar Radiation (Insolation) at the Earth's Surface

The intensity of solar radiation that ultimately reaches a place on the earth's surface depends upon the following factors:

(a) The value of the solar constant,
(b) The transparency of the atmosphere,
(c) The latitude of the place, and
(d) The seasonal and diurnal variation

8.6.1 The Solar Constant

It is remarkable that inspite of inconstant sun with large fluctuations in its radiation during solar flares and prominences, etc., the intensity of solar radiation reaching

8.6 Incoming Solar Radiation (Insolation) at the Earth's Surface

the outer boundary of the earth's atmosphere has remained more or less constant over the last few centuries. The reasons for this may be the following: the large fluctuations that often occur in solar activity mostly affect the extreme ultraviolet and X-ray part of the solar spectrum which contains a very small amount of solar energy as compared to the visible and longwave part which has more than 98% of the energy. Thus, in the mean, the variations are found to be almost negligible and the accepted mean value of the solar constant is about 1368 W m^{-2}.

Ångstrom, however, found from some measurements that the solar constant, S_0, could be connected with the sunspot activity by the formula (see, e.g., Saha and Srivastava, 1931, p 561)

$$S_0 = 1.903 + 0.011\sqrt{N} - 0.0006 N$$

where N is a co-efficient (known as the Wolf and Wolfer's number) characterizing the number and extent of sunspots, and S_0 is given in units of Cal/cm^2/min.

8.6.2 The Transparency of the Atmosphere – Effects of Clouds and Aerosols

In Sect. 8.4, we pointed out some of the factors which interfere with the passage of the incoming solar radiation through the atmosphere. Amongst these, the presence of the clouds is, perhaps, the most important, since it can cut out almost 70–80% of the incident radiation. Amongst the others involving particulate matter one may, perhaps, mention the periodic release of enormous amount of dust thrown up into the atmosphere by volcanic eruptions around the globe. The erupted material from such explosions remains suspended in the atmosphere sometimes for months and years. However, it is estimated that in the absence of clouds the total loss from the incoming radiation due to factors which cause scattering, reflection and absorption, may not normally exceed 6–7% of the incident radiation.

8.6.3 Distribution of Solar Radiation with Latitude – The Seasonal Cycle

It is well-known that as the earth moves around the sun in an elliptical orbit with the sun at one focus, its obliquity to the sun makes the zenithal position of the sun oscillate between 23.5°N and 23.5°S, in a seasonal cycle during the year. It is overhead at the equator at the equinoxes in March and September when days and nights are about equal. The earth is then at its mean distance of about 1.4968×10^8 km from the sun and receives solar radiation at a more or less constant rate. It is farthest from the sun in June-July when the sun is overhead at 23.5°N and the days are longer than nights and closest to it in December-January when the sun is overhead at 23.5°S and nights are longer than days.

The angle at which the sun's rays strike the earth's surface varies with time of the day, season and latitude. Since the length of a day or night varies with latitude, the insolation varies with latitude and season only. In the summer hemisphere, the length of the day gets longer and longer with increasing latitude till near the pole, it is almost 24 h of daylight. The reverse is the trend in the winter hemisphere where day gets shorter and shorter with increasing latitude and it is almost 24 h of night near the pole. Thus, the duration of sunlight is the longest over the earth's polar region which receives more radiation from the sun than any other place on earth at the time of the summer solstice, though the distance from the sun is greater and the sun's rays fall at slanting angles. Ignoring any loss in intensity due to passage through the atmosphere and using astronomical data, Milankovitch first computed the rate at which solar energy is received at the earth's surface at different latitudes in different seasons (see Fig. 8.4). In later work, he revised his computations by allowing a constant transmissivity of 70% for the atmosphere at all wavelengths and also allowing the sec θ effect (vide Eq. 8.5.5). The result of this later computation is shown in Fig. 8.5.

In the first case (Fig. 8.4), Milankovich found that there is little variation in the intensity of the solar radiation over the equatorial belt (the variation being limited to about 100 Ly per day between summer and winter), since the sun always shines overhead between 23.5°N and 23.5°S. However, large variations occur in polar latitudes where the intensity changes from nearly zero in winter to about 1100 Ly per day in summer. Thus, a secondary maximum in insolation is to be expected in high latitudes, which arises due to a combination of the effects of increasing duration of sunlight with increasing latitude and increasing radiation intensity with decreasing latitude. It is also observed that the maximum north polar radiation is 72 Ly per day less than the corresponding south polar radiation. This is because the earth is closest to the sun at the time of the southern hemisphere solstice and farthest away at the time of the northern hemisphere solstice.

However, it may be seen from Fig. 8.5 that large changes in computed values of the radiation reaching the earth's surface occur with an atmospheric transmissivity of 70%. For example, near the equator, instead of 790 Ly per day, the rate now is about 460 Ly per day. Very significant changes are noticed over the polar-regions, where a net cooling due to longer pathway shifts the maxima over the summer poles to latitudes near 35° at the solstice times.

8.6.4 Seasonal and Latitudinal Variations of Surface Temperature

Variations of the incoming solar radiation with season and latitude cause corresponding changes in surface temperatures which are observed all over the globe. An example is presented Fig. 8.6 for two stations in Asia; an equatorial station in Borneo and Beijing near 40°N latitude.

It is evident from Fig. 8.6 that the amplitude of the seasonal oscillation is close to 1 °C in Borneo, whereas it is about 15 °C at Beijing.

8.6 Incoming Solar Radiation (Insolation) at the Earth's Surface

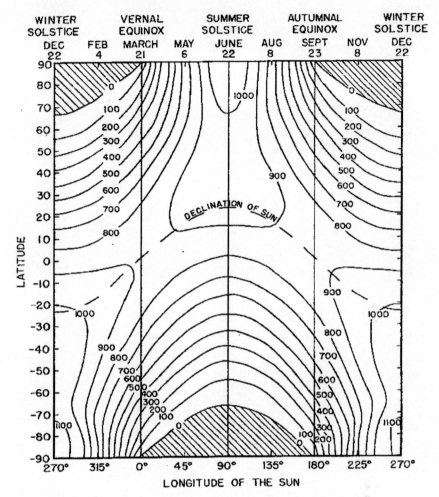

Fig. 8.4 Receipt of solar radiation at the earth's surface in Langley (Ly) per day with no loss in the atmosphere ($1\,\text{Ly day}^{-1} = 0.484\,\text{Wm}^{-2}$) (After Milankovich)

8.6.5 Diurnal Variation of Radiation with Clear and Cloudy Skies

We are all familiar with the diurnal cycle of temperature at a place. With the rising of the morning sun, air temperature goes up, reaches a maximum soon after local noon and then drops as the sun goes down in the evening. However, it is observed that more often than not this regular cycle is disturbed when there is excessive moisture in the atmosphere with clouds overhead. An example is given in Fig. 8.7 which shows the diurnal cycle of surface temperature at Washington, D.C. over a period of 5 days (from 1 to 5 July, 2003).

112   8 The Incoming Solar Radiation

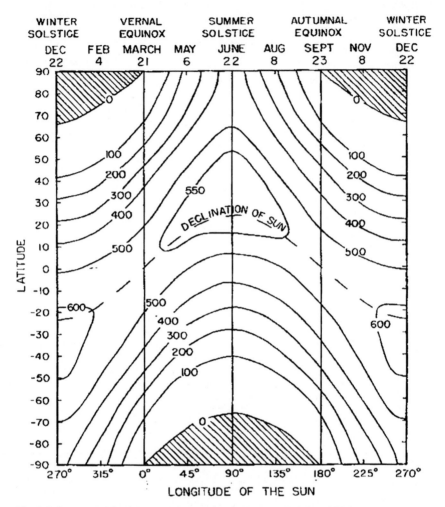

Fig. 8.5 Same as in Fig. 8.4, but with an atmospheric transmissivity of 70%.

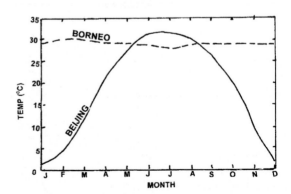

Fig. 8.6 The average daily maximum temperature (°C) in different months in Borneo and Beijing

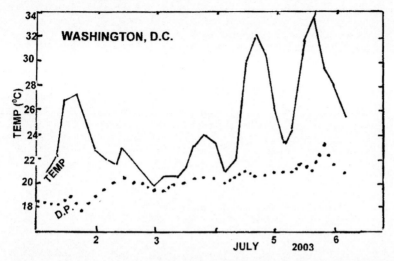

Fig. 8.7 Diurnal variation of surface temperature (TEMP) and dew-point (D.P) in °C at Washington, D.C., at 4-hourly intervals from 00 EST on 1 July to 24 EST on 5 July, 2003 (Daily maxima and minima are not marked)

It shows how the diurnal cycle of temperature was disrupted by an almost continuous rise of dewpoint from a value of about 17.7 °C late on 1 July to about 20.6 °C around noon the following day. The increase of the water vapour content of the atmosphere suppressed the amplitude of the diurnal cycle by causing the daytime maximum air temperature to come down and the nighttime minimum temperature to go up, while the dew-point rose rapidly. There was rainfall in the city during the period of moisture build-up between 1 and 2 July and the weather remained largely cloudy and disturbed till 4 July.

8.7 Reflection of Solar Radiation at the Earth's Surface – The Albedo

Not all the solar radiation that reaches the earth's surface is absorbed by it, since a part of the radiation is reflected back to space by what is known as the earth's surface albedo. The albedo is defined as the fraction of the incident radiation that is reflected by the surface. The earth's surface not being uniform, the albedo varies widely from place to place depending upon the nature and composition of the underlying surface. For example, the albedo of a dense forest is very different from that of a freshly-covered snow surface, or, for that matter from a given water surface at different solar elevations, as shown in the Table 8.1, which gives the approximate values of albedoes for a few selected types of surfaces (after Riehl, 1978).

Table 8.1 Typical values of earth's albedo, (After Riehl, 1978)

Surface type	Albedo, α (%)
Forests	3–10
Fields, green	3–15
Fields, dry, plowed	20–25
Grass	15–30
Bare ground	7–20
Sand	15–25
Snow, fresh	80
Snow (old)/ice	50–70
Water, solar elevation $> 40°$	2–4
Water, solar elevation 30–5°	6–40

Chapter 9
Heat Balance of the Earth's Surface – Upward and Downward Transfer of Heat

9.1 Introduction: General Considerations

It is estimated that of the total solar energy that is intercepted by the earth at the top of the atmosphere, only about half reaches the earth's surface and is absorbed by it as heat energy. When heated, the surface emits its own radiation. We showed in Fig. 6.2 that radiation from bodies at the temperatures of the earth's surface and atmosphere lies in the longwave part of the spectrum with maximum emissions at wavelengths between about 10μ and 15μ. It so happens that a fraction of this longwave radiation moving upward is absorbed by some gases present in the atmosphere, such as water vapour and carbon dioxide, which have strong absorption bands in the infrared part of the spectrum. The absorption leads to warming of the gases which then emit their characteristic longwave radiation, a part of which is sent back to the earth. Thus, the net radiative heating of the earth's surface depends on three factors:

(a) Heat gained from the incoming solar radiation, I,
(b) Heat lost or emitted by longwave radiation, E, and
(c) Heat gained from atmospheric gases as downward longwave radiation, G.

Concurrently, the earth's surface loses sensible heat (H_s) through vertical exchanges with the atmosphere above and the ground or water below. Over a water surface, considerable heat (H_e) is lost by evaporation.

So, the temperature of the earth's surface is determined by the following balance relation:

$$(I - E + G) - H_s - H_e = 0 \tag{9.1.1}$$

where

I is the net solar radiation absorbed by the earth's surface,
E the longwave radiation emitted by the surface,
G the longwave radiation returned to the earth's surface by the atmospheric gases (mainly water vapour and carbon dioxide),
H_s the sensible heat lost to the atmosphere above and soil or water below, and
H_e the heat lost by evaporation of water from the surface.

An accurate determination of the heat balance of the earth's surface using (9.1.1) is well-nigh impossible, because of the great inhomogeneity of the earth's surface and our incomplete knowledge and understanding of the various heat transfer processes.

About two-thirds of the earth's surface consists of oceans and one-third land. Though the major parts of the oceans appear as water surfaces, the polar regions are largely frozen and appear as solid ice surfaces. Then, there are the warm and cold ocean currents which cover wide areas and introduce large inhomogeneity in ocean surface temperatures. The land surface also is highly uneven, in fact more so than the ocean surface, with high snow-clad mountains over several areas flanked by warm valleys, vast deserts and lands co-existing with equatorial rainforests and rivers and in-land lakes, and so on. To date, inspite of numerous field experiments and laboratory and theoretical studies, the values of the co-efficients of heat exchanges with the environment under different stability conditions are only approximately known. So, when we talk of the heat balance of the earth's surface, the existing inhomogeneity of the surface and the above-mentioned uncertainties in the values of the exchange co-efficients must be borne in mind. However, notwithstanding these limitations, we may, perhaps, consider some broad aspects of the heat transfer processes outlined in the relation (9.1.1). We start off with the radiative processes.

But, before we consider the actual atmosphere, let us ask: What would have been the heat balance at the surface of our planet if it did not have an atmosphere? Although some may regard the question as hypothetical, it helps one to understand what role the existing atmosphere plays in the heat balance and what it means to life on earth.

9.2 Heat Balance on a Planet Without an Atmosphere

The hypothetical scenario is depicted in Fig. 9.1.

On an earth without an atmosphere, the solar radiation would impinge directly at the earth's surface and be absorbed by it to raise its temperature. Let S be a measure of the energy received per unit area of a surface held perpendicular to the incoming radiation per unit time, which, in this case, will be equal to the solar constant, S_0.

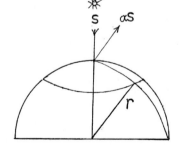

Fig. 9.1 Radiative heating and cooling of a planet without an atmosphere

Radiation intercepted by the earth then amounts to πr^2 S, where r is the mean radius of the earth. A fraction of this incident radiation, α, will be reflected from the surface and returned to space. The balance, $S(1-\alpha)$, will be absorbed by the surface of the earth. As a result of this absorption, let the surface temperature be raised to T. The surface will then give out longwave radiation which, according to Stefan-Boltzmann law, will amount to σT^4 per unit area of the earth's surface.

The balance between the incoming and the outgoing radiation may then be given by the relation

$$S(1-\alpha)\pi r^2 = \sigma T^4 \, 4\pi r^2 \qquad (9.2.1)$$

From (9.2.1), we get

$$T = \{S(1-\alpha)/4\sigma\}^{1/4} \qquad (9.2.2)$$

If it is assumed that the albedo of the earth, α, in the hypothetical case was no different from the present value of about 30%, then substituting the known values of S and σ in (9.2.2), we get a value of about 255 K for T, which is about 33 °C colder than the average temperature on the present earth. So, the question is: what made up this deficiency in the earth's surface temperature? The answer clearly is the present-day atmosphere with one of its remarkable property, popularly known as the greenhouse effect. In the following section, we look into some aspects of this property of our atmosphere.

9.3 Heat Balance on a Planet with an Atmosphere: The Greenhouse Effect

The presence of an atmosphere changes the scenario depicted in the foregoing section. In the atmosphere, there are some gases, such as water vapour, carbon dioxide and ozone, which, though they exist in very small and variable proportions, play very important roles in the heat budget of the earth's surface, because they are almost transparent to the incoming shortwave radiation but absorb a large fraction of the outgoing longwave radiation emitted by the earth's surface after the latter is heated by the shortwave radiation of the sun. The absorbed energy raises the temperature of the gases which then give out their own characteristic longwave radiation a part of which, moving downward, is returned to the earth. Thus, the presence of the gases prevents the earth's surface from being continually cooled by the outgoing longwave radiation. The process is somewhat similar to that which occurs in a greenhouse which allows the shortwave solar radiation to come in during daytime and warm up the enclosure, but does not allow the longwave radiation from inside to leave the enclosure with the result that the inside remains warm enough for the in-house plants to survive especially in cold climates where the outside may be extremely cold. For this reason, the atmospheric gases which absorb the outgoing longwave radiation from the earth's surface and re-radiate a part of it back to the earth's surface are called greenhouse gases and the process the greenhouse effect. The process is explained more fully in the next sub-section.

9.3.1 The Greenhouse Effect

The working of a layer of greenhouse gas in the atmosphere may be compared to that of a slab of glass placed horizontally a little above the earth's surface and exposed to the incoming solar radiation. It is well-known that glass is transparent to radiation of wavelengths shorter than about 4μ but partially opaque to radiation of longer wavelengths. Likewise, the gas layer is transparent to shortwave solar radiation moving downward but partially opaque to longwave radiation moving upward from the earth's surface (Fig. 9.2).

In Fig. 9.2, let I denote the net downward-moving solar radiation which after passing through the gas layer warms up the earth's surface to a temperature T. The earth's surface then emits longwave radiation at this temperature which is given by the Stefan-Boltzmann's law,

$$U = \sigma T^4 \tag{9.3.1}$$

Where U denotes the upward flux of the longwave radiation

Now, if a_λ is the absorption coefficient of the gas for wavelength λ, a fraction $a_\lambda U$ of the upward-moving flux is absorbed by the layer, while the remainder, $U(1-a_\lambda)$, leaves it. The absorbed energy warms up the gas which then emits its own radiation, approximately half of which moves upward and half downward. Let the part moving upward or downward be denoted by G. Then, for equilibrium, the upward- and downward-moving radiations must balance. Thus,

$$I = U(1-a_\lambda) + G = U - G \tag{9.3.2}$$

From (9.3.2), we get $G = a_\lambda U/2$, and $U = I/(1-a_\lambda/2)$.
Substituting the value of U from (9.3.1) in (9.3.2), we get

$$T = [I/\{\sigma(1-a_\lambda/2)\}]^{1/4} \tag{9.3.3}$$

If we assume that the layer absorbs all the longwave radiation that enters it from below, i.e., if the absorption co-efficient a_λ is unity, it is easy to see from (9.3.3) that

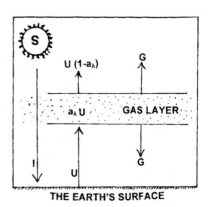

Fig. 9.2 Illustrating the greenhouse effect

the presence of the greenhouse gas enhances the ground temperature by about 19%. Laboratory experiments show that water vapour has a very strong absorption band in the longwave part of the spectrum. So if this gas is present in sufficient concentration in the atmosphere as with a cloudy sky, it would absorb practically all the outgoing longwave radiation and re-radiate it back to the earth's surface to warm it up or reduce its cooling. This may explain why on a cloudy day the warm and sultry weather that one experiences persists throughout the night when radiational cooling of the surface is prevented by downward longwave radiation from the clouds.

The greenhouse effect occurs in the atmosphere over both land and ocean and is not unique to earth alone. It occurs on other planets as well which have atmospheres with gases which can absorb longwave radiation emitted by the planet's surface. For example, on Mars which has a very tenuous atmosphere of carbon dioxide and dust particles, the greenhouse warming is about 5 K. On the planet Venus which has a much denser atmosphere of carbon dioxide, the greenhouse warming amounts to about 500 K, a temperature too high to support life of the kind we know on our planet. It is believed that when the planets formed, our earth had nearly as much CO_2 in its atmosphere as the planet Venus, but most of this CO_2 was absorbed by the earth's oceans and used up in the formation of rocks in the earth's crust, thus leaving its concentration to the present level.

In the present terrestrial atmosphere, greenhouse warming is caused naturally to the extent of about 90% by water vapour (H_2O) which exists in its three phases and the rest by carbon dioxide (CO_2) and ozone (O_3). According to Hettner's experiments, as quoted by Simpson (1928), water vapour absorbs longwave radiation strongly over a very wide range of wavelengths: for example, in bands centered at 1.37μ, 1.84μ and 2.66μ; a very intense band centered at 6.26μ; and a very wide band extending from about 14μ upwards to almost 80μ. Carbon dioxide has a narrow intense absorption band centered at 14.7μ and extending from about 12μ to 16.3μ. Ozone has a strong absorption band at $9-10\mu$. There is a growing concern among humankind, supported by several scientific studies, that the emission of increasing amount of CO_2 into the terrestrial atmosphere by industrial activity and burning of fossil fuels, etc., on earth may slowly enhance global warming to such an extent that it will one day produce catastrophic effects on this planet, such as melting of the arctic sea ice, flooding of all low-lying areas by raising the mean sea level, changing global climate, etc. Given this bleak scenario, it stands to reason that strict measures should be taken by all concerned to control the release of greenhouse gases especially CO_2 into the atmosphere to avert the catastrophe.

9.4 Vertical Transfer of Radiative Heating – Diurnal Temperature Wave

An interesting case of radiative transfer of heat in the atmosphere is the vertical movement of the diurnal cycle of solar heating from the earth's surface upward. The appropriate transfer or diffusion equation which may be found in any standard text-book on mathematical physics is

$$\partial T/\partial t = K_R \partial^2 T/\partial z^2 \qquad (9.4.1)$$

where T is temperature at height z and time t, and K_R is co-efficient of radiative diffusivity.

The boundary condition to solve (9.4.1) is given by

$$T = T_0 + A \sin pt \qquad (9.4.2)$$

where T_0 is the mean temperature at the earth's surface, A the amplitude and p the frequency of the temperature wave.

The solution of (9.4.1) consistent with the boundary equation (9.4.2) is

$$T = T_0 - \beta z + A \exp(-bz) \sin(pt - bz) \qquad (9.4.3)$$

where we have assumed a vertical lapse rate of temperature, β, and b is a constant which is given by

$$b^2 = p/2\, K_R, \text{ where } p = 2\pi/(24 \times 60 \times 60)$$

The diurnal variation of temperature at any height z is then given by (9.4.3) in which the amplitude A falls off exponentially with height and the time of the maximum temperature at height z occurs bz/p seconds later than at the surface. An estimated value of K_R in the atmosphere, as given by Brunt (1944), is 1.3×10^3.

9.5 Sensible Heat Flux

The earth's surface whether land or water always tends to remain in approximate thermal equilibrium with its environment, immediately above and below the surface. So, whenever there is a gradient of temperature between the surface and the environment above or below, there is a flux of heat flowing down the temperature gradient. In the case of land, heat may flow both upward into the atmosphere and downward into the soil or earth's crust. Over the ocean, heat may flow in the same way upward into the atmosphere and downward into the water. We first consider vertical transfer of sensible heat into the atmosphere, from both land and sea.

9.5.1 Vertical Transfer of Sensible Heat into the Atmosphere

The sensible heat flux, H_s, which gives the amount of heat flowing upward across unit area of a surface per unit time at a height z near surface, is usually governed by the flux-gradient relationship which may be stated as:

$$H_s = -\rho\, c_p\, K_H (\partial T/\partial z + \Gamma), \qquad (9.5.1)$$

9.5 Sensible Heat Flux

where ρ is air density, c_p is specific heat of air at constant pressure, K_H is co-efficient of diffusivity of heat, T is temperature, $-\partial T/\partial z$ is vertical gradient of temperature with height z, and Γ is dry adiabatic lapse rate of temperature.

The upward flux at height $z + \delta z$ is then given by

$$(H_s)_{z+\delta z} = H_s + (\partial H_s/\partial z)\, \delta z \tag{9.5.2}$$

where H_s denotes the sensible heat flux at height z.

The difference in flux between the two levels, $-(\partial H_s/\partial z)\,\delta z$, is then the heat which is retained in the layer δz and which warms it up at the rate $\rho\, c_p\, (\partial T/\partial t)\, \delta z$. Using the value of H_s from (9.5.1) and assuming that ρ and K_H do not vary appreciably with height, we arrive at the vertical heat transfer Eq. (9.5.3)

$$\partial T/\partial t = K_H\, \partial^2 T/\partial z^2 \tag{9.5.3}$$

which is known as the Fourier equation for heat transfer.

Equation (9.5.3), though derived under assumptions which may not strictly hold in the real atmosphere, is quite general and may be used to solve different types of heat transfer problems with suitable boundary conditions, provided we choose proper values for the co-efficient of thermal diffusivity, K_H.

In the atmosphere, heat transfer by molecular conduction or diffusion may be ruled out, since it is an extremely slow and inefficient process. So, heat is transferred mostly by radiation, turbulent motion and convection. The co-efficient of heat transfer by these latter processes is several orders of magnitude greater than that in molecular conduction. Brunt (1944) gives relative estimates for approximate values of the co-efficient of heat diffusivity in the different cases: 0.16 for molecular conduction, 1.3×10^3 for radiative diffusivity and 1×10^5 for convective or turbulent transfer. These are values derived under idealized conditions and may often vary in the real atmosphere. However, they give us an idea of the relative effectiveness of the different transfer processes. They appear to confirm that one of the most effective mechanism of heat transfer in the atmosphere is by turbulent motion or eddies, provided we use appropriate value for the co-efficient K_H.

In turbulence, K_H is written for $\overline{l'w'}$, where l' is a linear dimension of the eddy, w' the vertical component of the eddy velocity and the overbar denotes the mean value over a period of time. However, an important point in vertical transfer of heat by eddies needs to be emphasized. It may be seen from (9.5.1) that the upward heat flux across unit area of a horizontal surface at height z per unit time is

$$-\rho\, c_p\, K_H\, (\partial T/\partial z + \Gamma)$$

where Γ is the dry adiabatic lapse rate and $-\partial T/\partial z$ the prevailing lapse rate of temperature. The above expression is positive or negative, according as the actual lapse rate is greater or less than the dry adiabatic lapse rate. In other words, the heat flux is upward only when the atmosphere is vertically unstable. When the atmosphere is stable and the heat flux is downward, the eddy motion will warm up the lower layer and cool the upper layer till the lapse rate increases to the dry adiabatic lapse rate.

Taylor (1915) applied the heat transfer Eq. (9.4.3) to a case in which a mass of air flowing over a warm land at surface temperature T_0 moved over a cold sea with surface temperature T_s. Suppose the warm air had a vertical lapse rate β at the time it entered the sea and the drop in temperature $(T_0 - T_s)$ was all of a sudden at time $t = 0$. The problem was to solve the heat transfer equation with the specified boundary conditions in order to compute the height to which the effect of the surface cooling will extend after time t. The solution, which may be found in any standard text-book (e.g., Heat conduction by Ingersoll et al., 1948) or any other book on heat transfer, is given by

$$T(z, t) = T_0 - \beta z + (T_0 - T_s)\{1 - (2/\pi) \int_0^{z/\sqrt{4K_H t}} \exp(-\mu^2)d\mu\} \qquad (9.5.4)$$

In (9.5.4), the term within the second bracket is unity at $z = 0$, and falls to a value 0.1 at a height given by $z/\sqrt{(4 K_H t)} = 1.2$. Taylor assumed that the surface cooling had no appreciable effect beyond a height given by

$$z = 2\sqrt{K_H t} \qquad (9.5.5)$$

Taylor tested his theory of the vertical diffusion of heat by turbulence by measuring temperatures at different heights and at different distances from the coastline over the Great Banks of Newfoundland where a layer of marked temperature inversion existed over the cold sea with a temperature distribution approaching the dry adiabatic lapse rate at higher levels. Knowing the height of the inversion layer at a given distance and the time the air has flown over the sea to reach there, he computed the value of K_H to be of the order of $10^3 \text{ cm}^2 \text{ s}^{-1}$. He, however, noted that the turbulent transfer in the inversion layer which was vertically stable was highly subdued and felt that with a larger lapse rate of temperature and stronger turbulence, the value of K_H should increase.

Taylor derived (9.5.5) on the supposition that the surface temperature dropped suddenly to T_s at the coastline. Later, he changed this boundary condition and assumed that the surface temperature drops gradually over the sea at the uniform rate of n° degree per unit time, so that the surface temperature is given by $T_0 - nt$. Introducing the revised boundary condition, he found that the temperature at height z at time t is given by (9.5.6):

$$T(z,t) = T_0 - \beta z - nt[(1 + z^2/2 K_H t)\{1 - (2/\sqrt{\pi}) \int_0^{z/\sqrt{4 K_H t}} \exp(-\mu^2)d\mu\}$$
$$- (2/\sqrt{\pi})(z/\sqrt{4K_H t})\exp(-z^2/4K_H t)] \qquad (9.5.6)$$

The term which multiplies nt in (9.5.6) is unity at the surface and falls to a value of 0.1 at $z/(\sqrt{4K_H t}) = 0.8$. The height reached by the temperature drop in this case is, therefore, less than in the case of the sudden drop. However, an approximation can still be made and we may write

$$z^2 = 4 K_H t$$

This case also gives a value of K_H of the order of $10^3 \text{ cm}^2 \text{ s}^{-1}$. Thus, both (9.5.5) and (9.5.6) emphasize the fact that even in the stable atmosphere over the Great Banks, heat transfer by turbulent motion is much more effective than molecular motion. In more normal situation, or when a cold continental airmass enters a warm sea, such as the sea of Japan during the northern winter, where there is considerable turbulence present, the co-efficient of heat diffusivity may be of the order of $10^5 \text{ cm}^2 \text{ s}^{-1}$ or even greater.

Another rather complex case of airmass modification through vertical heat transfer is observed over the Arabian sea during the northern summer, in which a hot continental airmass from east Africa with a surface temperature of over 40 °C first flows over a cold sea swept by the Somali current where the temperature often drops to about 20 °C or below and then enters a warm region of the eastern Arabian sea with temperatures around 30 °C. A strong low-level temperature inversion develops over the western part of the sea and makes the atmosphere vertically very stable with little or no clouds in the sky. But as the airmass flows over the warmer part of the sea, considerable turbulence and convection currents develop which weaken the inversion and transfer heat and moisture rapidly from the sea surface to a height of 2–3 km. With inversion gone, towering clouds develop and a lot of precipitation occurs over the eastern part of the sea.

9.6 Evaporation and Evaporative Heat Flux from a Surface

The process of loss of water in the form of vapour from a wetted land surface, or from a water surface, such as that of ponds, rivers, lakes or open oceans, whenever the vapour pressure of the air above the surface falls short of its saturation value at the temperature of the underlying surface is called evaporation. The evaporative flux, E, is the amount of water lost per unit area of the surface per unit time and may be easily derived using the equation of state for water vapour (4.3.1) and expressed by the relation

$$E = -\rho K_D (\in /p) \partial e / \partial z \tag{9.6.1}$$

where K_D is the co-efficient of eddy diffusivity for water vapour, ρ is the density of air, $\in$ is the ratio of the molecular weights of water vapour to dry air ($\in = 0.625$), p is the atmospheric pressure, e is the vapour pressure of water, and z is a small height above the surface.

The evaporative heat flux, H_e, is then given by

$$H_e = -L\rho\, K_D (\varepsilon/p) \partial e / \partial z = LE \tag{9.6.2}$$

where L is the latent heat of vaporization of water at the temperature of the underlying surface.

9.6.1 Bowen's Ratio

The earth's surface loses heat to the atmosphere by both sensible heat flux and evaporative heat flux. It is of interest to know the relative importance of one over the other. The information is given by the values of the ratio, H_s/H_e, which is known as the Bowen's ratio (Bowen, 1926). If we denote this ratio by R, then, from (9.6.1) and (9.6.2) we get

$$R = \rho\, c_p (K_H/K_D)(p/\in L)[\{(\partial T/\partial z) + \Gamma\}/(\partial e/\partial z)]$$

If in the above expression for R, we assume $K_H = K_D$, and neglect Γ, we get

$$R = c_p\, (p/\in L)(\partial T/\partial z)/(\partial e/\partial z) \tag{9.6.3}$$

It is difficult to evaluate R over a land surface because of the highly varying temperature and moisture fluxes near the surface. Steadier conditions are found over the ocean surface. But, even there, not many comprehensive studies have been made. Those made suggest that as a rule, the ratio is small in low latitudes where it remains constant throughout the year, but is much greater in middle latitudes where it attains values near 0.5 in winter and falls to about −0.2 in summer. A negative value suggests sensible heat flowing from the atmosphere to the ocean. The average value of the ratio over the global ocean appears to be around 0.1, indicating that sensible heat flux from the ocean surface is only about 10%, while the evaporative heat flux constitutes about 90% of the total heat flux. These values point to the great importance of the world's oceans as the steady supplier of heat energy in the form of evaporative flux for the global atmospheric circulation.

9.6.2 Evaporative Cooling

Evaporation of water from the surface of a body whether a gas, liquid or solid leads to the cooling of the body. Numerous examples can be cited from everyday experience where this property of evaporation has been observed and utilized by humankind. The following are a few of them:

(i) Desert Coolers

These are mechanical devices used in several tropical countries to cool air in living rooms by drawing outside dry hot air (which may be at a temperature of about 45 °C) through a rectangular wooden frame which is stuffed with a mesh of sawdust or sweet-smelling dried straw popularly called khas-khas, mounted on the door or window of a room and kept wet by letting streams of water fall through the mesh. During its passage through the wetted mesh, the hot air parts with its heat in evaporating water and is thereby cooled to its wet-bulb temperature. It is estimated that a device of this type can lower the air temperature inside the room from 40°C to almost 25–30°C. However, since evaporation depends on the degree of saturation of the air, a rise in the moisture content of the outside air makes air coolers less efficient.

(ii) Air cooling during rainfall

An afternoon shower in summer brings welcome relief from oppressive heat because the air near the ground loses heat to the falling drops which get evaporated. However, relief most often is only temporary, since after the shower, the temperature tends to go up and humidity also increases, making the weather uncomfortable again.

(iii) Ground cooling/heating after night rainshower

It is a common experience that after a heavy rainshower during the night, the surface temperature does not rise appreciably even in the presence of a hot sun next day, because most of the solar heating goes to evaporate water from the ground. The temperature rises rapidly again only when evaporative cooling comes to an end.

(iv) Windchill effect

People moving out in wintry weather feel colder when a wind blows than when it is calm. The colder and drier the air, the chillier is the feeling. The reason is simple. Normally, in calm conditions, the body being warmer than the outside cold air loses sensible heat to the air and one feels cold. But when the wind blows, it removes extra heat from the body through evaporation of water perspired from the exposed parts, like the face or the hands. So, one feels colder.

9.7 Exchange of Heat Between the Earth's Surface and the Underground Soil

The fundamental equation of heat transfer into the soil or any solid structure below the surface is the Fourier equation

$$\partial T/\partial t = K_H \partial^2 T/\partial z^2 \tag{9.7.1}$$

where K_H is the co-efficient of conductivity of heat and the vertical co-ordinate z increases from 0 at the surface downward.

Since the earth's surface temperature varies periodically in daily and annual cycles, we solve the problem generally for a periodic wave at the boundary and then apply the results to diurnal and annual waves.

The periodic wave at the boundary $(z = 0)$ may be expressed as

$$T = T_0 \sin pt$$

where T_0 is the amplitude and p the frequency of the thermal wave.

A solution of (9.7.1) consistent with the prescribed boundary condition is

$$T = T_0 \exp\{-z\sqrt{(p/2K_H)}\} \sin \{pt - z\sqrt{(p/2 K_H)}\} \qquad (9.7.2)$$

which gives the temperature analytically at any time t and at any depth z.

9.7.1 Amplitude and Range

It may be seen from (9.7.2) that the amplitude of the thermal wave falls off exponentially with increasing z. Thus, the range or the maximum variation of temperature at any depth z below the surface is given by T_R, where

$$T_R = 2T_0 \exp\{-z\sqrt{(p/2 K_H)}\} = 2T_0 \exp\{-z\sqrt{(\pi/PK_H)}\} \qquad (9.7.3)$$

where P is the period of the wave and related to p by the relation, $p = 2\pi/P$.

It follows from (9.7.3) that the range of temperature in the interior of the land increases with an increase in the period of the wave at the surface.

9.7.2 Time Lag

The expression (9.7.2) also enables us to calculate the time lag in the occurrence of the maximum or minimum temperature at any depth z after its occurrence at the surface, as well as the velocity and wavelenth of the wave.

Since the temperature at the surface is given by a sine-curve, the maximum occurs at time t given by

$$Pt = (2n+1)\pi/2$$

where n has zero or even values. The minima occur at odd values of n.

Let us take the maximum temperature to calculate lag, velocity and wavelength.

Let the maximum which occurs at $z = 0$ at time given by $pt = \pi/2$ travel to a depth z where it occurs at time t_1. Then,

$$Pt = \pi/2,$$
$$pt_1 = \pi/2 + z\sqrt{(p/2K_H)}$$

9.7 Exchange of Heat Between the Earth's Surface and the Underground Soil

The time lag $(t_1 - t)$ is then given by subtraction

$$t_1 - t = (z/p)\sqrt{(p/2\,K_H)} = (z/2)\sqrt{(P/\pi K_H)} \tag{9.7.4}$$

The same reasoning may be applied to calculate the time lag for the temperature minimum or any other phase of the wave.

9.7.3 Velocity

The downward velocity, v, of the wave, using (9.7.4), is given by

$$v = z/(t_1 - t) = 2\sqrt{(\pi K_H/P)} \tag{9.7.5}$$

9.7.4 Wavelength

The wavelength of the downward-moving thermal wave may be calculated from the relation

$$\lambda = v\,P = 2\sqrt{(\pi\,P\,K_H)} \tag{9.7.6}$$

Equations (9.7.4–9.7.6) may be used to compute the value of K_H, the thermal diffusivity, of any underground medium from measurement of the values of the time lag, velocity and wavelength of the thermal wave.

9.7.5 Diurnal Wave

In the case of a diurnal wave, $P = 86400\,\text{s}$. Taking $K_H = 4.8 \times 10^{-3}\,\text{cm}^2\,\text{s}^{-1}$, an appropriate value for wet soil (see Appendix 4), we get from (9.7.5) and (9.7.6) the following values for the velocity and the wavelength:

$$v = 8.357 \times 10^{-4}\,\text{cm}\,\text{s}^{-1}; \qquad \lambda = 72.2\,\text{cm}$$

During the day, a maximum temperature of about 45 °C is reached around 2 p.m. and minimum of about 25 °C a little before sunrise at a land station in the tropics. These give the amplitude of the diurnal wave at the surface to be about 10 °C (half of the daily range) which will decrease exponentially with depth at a rate given by the expression

$$T_0 \exp\{-z\sqrt{(\pi/K_H\,P)}\}$$

Using this expression, it is easy to show that the amplitude will fall off to a fraction of 0.42 at 10 cm, 0.076 at 30 cm, 1.8×10^{-4} at 100 cm, and so on. Thus, the amplitude of 10 °C at the surface will be reduced to 4.2 °C at 10 cm, 0.76 °C at 30 cm and about 0.0018 °C at 1 m. It will be almost imperceptible at greater depths.

Also, the time lag between the time of occurrence of a maximum or a minimum at the surface and that at a level below will increase with depth, according to (9.7.4), to 3.32 h at 10 cm, about 10 h at 30 cm and about 33 h at 1 m.

People living in brick houses with concrete roofs must be aware of the diurnal variation of temperature in their rooms. The temperature is maximum at the roof top in the afternoon around 2 p.m., but it takes approximately 8 h or so for the wave maximum to travel through the roof slab which is usually about 25 cm thick to reach the rooms down below. This means that the temperature in the rooms is maximum late in the evening around 10 p.m. when outside is cooling down. Similarly, due to lag in the timing of the minimum, the coolest temperature inside the rooms is around noon when the outside is blazing hot. People in the tropics who are familiar with this diurnal cycle of temperatures in their homes take advantage of it by sleeping outside on the lawn at night and remaining indoors around noon during the hot summer months.

9.7.6 Annual Wave

In the annual cycle of temperatures at the earth's surface, the maximum is experienced during the summer and the minimum during the winter. For example, at New Delhi in India, the summer temperatures reach almost 50 °C in May–June, while the winter temperatures often drop to near freezing point in December–January. This gives amplitude of about 25 °C at the surface and a period of $365\frac{1}{4}$ days. Taking a value of $0.0048 \text{ cm}^2 \text{ s}^{-1}$ for the co-efficient of thermal diffusivity for wet soil we obtain the following values for the velocity and the wavelength of the annual wave from (9.7.5) and (9.7.6):

$$v = 3.78 \text{ cm } (\text{day})^{-1}; \quad \lambda = 13.8 \text{ m}$$

The results show that at a depth $z = \lambda/2 = 6.9$ m, the annual wave appears in the opposite phase, i.e., when it is summer at the surface, it is winter at this depth or vice versa. The wave appears in the same phase as at the surface at a depth of 13.8 m after 1 year. The amplitude and the time lag at different depths are given in Table 9.1.

Table 9.1 shows that the amplitude of the annual wave falls off rapidly with depth below a depth of about 10 m where it arrives about 8 days after occurrence at the

Table 9.1 The amplitude and time lag of the annual wave at different depths in wet soil at New Delhi (India)

Depth (m)	1	10	30
Amplitude (C)	15.8	0.26	29×10^{-6}
Time lag (days)	2.65	7.95	26.5

surface. It should, however, be clarified that v as given by (9.7.5) is merely the velocity of the thermal wave and not the rate of actual penetration of heat energy into the soil or underground medium, which, inter alia, depends upon the physical and chemical properties of the underground material and its water content. In Appendix 6, we tabulate the experimental values of the density, specific heat, the co-efficient of thermal conductivity and the co-efficient of thermal diffusivity of a few common types of underground materials.

The Fourier equation (9.7.1) has been applied widely to study heat transfer in such diverse practical problems as determining the depth at which underground water pipes, electrical cables, etc., should be laid in areas subjected to frequent occurrence of severe heat and cold waves in order to protect them from damage and other undesirable effects, finding the thickness of different insulation materials in the manufacture of cold storages, furnaces, etc., estimating the age of the earth counting time from the moment its surface first started solidifying from fluid magma, and dealing with several other problems of geological, geophysical and astrophysical interest. Several of these applications are discussed in standard text-books on heat transfer.

9.8 Radiative Heat Flux into the Ocean

In the previous section, we studied heat transfer into the soil or crust below a land surface. We now look at the oceans which occupy more than two-thirds of the earth's surface. A difference to note here is that the solar radiation is incident at an interface between a gaseous and a liquid medium, in contrast to that between a gaseous and a solid medium. Physically, the difference is important, because while in the case of the land, only a small part of the radiation penetrates a thin layer of earth, in the oceans the incident radiation can penetrate to a much deeper level, producing diverse physical, chemical and biological effects for aquatic plant and animal life.

Numerous voyages of discovery and ocean explorations by scientists during the last several centuries have revealed a wealth of information about the oceans, a detailed description of which is beyond the scope of this book. For such information, the reader should consult a standard text-book on oceans. In what follows, we simply review a few salient aspects of the properties of ocean water and how it responds to the incoming solar radiation.

9.8.1 General Properties of Ocean Water

(a) Salinity and Density

Ocean water always contains a certain amount of dissolved salt, the proportion of which in a given sample of sea water, called salinity, usually varies between 33 and 37 parts per thousand with an average of about 35 parts per thousand. The salinity alters the usual properties of fresh water in several ways. For example, (i) it increases the density of fresh water by about 2.5%; (ii) the presence of salt and impurities in ocean water lowers the freezing point of pure water; (iii) the osmotic pressure and refractive index of water increases with salinity; and so on. In some coastal regions where sea water is diluted by influx of river water, or where precipitation exceeds evaporation, salinity can decrease considerably. However, it is the striking contrast between the properties of the atmosphere and the ocean that is of prime importance. Ocean water is about 800 times denser than air which has a density of about 1.2–$1.3\,\mathrm{kg\,m^{-3}}$ near surface. Further, even though the oceans cover about 70% of the earth's surface and its average depth is only about 4 km, the global ocean has a mass which is about 280 times greater than that of the atmosphere which covers the whole surface of the earth and extends to almost unlimited height into space. The great difference between the densities of the ocean and the atmosphere makes the common boundary between them vertically very stable. The stability greatly restricts the vertical movement in the ocean as compared to that in the atmosphere. For example, the amplitude of an ocean wave hardly exceeds 1 m, whereas the buoyancy plumes in the atmosphere can rise to hundreds of metres. The density of ocean water varies with pressure, temperature and salinity.

(b) Pressure of ocean water

At the surface of the ocean, pressure is that of the overlying atmosphere which measures about $10^4\,\mathrm{kg\,m^{-2}}$ and is called a bar. Since the mass of ocean water is nearly $10^3\,\mathrm{kg\,m^{-3}}$ and the acceleration due to gravity is approximately $10\,\mathrm{m\,s^{-2}}$, it follows from the hydrostatic approximation (2.3.4) that a column of ocean water 10 m deep would exert the same pressure as the whole atmosphere, i.e., 1 bar or 10 decibars (db). In other words, pressure increases with depth at the rate of 1 db per meter. At this rate, the pressure at the bottom of the ocean may be as high as 4000 db or even higher, if the ocean is deeper at a place. However, as we shall see later in this section, the density of sea water at times varies considerably with depth, especially in the surface layer.

(c) Specific heat and heat capacity

The specific heat of water is about 4 times that of air. This multiplied by high density endows the ocean water with a heat capacity which is about 1100 times greater than that of the atmosphere. So, a mere 2.5 m deep layer of the ocean has the same heat capacity as the whole depth of the atmosphere. In other words, the heat required to raise the temperature of 2.5 m of water by 1 K is the same as that required to raise the temperature of the whole atmosphere by 1 K. It is this great heat storage capacity of the ocean that maintains it more or less as a natural thermostat and allows only small diurnal and annual variations in the temperature of the air in direct contact with the ocean surface, as compared to the large variations that are observed over the continents (Monin, 1975).

9.8.2 Optical Properties of Ocean Water – Reflection and Refraction

When a beam of light of wavelength λ and intensity I is incident upon an ocean surface at a point O (see Fig. 9.3) from a direction which makes an angle θ_a with the outward normal to the surface, a fraction F of it is reflected from the surface, while the remainder R enters the body of the ocean where it is partly absorbed and partly transmitted to lower layers (see Fig. 9.3).

Now, if for a moment we neglect absorption and consider transmission only, the beam entering the water is refracted due to change in the refractive index of the medium from air to water in a direction which makes an angle θ_w with the normal to the surface. Suppose, at some small depth at P, the refracted beam R meets a layer of particulate matter of different density and refractive index held in suspension in sea water. At P, a fraction I' is reflected or scattered in a direction which makes an angle θ_w with the normal to the surface at P, while the remainder is transmitted to lower layers. The reflected beam I' on reaching the interface at O' from below will suffer reflection as well as refraction. The reflected part F' re-enters the water making an angle θ_w with the normal to the surface at O', while the refracted beam R' emerges into the air making an angle θ_a with the local normal. Similar reflection and scattering may be expected to occur at myriads of points inside the water wherever particles intercept the radiation, followed by multiple total internal reflection at the interface. In this way, a large part of the solar energy is retained inside the body of the ocean to warm it up. The trapping of solar radiation in this manner leads to continual warming of the surface layer of the ocean.

However, it must be stated that the process of reflection, refraction and absorption of light radiation in the ocean is highly complex. The laws stated below follow from the electromagnetic theory of light (for details of the theory, reference may be made to any standard text-book on optics):

(i) The reflected and the refracted rays at a surface lie in the same plane as the incident ray and the normal to the surface;
(ii) The angle of reflection is equal to the angle of incidence; and

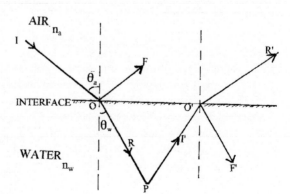

Fig. 9.3 Processes of reflection and refraction of solar radiation in the ocean

(iii) The product of refractive index of the medium of the incident ray and the sine of the angle of incidence is equal to the product of the refractive index of the medium of the refracted ray and the sine of the angle of refraction. In other words, $n_w \sin \theta_w = n_a \sin \theta_a$, where n denotes the refractive index, θ the angle and the subscripts w and a refer to water and air respectively. This is known as Snell's law.

The important point to note here is that as the angle of incidence increases, there is more of reflection into the same medium and less of transmission into the other medium. When this principle is applied to the interface at O' (Fig. 9.3), it means that for an angle of incidence θ_w greater than a certain critical value θ_w^*, the beam I' will be totally reflected back into water. Such total internal reflection helps in retaining the solar radiation inside the water at O' and will not occur at O where, though the angle of incidence θ_a can increase to $\pi/2$, the angle of refraction θ_w has always a value less than θ_w^*.

Total internal reflection of light radiation of the kind described in the preceding para at multiple points along the ocean-atmosphere interface traps solar energy in the upper part of the ocean which leads to its warming This warming is in addition to that which results from direct absorption of the solar beam that enters the ocean from above. The increased warming of the ocean by total internal reflection of solar radiation in this manner appears to be somewhat analogous to the Greenhouse effect in the atmosphere, a brief description of which was given earlier in this chapter.

9.8.3 *Absorption and Downward Penetration of Solar Radiation in the Ocean*

Sverdrup et al. (1942) have shown that the depth to which light radiation penetrates into the ocean depends upon the wavelength of the radiation. Infrared part of the radiation is strongly absorbed by water and dissolved gases like CO_2 in the surface layer allowing shorter waves only to penetrate further down. This is evident from a schematic representation of the energy spectrum of the radiation from the sun and the sky penetrating the sea surface, and of the energy spectra in pure water at depths of 0.1, 1, 10, and 100 m, presented by them in Fig. 9.4 of their book (not reproduced). Inset, they gave curves of percentages of total energy and of the energy in the visible part of the spectrum reaching different depths.

They show that radiation of wavelengths greater than 1μ is almost totally absorbed by water within the first 10 cm of the surface. According to them, pure water is transparent for visible radiation between 0.35μ and 0.75μ only. The shorter the wavelength, the greater is the depth. In the visible part, it is only the shorter waves in the blue-green-yellow region between 0.4μ and 0.6μ which can reach a depth of about 100 m. The intensity of the transmitted beam, however, continually diminishes with depth.

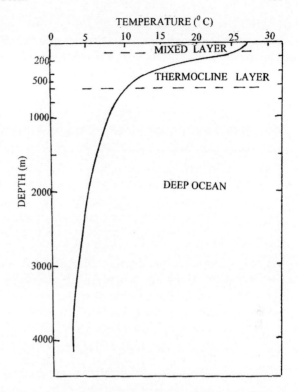

Fig. 9.4 Schematic showing vertical distribution of temperature in tropical oceans

9.8.4 Vertical Distribution of Temperature in the Ocean

The near-total absorption of solar radiation in the upper strata of the ocean together with a certain degree of mixing caused by wind-driven ocean currents near the surface produces a vertical profile of temperature in tropical oceans, which shows the following three distinct layers (see Fig. 9.4):

(i) A mixed or surface layer upto a depth of about 100 m in which there is little variation of temperature with depth,
(ii) A thermocline layer in which temperature drops steeply from a value of about 21 °C to 26 °C at 10 m to about 5 °C at about 1000 m, and
(iii) A deep ocean layer, below about 1 km, in which there is little fall of temperature with depth. This layer constitutes the cold and dark region of the ocean.

Further, the depth of all the layers varies with latitude. For example, the mixed layer is rather shallow ($\leq$ 100m) in low latitudes, compared to the midlatitudes, between about 30° and 50°, where it may extend to a depth of about 300 m. The layer disappears beyond latitude about 50°. The depth of the main thermocline layer and the deep layer also appears to follow the same type of variation with latitude as the mixed layer.

While comparing the thermal structure of the ocean with that of the atmosphere, Defant (1961) has called the combined mixed-thermocline layer as the troposphere and the deep ocean as the stratosphere of the ocean. This analogy is often a useful one, especially when dealing with conditions in low latitudes.

9.9 The Thermohaline Circulation – Buoyancy Flux

Earlier in this chapter, we mentioned that the density of ocean water at a given pressure depends on both temperature and salinity. The surface layers of the ocean are subject to such physical processes as evaporation, precipitation, discharge from rivers, etc., as well as atmospheric temperature and the net effect of these processes may be either an increase or decrease of density. In the open ocean, the salinity increases when evaporation exceeds precipitation. The effect of a change in salinity on density may be in the same direction as that due to change of temperature, or in the opposite direction. When both the effects are in the same direction, they support each other in causing a large increase or decrease of density, but when they oppose each other the net effect may be no change or only a small change in density.

Now, it is well-known that both temperature and salinity usually vary with latitude. The temperature is maximum at the equator and decreases towards the poles. The temperature-induced density will, therefore, be minimum at the equator and increase towards the poles. Normally, a high temperature at the equator will produce high evaporation but its effect is offset by precipitation exceeding evaporation in equatorial latitudes. So the net effect of both temperature and salinity at the equator is to produce a lowering of surface water density. Compare this with conditions over the subtropical belt where the temperature may be lower than at the equator but where evaporation exceeds precipitation. Here both temperature and salinity combine to produce a relatively large increase in density. The effect of these changes of density between the equator and the subtropical belt is to force a meridional circulation in the ocean called the thermohaline circulation, in which denser water over the subtropical belt will sink under the force of gravity to a depth where the density difference with the environment disappears and then move equatorward at a subsurface level to rise at the equator. At the surface, the equatorial waters will move towards the subtropical belt to replace the sinking water.

Poleward of the subtropical belt also there will be a thermohaline circulation but it will be a weak one, because in the higher latitudes while density will increase on account of lower temperature, it will decrease due to precipitation being in excess of evaporation in the polar belt.

Agafonova and Monin (1972) have constructed a quasi-climatic map of buoyancy flux (divided by g) at the ocean surface (generated by changes in temperature and salinity) as a driving term for the thermohaline circulation of the ocean.

This map is shown in Fig. 9.5 (their Fig. 9.1).

A pure thermohaline circulation is, however, difficult to observe in the real ocean on account of the prevalence of strong wind-driven ocean currents which most often

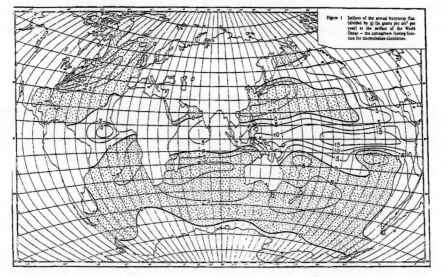

Fig. 9.5 Isolines of the annual buoyancy flux (divided by g) (in grams per cm^2 per year) at the surface of the World Ocean-the atmospheric forcing function for thermohaline circulation (Agafonova and Monin, 1972, their Fig. 1)

are too strong for it to develop. Density changes on account of net heating or cooling and excess evaporation or precipitation, however, do occur which lead to buoyancy fluxes in the upper ocean. Surface water the density of which has been increased by excess cooling or evaporation will then sink until it meets water of the same density. Reversely, when surface density decreases on account of excess heating or precipitation, water from a subsurface level will rise to the surface.

9.10 Photosynthesis in the Ocean: Chemical and Biological Processes

The sunlight that enters the ocean plays a very important role for life in the ocean, by promoting what is known as greenplant photosynthesis in which molecules of carbon dioxide dissolved in sea water chemically combine with molecules of water in the presence of sunlight to produce carbohydrates which form the building blocks of plant bodies and release oxygen. The process is reversible and in the respiratory cycle in which the released oxygen is consumed, the energy is released. The chemical reaction may be expressed by the reversible equation:

$$CO_2 + H_2O + \text{Light energy} \rightleftarrows CH_2O + 6O_2 \qquad (9.10.1)$$

Thus, photosynthesis is the chemical-biological process in which water molecules are split with the energy of sunlight and oxygen is released. In to-day's ocean, green plants, single-celled phytoplankton (free-floating organisms), especially chlorophyll – containing bacteria, all perform oxygenic photosynthesis. However, it is well-known that the reverse process of respiration does not remove all the oxygen produced, since a small fraction (0.1%) of carbon gets buried in sediments and escapes oxidation. It follows from Eq. (9.10.1) that a burial of 1 mole of organic carbon will generate one mole of oxygen. Oxygen is also generated by several other chemical and biological processes in the ocean. It is believed that the transition from the ancient atmosphere which had no oxygen to the oxygen-rich modern atmosphere commenced only when life appeared on earth more than 2.3 billions of years ago and that in this transformation, greenplant and bacterial photosynthesis played a key role.

Chapter 10
Heat Balance of the Earth-Atmosphere System – Heat Sources and Sinks

10.1 Introduction – definition of heat sources and sinks

We showed in Chap. 8 that the intensity of the incoming solar radiation at the earth's surface varies widely with latitude and the transmissivity of the overlying atmosphere, being, in the annual mean, maximum at the equator and minimum at the poles. As against this and as shown by measurements and computations, the intensity of the outgoing longwave radiation varies only slightly with latitude. Thus, the difference between the incoming and the outgoing radiation creates a diabatic heat source over the equatorial latitudes and a heat sink over the polar-regions. In order that the equatorial region may not continually warm up and the polar region continually cool down, excess heat from the equatorial region must flow out to the polar region for a heat balance between the two regions. In the present text, we define a diabatic heat source or heat sink by the following criteria:

$$dH/dt > 0, \; \nabla^2(dH/dt) < 0 \quad \text{(Heat source)}$$
$$dH/dt < 0, \; \nabla^2(dH/dt) > 0 \quad \text{(Heat sink)}$$

where H is the mean heat content of a unit mass of air over a region at a given time t and ∇ is Del operator.

Simpson (1928, 1929) in England appears to have been one of the first to have carried out a detailed computation of the heat budget of the earth-atmosphere system, using method and data which produced surprisingly realistic results. Some details of his method and findings will be reviewed in this chapter.

After the introduction of the earth-observing satellites in the sixties and seventies of the last century there was renewed interest in the subject and direct measurements were made from space platforms of the intensities of the incoming solar radiation, the albedo and the outgoing longwave radiation. The analysis of the satellite data enabled direct computation to be made of a heat budget of the earth-atmosphere system. Such computations have been made by several workers (e.g. Raschke et al., 1973; Winston et al., 1979). Some aspects of their work and findings will be reviewed.

An energy balance method in which net heating or cooling of the atmosphere is computed from vertically-integrated radiative heat flux divergence and sensible and latent heat fluxes has been widely used. After the Global Weather Experiment (GWE) in 1979, several workers (e.g., Yeh and Gao, 1979; Nitta, 1983; Luo and Yanai, 1984; Chen et al., 1985; Murakami, 1987) used the energy balance method to compute heat sources and sinks over the Tibetan plateau. Gutman and Schwerdtfeger (1965) and Rao and Erdogan (1989) used the method to compute heat sources and sinks over the Bolivian plateau in South America. The first law of thermodynamics has also been applied to compute the diabatic heating of the atmosphere and locate atmospheric heat sources and sinks from observed atmospheric parameters.

In recent years, a form of the mass continuity equation in isentropic co-ordinates has been integrated by Johnson and his co-workers in the university of Wisconsin (e.g., Johnson, 1980; Wei et al., 1983; Johnson et al., 1985; Schaack et al., 1990; Schaack and Johnson, 1994) to estimate atmospheric heating. They applied the method to locate three-dimensional heat sources and sinks over different parts of the globe in different seasons. Some aspects of these studies will also be reviewed.

Atmospheric heat sources and sinks at the earth's surface are qualitatively identified by meteorological temperature and pressure systems. A 'heat low' or 'a trough of warm low pressure' at the surface, for example, is usually a heat source and associated with penetrative convection, cloudiness and precipitation. On the other hand, a 'cold high' or 'a ridge of cold high pressure' at surface is to be identified as a heat sink where normally air subsides and the process inhibits formation of cloud and rain. Defined in this way, it should be possible to qualitatively verify the accuracy of any computation of heat balance or atmospheric heating over any part of the globe from actual meteorological observations of temperature and pressure and other relevant data. Remote sensing of temperature, moisture and cloudiness by satellites is of great help in this regard.

But, before we take up the study of the heat balance of the earth-atmosphere system, let us first look at the various physical processes known to be involved in the balance.

10.2 Physical Processes Involved in Heat Balance

Physical processes involved in the heat balance of the earth-atmosphere system along with a rough estimate of their contributions to the annual budget can be qualitatively understood as follows: If we take the intensity of the incoming solar radiation at the outer boundary of the atmosphere as 100 units, out of which a total of 30 units is returned to space by way of reflection from the ground and cloud surfaces and backscatter from air molecules, 16 units are absorbed by atmospheric gases such as water vapour and ozone and suspended dust particles, and 3 units are absorbed by clouds, the remaining 51 units are absorbed by the earth's surface. There is also downward longwave radiation amounting to about 98 units

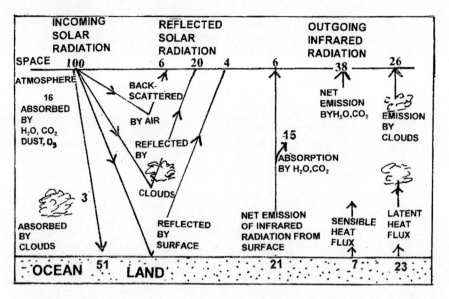

Fig. 10.1 Heat balance of the earth-atmosphere system (Reproduced with permission from a report of the National Academy of Sciences panel on Understanding Climate Change', 1975) (Courtesy: National Academies Press, Washington, D.C.)

from the greenhouse gases in the atmosphere, according to an estimate made by London and Sasamori (1971), which, at times, may even exceed the incoming solar radiation. The surface then emits longwave radiation. The net surface emission (excess of upward over downward longwave radiation) is 21 units of which 15 units are absorbed by water vapour and carbon dioxide in the atmosphere, while 6 units escape to space. The other outgoing longwave radiation includes 38 units from water vapour and carbon dioxide and 26 units from clouds. Thus, a total of 70 units of the incoming solar radiation which are absorbed by the earth-atmosphere system are balanced by 70 units of the outgoing longwave radiation.

A schematic in Fig. 10.1, reproduced with permission from a report of the U.S. National Academy of Sciences panel on 'Understanding climate change' (1975), shows these processes.

10.3 Simpson's Computation of Heat Budget

Simpson (1928) computed the annual heat budget of the earth-atmosphere system, making several simplifying assumptions regarding the radiative properties of the earth and its atmosphere. It is not possible to give the full details of his work in this brief survey. Those interested in them may look up his original memoirs (Simpson, 1928 and 1929). Only some salient points of his work are presented here.

After allowing for reflection from the atmosphere, the earth's surface and average cloud amount at different latitudes, Simpson used a value of 0.278 gm-cals per cm^2 per minute for the effective incoming solar radiation. He computed the total amount of outgoing radiation with clear and cloudy skies at all latitudes. For the purpose of these computations, he divided the radiation at terrestrial and atmospheric temperatures into the following three categories:

(a) Wavelengths at which water vapour is transparent to radiation, $8^1/_2$–11μ;
(b) Wavelengths in which water vapour amounting to 0.3 mm of precipitable water will completely absorb all the radiation: $5^1/_2$–7μ, and > 14μ;
(c) Wave-lengths in which water vapour absorption is intermediate between (a) and (b): 7–$8^1/_2$ μ and 11–14μ.

For a clear sky, Simpson assumed that the radiation of category (a) originated at the earth's surface with a temperature of 280 K, and that of (b) originated in the stratosphere which contains 0.3 mm of precipitable water and has a temperature of 218 K. For (c), he assumed intermediate values between the radiations emitted at the surface and the stratospheric temperatures. For an overcast sky, Simpson used the same procedure but replaced the temperature of the earth's surface by that of the cloud top which he assumes to be 261 K in all latitudes. In his later work, Simpson uses a mean cloud amount of 5/10 for all latitudes and obtains a final figure of 0.271 gm-cals per cm^2 per min for the total outgoing radiation from the earth-atmosphere system, a value which is pretty close to that of the net incoming solar radiation. Simpson's results are shown in Fig. 10.2.

In Fig. 10.2 (reproduced from Brunt, 1944, © Cambridge University Press, with permission), Curve I gives the intensity of the net incoming solar radiation and Curve II that of the total outgoing radiation at different latitudes. The difference

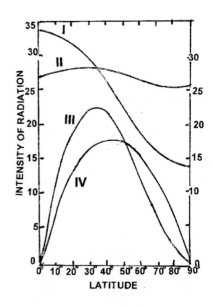

Fig. 10.2 Solar and terrestrial radiation

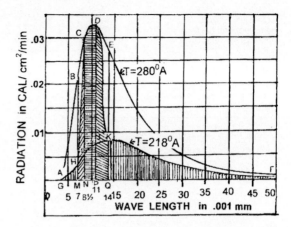

Fig. 10.3 Simpson's graphical representation of the outgoing terrestrial radiation

between the two curves shows that the incoming radiation exceeds the outgoing radiation between the equator and about latitude 35°, but in higher latitudes the outgoing radiation exceeds the incoming radiation. In other words, there is an accumulation of heat in low latitudes which will cause continuous warming of the tropics and depletion in high latitudes which will continuously cool those latitudes. Since this undue heating and cooling are not observed, it follows that a heat balance is reached between the source and the sink through the general circulation of the atmosphere. Curve III in Fig. 10.2 gives the total horizontal transfer of heat across circles of latitude and Curve IV the horizontal heat transfer per cm of circle of latitude.

Fig. 10.3 depicts the method used by Simpson to compute the total outgoing radiation from the earth-atmosphere system, as explained above.

Considering the uncertainties involved in his various assumptions, the results obtained by him were found to be surprisingly accurate and realistic when he (Simpson, 1929), worked out similar data, using monthly means of temperatures.

10.4 Heat Balance from Satellite Radiation Data

The earth-observing satellites which carried scanning radiometers on board in the late sixties and early seventies of the last century enabled direct measurements of incoming and reflected solar radiation and outgoing longwave radiation to be made from a space platform. Raschke et al. (1973) utilized the radiation measurements of the Nimbus 3 satellite flown in 1969 and 1970 to compute the heat budget of the earth-atmosphere system. Their results are shown in Fig. 10.4.

Raschke et al. (1973) computed the budget using the relation

$$Q_N = S_0 (1 - \alpha) - Q_R \qquad (10.4.1)$$

where, Q_N is the net radiation, S_0 is the solar constant, α is the albedo of the earth-atmosphere system, and Q_R is the total outgoing longwave radiation.

Fig. 10.4 Global (G) and hemispherical (N = north, S = south) averages of the radiation budget and its components, computed from Nimbus-3 radiation measurements in 1969 and 1970 (From Raschke et al., 1973, with permission of American Meteorological Society)

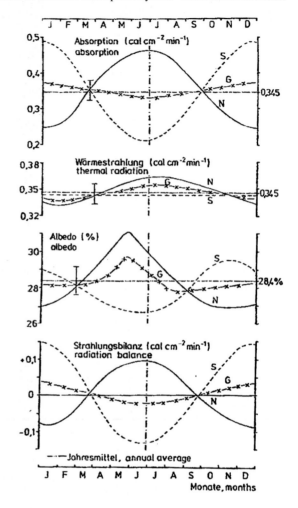

The curves in Fig. 10.4 were constructed from 15-day averages of each of the radiation components given in (10.4.1). The seasonal migration of the heat sources and sinks between the hemispheres is well brought out by them.

Figure 10.5 shows the distribution of the zonally-averaged annual mean net radiation, Q_N, which is the result of absorbed incoming solar radiation minus outgoing longwave radiation, with latitude, computed from data presented by Winston et al. (1979).

Figure 10.5 confirms that in the annual mean there is a heat source over the equatorial region extending to about 30N and S, and a heat sink over the higher latitudes. Since no undue accumulation or depletion of heat takes place in any part of the earth-atmosphere system, it follows that the atmosphere and the ocean must be so circulating as to achieve a heat budget of the system, i.e., remove excess heat from the source and deliver it to the sink. In this task, the oceanic transfers, though slow, may, in the long term, be considerable. Most of the atmospheric transfers are

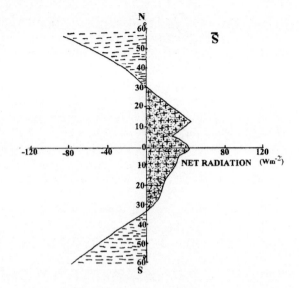

Fig. 10.5 Latitudinal distribution of zonally-averaged annual mean net radiation, Q_N (denoted by $\bar{S}$ in the Figure), in Wm^{-2}, based on data of Winston et al. (1979). Positive values indicate heat source, negative heat sink

carried out by the general circulation and its perturbations, while the oceanic ones by wind-driven ocean currents.

The seasonal movement of the sun across the equator, however, continually redistributes the computed zonally-averaged net radiation of the earth-atmosphere system, as indicated by Fig. 10.6, which is self-explanatory.

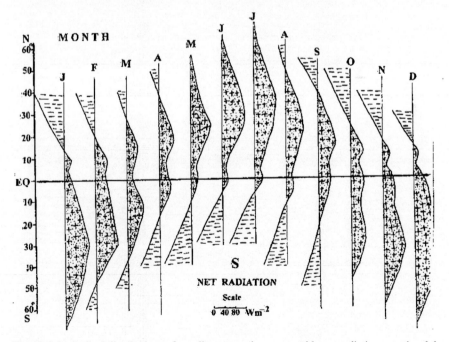

Fig. 10.6 Latitudinal distributions of zonally-averaged mean monthly net radiation over the globe in different months (Based on radiation data of Winston et al., 1979)

Fig. 10.6, which presents the latitudinal distribution of the zonally-averaged heat sources (positive values) and sinks (negative values) for the different months of the year, testfies that the equatorial belt between 10S and 15N remains a heat source almost throughout the year. The same features are also revealed by their geographical distribution presented in Fig. 10.7 (a, b).

In summer (Fig. 10.7a), a heat source with positive values of net radiation appears over a wide belt of the northern hemisphere, extending from about 10S northward to about 60N, while a heat sink appears over the whole southern hemisphere south of 10S. In general, there is a tendency for higher positive values of net radiation to be located over the continents of Asia, Africa, and North America than over the neighboring oceans. High positive values also appear over the warmer parts of the northern oceans.

In winter (Fig. 10.7b), the situation is reversed. The northern hemisphere, north of about 15N, now appears as a heat sink. Positive values of net radiation appear over the whole latitude belt south of this latitude with maxima located over the

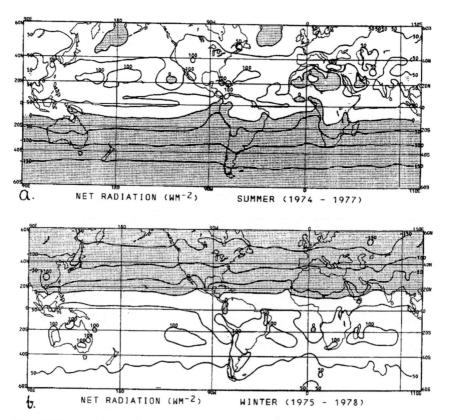

Fig. 10.7 Geographical distribution of net radiation (Wm^{-2}) over the globe during the northern hemisphere: **(a)** Summer (June–August); **(b)** Winter (December–February) Areas with negative values are shaded (After Winston et al., 1979)

subtropical belt of the southern hemisphere, especially over the continents of South America, South Africa and Australia and some warmer parts of the southern oceans. Heat sinks appear over the vast desert areas of Africa and Asia and also over ocean areas dominated by cold ocean currents in both the seasons.

10.5 Heat Sources and Sinks from the Energy Balance Equation

Heat sources and sinks have been estimated from values of net radiation and sensible and evaporative heat fluxes, using a relation of the form

$$\text{Heat source (or sink)} = (R_\infty - R_0) + Q_s + Q_e \qquad (10.5.1)$$

where

R_∞ is the net radiation at the outer limit of the atmosphere,
R_0 is the net radiation at the earth's surface, and
Q_s and Q_e are the sensible and evaporative heat fluxes respectively.

The relation (10.5.1) has been used to compute heat sources and sinks over different regions of the globe, notably the Tibetan plateau during the northern summer by several workers (e.g., Yeh and Gao,1979; Nitta,1983; Luo and Yanai, 1983, 1984; and Chen et al., 1985) and the Bolivian plateau in South America during January by Rao and Erdogan (1989) and others.These computations were made using conventional meteorological data of net radiation, temperature, wind and moisture content and surface vegetation over the respective regions, and, in some cases, satellite-observed net radiation. We give here a brief description of the work of the above-mentioned workers in respect of the Tibetan plateau.

Yeh and Gao (1979) in their book on the meteorology of the Tibetan plateau summarizes the results of early Chinese research on the heat balance of the western and the eastern parts of the plateau as well as the whole plateau. They tabulated the monthly mean values of the sensible and latent heat fluxes as well as the precipitation, using the long-term surface records of the available meteorological stations at a mean height of about 4 km, the line of demarcation between the two parts being about 85E. From the tabulated data, they estimated the monthly mean values of net atmospheric heat source over the two parts of the plateau as well as the plateau as a whole. Their results are shown in Table 10.1.

The values of the monthly mean sensible and latent heat fluxes and the net heat source over the two parts of the plateau, as computed by these workers, are shown in Fig. 10.8.

From Fig. 10.8, one may immediately note an extremely large sensible heat flux over the arid western part of the plateau (where it reaches a maximum value of about 450 units in June), compared to the eastern part where the values are much lower and a maximum of about 250 units is reached in May, a month earlier than over the western part. The distribution of latent heat flux between the two parts, however, varies in the reverse direction, being much larger over the eastern than over the

Table 10.1 The net atmospheric heat source (Wm^{-2}) over the Tibetan plateau (after Yeh and Gao, 1979).

	Month												Annual mean
	1	2	3	4	5	6	7	8	9	10	11	12	
Eastern Plateau	−72	−48	−8	38	73	89	92	67	40	−17	−53	−78	10
Western Plateau	−71	−31	59	106	134	138	120	90	51	6	−55	−76	39
Weighted Mean	−72	−42	25	60	94	109	101	74	44	−10	−54	−77	21

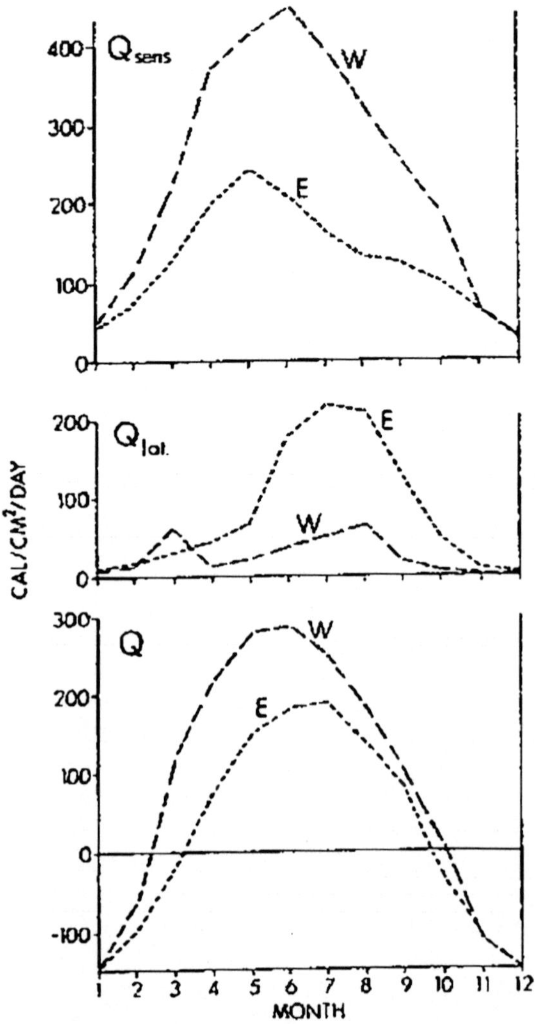

Fig. 10.8 Ten-year (1961–1970) means for sensible heat flux at the surface(Q_{sens} denoting Q_s), latent heat of precipitation(Q_{lat} denoting Q_e) and net atmospheric heat source (Q), over the western(W) and eastern(E) Tibetan plateau (After Yeh and Gao, 1979)

10.5 Heat Sources and Sinks from the Energy Balance Equation

western part. Further, the latent heat flux over the eastern part reaches its maximum value in July, a month earlier than over the western part. It is also noteworthy that a feeble maximum appears in latent heat flux curve over the western part of the plateau in March, while no such maximum appears then over the eastern part. Because of the predominance of sensible heat flux over latent heat flux, the western part of the plateau contributes more to the net heat source over the plateau than the eastern part, as shown by the distribution of net atmospheric heat source (Q) in Fig. 10.8.

There was renewed interest in the study of the heat budget of the Tibetan plateau during the Global Weather Experiment (GWE) in 1979 when there was an improved network of surface and upper-air observing stations over the plateau. Several studies were conducted using the special data that became available during this experiment. For details of data used by these studies, the original papers should be consulted. Here we give only a brief description of the method that was used by Nitta (1983) and Luo and Yanai (1983, 1984) and also by Chen et al. (1985) and some of the results reported by them. In addition to heat budget, they also computed the moisture budget, using the following expressions for Q1 which they called the apparent heat source, and Q2 the moisture sink

$$(1/g) \int_{p_T}^{p_s} Q1 \, dp \approx (1/g) \int_{p_T}^{p_s} Q_R \, dp + LP + SH \qquad (10.5.2)$$

$$(1/g) \int_{p_T}^{p_s} Q2 \, dp \approx LP - LE \qquad (10.5.3)$$

where p_s and p_T are pressures at the surface and top of the atmosphere respectively, Q_R stands for radiative heating, LP is the latent heat released to the atmosphere by precipitation, SH is sensible heat exchange with the ground and LE is the heat used for evaporation of water from the lower boundary.

If we denote the left-hand sides of (10.5.2) and (10.5.3) by $<Q1>$ and $<Q2>$ respectively and assume that

$$(1/g) \int_{p_T}^{p_s} Q_R \, dp = R_\infty - R_0 \qquad (10.5.4)$$

where R_∞ and R_0 are the net radiation at the top of the atmosphere and the earth's surface respectively and both values are treated as positive if directed downward, (10.5.2) and (10.5.3) may finally be written as

$$<Q1> = (R_\infty - R_0) + LP + SH = RC + LP + SH \qquad (10.5.5)$$
$$<Q2> = LP - LE \qquad (10.5.6)$$

where RC in (10.5.5) stands for $(R_\infty - R_0)$ and signifies radiative cooling.

While Nitta (1983) and Luo and Yanai (1983, 1984) evaluated Q1 and Q2 at individual pressure surfaces and then carried out the vertical integration, Chen et al.

(1985) adopted a slightly different procedure and evaluated the apparent heat source and moisture sink over the plateau directly from observations. R_∞ was evaluated from Nimbus-7 radiation data and R_0 was calculated from an empirical relationship which took into consideration, amongst others, the intensities of the incoming solar radiation and the outgoing longwave radiation at the ground, the surface albedo, the declination of the sun, the latitude of the place, the mean cloud amount and water vapour pressure in the atmosphere and the state of vegetation at the surface.

The sensible heat flux, SH, was calculated from the standard relation

$$SH = \rho\ c_p\ C_D\ V_0\ (T_g - T_a) \tag{10.5.7}$$

where ρ is air density, c_p is the specific heat of air at constant pressure, C_D is drag co-efficient, V_0 is the mean wind speed at about 10 m above the ground, T_g is the ground temperature and T_a is the air temperature immediately above the ground. The latent heat released into the atmosphere, LP, is calculated easily from precipitation data. LE is difficult to calculate over the plateau because of the paucity of evaporation data and was obtained as a residual in the heat balance equation.

Table 10.2 presents the mean values of the apparent heat source and moisture sink and their components (SH, LP, RC, LE) over the whole plateau and its eastern and the western parts during the period June-August 1979, as computed by Chen et al. (1985), along with those of Yeh and Gao (1979), for comparison. The values computed by Nitta (1983) for the eastern part of the plateau are also included. The values computed by Luo and Yanai (1984) for the eastern and the western parts of the plateau were available for the month of June only and are, therefore, not included in the Table.

It is clear from Table 10.2 that a net heat source resides over the Tibetan plateau during the northern summer. However, the extent of atmospheric heating varies widely amongst the different computations, being minimum in CRF. Strong radiative cooling is found over the plateau by all the computations. It is also found that while sensible heating dominates over precipitation heating over the arid western part, precipitation heating dominates over sensible heating over the more humid eastern part.

Table 10.2 Mean values of the atmospheric heat source and its components, Wm^{-2}, over Tibet during June-August 1979, computed by Chen, Reiter and Feng (CRF), 1985; Yeh and Gao (Y&G), 1979; and Nitta, 1983. (After Chen et al., 1985)

	Whole plateau		Eastern part			Western part	
	CRF	Y&G	CRF	Y&G	Nitta	CRF	Y&G
< Q1 >	64	94	77	82	120	46	116
< Q2 >	19		31		25	1	
SH	59	118	50	82	(105)	70	189
LP	65	73	87	97	90	35	24
RC	−60	−96	−60	−97	−75	−59	−97
LE	47		56		65	34	

10.6 Computation of Atmospheric Heating from Mass Continuity Equation

Johnson et al. (1987) reviewed the method to estimate three-dimensional atmospheric heating over different parts of the globe by integrating the time-averaged mass continuity equation in isentropic co-ordinates in the following form

$$\partial \overline{(\rho\, J_\theta)}/\partial t_\theta + \nabla_\theta \cdot \overline{(\rho J_\theta U)} + \partial \overline{(\rho\, J_\theta \dot\theta)}/\partial \theta = 0 \tag{10.6.1}$$

where ρ is density, θ is potential temperature, z is height, J_θ is the transformation Jacobian $|\partial z/\partial \theta|$ (from z to θ surface), U is velocity, t is time and the overbar denotes a time average.

The atmospheric mass within an isentropic layer is determined by the difference in pressure between the lower and the upper isentropic levels, as given by the hydrostatic equation

$$\rho\, J_\theta = -(1/g)\, \partial p/\partial \theta \tag{10.6.2}$$

where p is pressure and g is acceleration due to gravity.

The diabatic mass flux through an isentropic surface θ is estimated by vertical integration of (10.6.1)

$$\overline{\rho\, J_\theta\, \dot\theta} = \int_\theta^{\theta_T} [\partial \overline{(\rho J_\theta)}/\partial t_\theta + \nabla_\theta \cdot \overline{(\rho J_\theta U)}]\, d\theta \tag{10.6.3}$$

where the diabatic mass flux is assumed to vanish at the isentropic surface (θ_T) at the top (T) of the atmosphere.

This form of the continuity equation gives the relation between the diabatic mass flux at an isentropic surface and the mass tendency and mass divergence within the overlying atmosphere.

The heating rate, $\dot\theta$, obtained from (10.6.3), is equivalent to the thermodynamic relation

$$\dot\theta = Q/\{c_p\, (p/p_{oo})^\kappa\} \tag{10.6.4}$$

where Q is the rate of heating and the denominator within the second bracket on the right-hand side is the Exner function with p_{oo} as standard pressure at 1000 mb surface.

The heating rate calculated from (10.6.3) is open to several sources of error, due to inaccuracies likely in the specification of the vertical profile of the mass tendency or the horizontal mass divergence. A systematic mass-weighted adjustment is therefore applied to the computed diabatic mass flux based on the integral constraint that the sum of the vertically-integrated mass tendency and mass divergence must reduce to the diabatic mass flux at the earth's surface. The following is a brief description of the adjustment procedure.

The vertical integral of the isentropic mass continuity equation (10.6.3) when integrated from surface to top of the layer may be written as

$$(\rho J_\theta \, \dot\theta)|_{\theta s} = \int_{\theta_s}^{\theta_T} [\partial(\rho J_\theta)/\partial t_\theta + \nabla_\theta \cdot (\rho \, J_\theta \, \mathbf{U})] \, d\theta \qquad (10.6.5)$$

where θ_T and θ_S are the potential temperatures at the top of the layer and the earth's surface respectively. In (10.6.5), the diabatic mass flux at the top of the layer is assumed to vanish.

Due to inaccuracies from various sources in the computation of the right-hand side of (10.6.5), the computed diabatic mass flux at the earth's surface using (10.6.5) will differ from that estimated from the actual variation of potential temperature at the surface which may be expressed as

$$\dot\theta_s \, (\lambda, \phi, t) = \partial \theta_s / \partial t_\theta + \mathbf{U}_s \cdot \nabla_\theta \, \theta_s \qquad (10.6.6)$$

where λ and ϕ are the longitude and latitude of the place respectively.

The true value of the diabatic mass flux at the earth's surface is, therefore, assumed to be given by

$$\rho J_\theta \, \dot\theta^*|_{\theta s} = \rho \, J_\theta \, \dot\theta_s(\lambda, \phi, t) = \rho \, J_\theta \, (\partial \theta_s/\partial t_\theta + \mathbf{U}_s \cdot \nabla_\theta \, \theta_s) \qquad (10.6.7)$$

The difference between (10.6.5) and (10.6.7) then represents the vertically-integrated systematic error, δ, given by

$$\delta = (\rho \, J_\theta \, \dot\theta)|_{\theta s} - (\rho \, J_\theta \, \dot\theta)^*|_{\theta s} \qquad (10.6.8)$$

If $\acute{\epsilon}$ represents error per unit mass of the atmospheric column, it is given by

$$\acute{\epsilon} = \delta / \int_{\theta_s}^{\theta_T} \rho \, J_\theta \, d\theta = -g \, \delta/[p(\theta_T) - p(\theta_s)] \qquad (10.6.9)$$

The adjusted diabatic mass flux at an isentropic surface (θ) is then given by

$$(\rho \, J_\theta \, \dot\theta) = \int_\theta^{\theta_T} [\partial(\rho \, J_\theta)/\partial t_\theta + \nabla_\theta \cdot (\rho \, J_\theta \, \mathbf{U}) - \rho \, J_\theta \, \acute{\epsilon}] \, d\theta \qquad (10.6.10)$$

In this method, the integrated error is distributed through the vertical column in proportion to the mass within an isentropic layer.

Johnson and his co-workers used the adjusted diabatic mass flux to estimate mass-weighted vertically-averaged heating rates over the whole globe over different periods of time. They computed the distributions of atmospheric heating by using the Global Weather Experiment (GWE) level IIIb data set prepared by the European Center for Medium-Range Weather Forecast (ECMWF) for the four seasons of 1979. Their results reveal several interesting details regarding the locations of

10.6 Computation of Atmospheric Heating from Mass Continuity Equation

heat sources and sinks over different parts of the globe in different seasons, as summarized below.

In January, with the sun in the southern hemisphere, the principal heat sources appear over land areas south of the equator, with maximum heating over the continents of South America, Southern Africa and Australia, and over warm equatorial oceans, not directly affected by cold ocean currents. The heat source over the Australian region appears to be a southward extension of an extensive and intense heat source that resides over the maritime continent and ocean areas lying to the northeast of Australia and extending southeastward deep into southwestern Pacific along the South Pacific Convergence Zone (SPCZ). Principal heat sinks appear over the southern oceans dominated by cold ocean currents. In general, the land areas to the north of the equator appear as heat sinks, except the elevated Tibetan plateau where a weak heat source appears to exist. By contrast, large parts of the northern oceans lying to the east of the continents of Asia and North America appear as prominent heat sources in January.

In April, with maximum solar heating now being over the equatorial zone, there is a general northward movement of the heat sources and sinks from their January locations. The whole equatorial belt now appears as a heat source. A continuous and well-marked heat source appears all along the equatorial Pacific which was earlier interrupted by cold ocean currents over the eastern Pacific. The intense heat source over the maritime continent and adjoining southwestern Pacific to the northeast of Australia with extension along the SPCZ appears to persist. The heat source over the Tibetan plateau appears to have intensified.

In July, with the sun in the northern hemisphere, heat sources and sinks appear to have moved further northward. In the northern hemisphere, all the continents appear as heat sources while adjoining ocean areas to the south appear as heat sinks. The heat source over the Tibetan plateau appears to have further intensified to a value exceeding 3 K per day. It now appears to be directly connected with the heat source over the maritime continent and the SPCZ which extends into the southwestern Pacific. The equatorial heat source along the intertropical convergence zone (ITCZ) in the Pacific appears to be interrupted by cold ocean currents in the mid-Pacific. In general, most of the land and ocean areas south of the equator appear as heat sinks.

The October distribution largely resembles that of April, except that the heat source maximum over the Tibetan plateau which was conspicuous in July has considerably weakened and shows sign of retreat southeastward, thereby strengthening the heat source over equatorial western Pacific and the SPCZ.

Besides the vertically-averaged estimates of heating, they also worked out layer-averaged heating rates for the isobaric layers, surface to 800 mb, 800–600 mb, 600–400 mb, and 400–200 mb. The heating rates in isobaric layers were obtained by interpolation from isentropic to isobaric surfaces.

Part II
Dynamics of the Earth's Atmosphere – The General Circulation

Chapter 11
Winds on a Rotating Earth – The Dynamical Equations and the Conservation Laws

11.1 Introduction

Nature as a whole tends to remain in a quasi-balanced state and nullify any imbalance that may arise or is created between its different parts or regions on account of differences of pressure, temperature or density at any time. The principle of heat balance requires that any imbalance in the distribution of heat in the earth- atmosphere system, such as that between a heat source over a warm low pressure area and a heat sink over a neighboring relatively cold high pressure area will force the atmosphere to circulate so as to remove the imbalance and restore the original heat balance. In the balancing process, the oceans also take part, but because of much lower density and faster movement of air, the atmosphere plays a much more active and effective role in the required heat transfer process than the oceans in shorter time scales. On longer time scales, say months or seasons, however, the contributions of oceans in this regard may become just as important, if not greater.

The forces that cause air motion are basically three: (i) the pressure gradient force, (ii) the force of gravity, and (iii) the force of friction or viscosity. In deriving the equations of motion involving these forces, we shall make use of the Newton's second law of motion which relates the motion to the forces acting on the moving body. The law of conservation of mass will be used in deriving the equation of continuity and the first law of thermodynamics which is the law of conservation of energy will be used in deriving the thermodynamic energy equation. Further, while applying the equations of motion to the real atmosphere, it is important to recognize that motions occur on varieties of space and time scales ranging from the smallest and fastest molecular vibrations to the largest and slowest-moving planetary-scale waves and that it is sometimes realistic to simplify the equations to adapt them to the scale of motion under consideration.

11.2 Forces Acting on a Parcel of Air

As already mentioned, the forces acting on a parcel of air in the atmosphere are:

(i) Pressure gradient force,
(ii) Gravity force, and
(iii) Frictional force

11.2.1 Pressure Gradient Force

Since pressure on a surface is defined as the force exerted on unit area of the surface and it is a continuous function of space variables, we take p_0 as pressure at a point P (x_0, y_0, z_0) which lies at the center of an infinitesimally small rectangular volume with sides, δx, δy, δz in a rectangular co-ordinate system (x, y, z) (see Fig. 11.1). It is assumed that the pressure increases eastward towards the positive x-direction, northward towards the positive y-direction, and upward towards the positive z-direction.

Taking the x-direction first, the force of the outside air on face ABCD of the volume is $\{p_0 - (\partial p/\partial x)(\delta x/2)\}\,\delta y\,\delta z$, whereas that on the opposite face PQRS is $-\{p_0 + (\partial p/\partial x)(\delta x/2)\}\,\delta y\,\delta z$.

The net force in the x-direction is then given by the vector addition of the forces on the two faces, i.e., by

$$-(\partial p/\partial x)\,\delta x\,\delta y\,\delta z$$

Similarly, the net forces in the y and z direction are respectively,

$$-(\partial p/\partial y)\,\delta x\,\delta y\,\delta z$$
$$\text{and} \quad -(\partial p/\partial z)\,\delta x\,\delta y\,\delta z$$

The total net force per unit mass on the volume element is then obtained by adding the above three forces and dividing the sum by the mass of the volume which is given by $\rho\,\delta x\,\delta y\,\delta z$. Thus we get

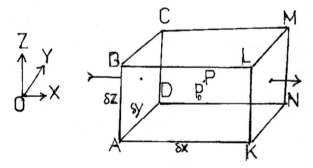

Fig. 11.1 Computation of pressure gradient force

11.2 Forces Acting on a Parcel of Air

$$-(1/\rho)\{(\partial p/\partial x)+(\partial p/\partial y)+(\partial p/\partial z)\} = -(1/\rho)\nabla p \qquad (11.2.1)$$

as the force per unit mass due to pressure gradient. Here ∇ denotes the three-dimensional gradient operator with components, $\partial/\partial x$, $\partial/\partial y$, $\partial/\partial z$ along the three co-ordinate axes x, y, z respectively, and the minus sign signifies that the force is acting in the direction from high to low values of pressure.

11.2.2 Gravity Force

We have seen in Chap. 1 that the field of earth's gravity at any height may be expressed in terms of the geopotential Φ at that height. So, if a parcel of air of unit mass is displaced through a vector distance $\delta\mathbf{r}$, the work done against gravity, δW, is given by

$$\delta W = -\nabla\Phi \cdot \delta\mathbf{r}$$

where $-\nabla\Phi$ is the gradient of the geopotential.

But, since $-\delta W = \mathbf{F} \cdot \delta\mathbf{r}$, where $\mathbf{F}$ is the force acting on the parcel, we get

$$\text{Force due to gravity per unit mass} = -\nabla\Phi \qquad (11.2.2)$$

11.2.3 Force of Friction or Viscosity

Like any real fluid, air possesses the property of viscosity by which it resists motion of any part of its medium relative to the other. So, whenever air at a horizontal level in the atmosphere moves relative to a lower level in the atmosphere or the ground below, a vertical gradient of horizontal velocity is created in proportion to the shearing stress between the upper and the lower levels. Since the air at the upper surface is moving faster, there is a net downward transfer of the horizontal momentum by molecular motion and the lower level experiences the same shearing stress as the upper. So, there is no net horizontal force acting on the atmospheric layer between the two levels. In the steady state (see Fig. 11.2), a measure of the shearing stress is given by the expression

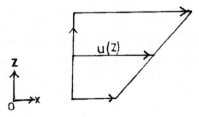

Fig. 11.2 Illustrating frictionally-generated steady-state shear flow in the vertical

Fig. 11.3 Computation of the frictional force

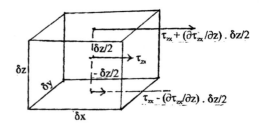

$$\tau_{zx} = \mu \, \partial u / \partial z \qquad (11.2.3)$$

where τ_{zx} denotes the shearing stress per unit area in the x-direction, u is the x-component of the wind vector at a height z above the lower boundary, and μ is called the co-efficient of molecular or dynamic viscosity.

However, in a more general case, where the shearing stress varies with height, a horizontal frictional force acts on the layer (see Fig. 11.3).

The magnitude of frictional force may be calculated as follows: Let τ_{zx} be the shearing stress at a horizontal surface at height z at the center of a rectangular volume element with sides δx, δy, δz along the rectangular co-ordinate axes, x, y, z respectively. Then the shearing stress in the x-direction across unit area of the lower boundary surface is $\tau_{zx} - (\partial \tau_{zx}/\partial z) \, \delta z/2$, while that across the upper boundary surface is $\tau_{zx} + (\partial \tau_{zx}/\partial z) \, \delta z/2$. The net frictional force on the volume element acting in the x-direction is then

$$[\{\tau_{zx} + (\partial \tau_{zx}/\partial z) \, \delta z/2\} - \{\tau_{zx} - (\partial \tau_{zx}/\partial z)\delta z/2\}] \, \delta x \, \delta y,$$

or $(\partial \tau_{zx}/\partial z)\delta x \, \delta y \, \delta z$. On substituting the value of τ_{zx} from (11.2.3) and dividing by the mass, $\rho \delta x \delta y \delta z$, we get for the x-component of the frictional force per unit mass

$$F_x = (1/\rho)\partial(\tau_{zx})/\partial z = (1/\rho)\partial(\mu \, \partial u/\partial z)/\partial z$$

For constant μ, F_x may be written as $F_x = \nu \, \partial^2 u/\partial z^2$
where $\nu (= \mu/\rho)$ is called the co-efficient of kinematic viscosity.

Similarly, the y-component of the frictional force per unit mass may be written

$$F_y = \nu \partial^2 v/\partial z^2 \qquad (11.2.4)$$

Thus, the frictional force per unit mass, $\mathbf{F} = \nu \, \partial^2 \mathbf{V}/\partial z^2$

11.3 Acceleration of Absolute Motion

Newton's second law of motion applies to absolute motion which is motion measured relative to a fixed frame of reference (fixed relative to space or distant stars). It states that the rate of change of momentum of a body is equal, in magnitude and direction, to the vector sum of the forces acting on the body. If we consider that the forces act on an infinitesimal parcel of air of constant density ρ, the absolute acceleration of the parcel following motion may be expressed in the vector form

$$\rho(dV_a/dt)_a = -\nabla p - \rho\nabla\Phi_a + \rho F \qquad (11.3.1)$$

where $(d/dt)_a$ denotes differentiation following absolute motion,

V_a is the absolute velocity vector,
p is pressure,
Φ_a is geopotential in absolute co-ordinate system,
F is the frictional force vector,
∇ is Del operator,
t is time,
and the subscript 'a' refers to the fixed frame.

In (11.3.1), the first term on the right-hand side gives the pressure force, the second the gravitational force and the third the force of friction.

It is sometimes convenient to consider acceleration per unit mass, instead of per unit volume. In that case, we divide (11.3.1) by ρ, and obtain

$$(dV_a/dt)_a = -\alpha\nabla p - \nabla\Phi_a + F \qquad (11.3.2)$$

where α is called the specific volume, or volume per unit mass $(= 1/\rho)$.

Equation (11.3.2) is then the equation in vector form for acceleration of absolute motion per unit mass.

11.4 Acceleration of Relative Motion

In practice, however, it is convenient to measure the velocity or acceleration of a parcel relative to a co-ordinate system which is fixed to the earth and call it relative velocity or acceleration. For this, we allow for the rotation of the earth's surface by adding an extra term V_E, which we may call the earth velocity, to the relative velocity. A measure of this extra term is given by the vector product, $V_E = \Omega \times r$, where Ω is the angular velocity of the earth and r the radial vector position of a point P on the surface of the spherical earth (see Fig. 11.4).

The absolute velocity V_a is then related to the relative velocity V_r by the vector equation

$$V_a = V_r + V_E = V_r + \Omega \times r \qquad (11.4.1)$$

Since r is the position vector of the point P, we may write

$$V_a = (dr/dt)_a, \text{ and } V_r = (d\,r/dt)_r$$

where the subscripts 'a' and 'r' refer to the absolute and the relative motion respectively.

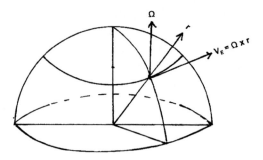

Fig. 11.4 Illustrating the Earth velocity vector, V_E

Equation (11.4.1) may then be written in the form

$$(d\mathbf{r}/dt)_a = (d\mathbf{r}/dt)_r + \mathbf{\Omega} \times \mathbf{r} \qquad (11.4.2)$$

It follows from (11.4.2) that the differential operator

$$(d/dt)_a = (d/dt)_r + \mathbf{\Omega} \times$$

is quite general and may be applied to any vector.

When applied to the absolute velocity vector $\mathbf{V}_a$, we have the equation

$$(d\mathbf{V}_a/dt)_a = (d\mathbf{V}_a/dt)_r + \mathbf{\Omega} \times \mathbf{V}_a \qquad (11.4.3)$$

Substituting for $\mathbf{V}_a$ from (11.4.1) on the right-hand side of (11.4.3), we get

$$(d\mathbf{V}_a/dt)_a = (d\mathbf{V}_r/dt)_r + 2\mathbf{\Omega} \times \mathbf{V}_r + \mathbf{\Omega} \times (\mathbf{\Omega} \times \mathbf{r}) \qquad (11.4.4)$$

It can be shown that the triple vector product on the right-hand side of (11.4.4) is equal to $-\Omega^2 \mathbf{R}$, where $\mathbf{R}$ is the position vector of the point P issuing perpendicularly from the axis of the earth (see Fig. 11.5).

Equation (11.4.4) may, therefore, be written in the final form

$$(d\mathbf{V}_a/dt)_a = (d\mathbf{V}_r/dt)_r + 2\mathbf{\Omega} \times \mathbf{V}_r - \Omega^2 \mathbf{R} \qquad (11.4.5)$$

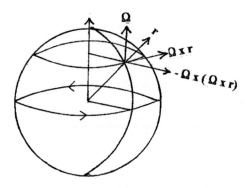

Fig. 11.5 Direction of the triple vector product, $\mathbf{\Omega} \times (\mathbf{\Omega} \times \mathbf{r})$

11.4 Acceleration of Relative Motion

Equation (11.4.5) is then the relationship between the accelerations in the two co-ordinate systems.

The equation for relative acceleration in the earth co-ordinate system is obtained by substituting for $(dV_a/dt)_a$ from (11.4.5) in (11.3.2). This gives

$$(dV_r/dt)_r = -\alpha \nabla p - 2\Omega \times V_r + F + (-\nabla \Phi_a + \Omega^2 R) \qquad (11.4.6)$$

It is immediately clear from (11.4.1) and (11.4.6), that the rotation of the earth introduces two additional forces, viz., $-2\Omega \times V_r$ and $\Omega^2 R$ in the expression for the relative acceleration. These are fictitious or apparent forces and depend upon the rotation of the earth. The first is called the Coriolis force and the second the centrifugal force. Also, the two terms within the bracket on the right-hand side of (11.4.6) constitute **g**, the earth's effective gravity, which is equal to $-\nabla \Phi$, vide (1.2.3) (see Chap. 1). Henceforth, we drop the subscript 'r', since, unless otherwise stated, we shall always deal with the relative motion, i.e., motion relative to the earth's surface. The final form of the equation of motion in vector notation is then

$$dV/dt = -\alpha \nabla p - 2\Omega \times V - \nabla \Phi + F \qquad (11.4.7)$$

11.4.1 Coriolis Force

The Coriolis force per unit mass, $-2\Omega \times V$, which is named after the French mathematician, G.G. de Coriolis (1792–1843), who first derived it, acts at right angle to the plane containing the earth's rotational axis and the relative velocity in the horizontal plane. It influences the direction of motion but not the speed of the wind. Since it changes the direction of the wind, it is often called the deviating force due to earth's rotation. The Coriolis force acts to the right of the wind direction in the northern hemisphere (NH) and to the left in the southern hemisphere (SH), as shown in Fig. 11.6.

The term giving the Coriolis force in (11.4.7) is in vector form. It is sometimes convenient to resolve it in a rectangular co-ordinate system with the x-axis pointing eastward, the y-axis pointing northward and the z-axis pointing vertically upward. In

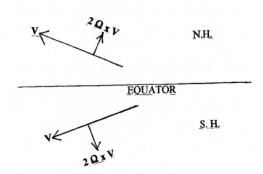

Fig. 11.6 The direction of the Coriolis force relative to wind direction

Fig. 11.7 Components of the earth's angular velocity vector, Ω, at a given latitude, ϕ

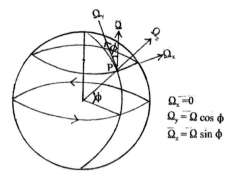

this co-ordinate system, the earth's angular velocity, Ω, has the components, $\Omega_x = 0$; $\Omega_y = \Omega\cos\phi$; $\Omega_z = \Omega\sin\phi$, where ϕ is the latitude, shown in Fig. 11.7.

If we take u, v, w as the three components of the wind vector **V** along the x, y, z axes respectively, the resolved components of the Coriolis force, $-2\Omega \times \mathbf{V}$, along these axes, are respectively,

$$-2\Omega(w\cos\phi - v\sin\phi),\ -2\Omega u\sin\phi,\ 2\Omega u\cos\phi$$

11.5 The Equations of Motion in a Rectangular Co-ordinate System

The components of the equation of motion (11.4.7) in a rectangular co-ordinate system (x, y, z) are then

$$du/dt = -\alpha\,\partial p/\partial x - 2\Omega(w\cos\phi - v\sin\phi) + \nu\partial^2 u/\partial z^2$$
$$dv/dt = -\alpha\,\partial p/\partial y - 2\Omega u\sin\phi + \nu\partial^2 v/\partial z^2 \quad (11.5.1)$$
$$dw/dt = -\alpha\,\partial p/\partial z + 2\Omega u\cos\phi - g$$

In (11.5.1), no vertical component of the frictional force has been added, since it is not regarded as of any consequence for the types of motion we usually consider in the atmosphere, in which motions are largely horizontal.

11.6 A System of Generalized Vertical Co-ordinates

It is sometimes useful and advantageous to use a vertical co-ordinate which is an independent single-valued monotonous function of height. Such a vertical coordinate can be pressure, potential temperature, or any other variable which can be expressed as a function of height. A vertical co-ordinate called sigma σ ($= p/p_s$, where p_s is surface pressure) has been particularly useful to take care of the variable topography

11.6 A System of Generalized Vertical Co-ordinates

of the earth's surface. Let us take 'c' as the generalized vertical co-ordinate given by $c = c(x, y, z, t)$. The word 'co-ordinate' here simply means a re-labelling of the vertical axis. It does not change the direction when changing from z to p (for example). Our task is then to derive an expression for the gradient of an arbitrary function α along the c = constant surface, when its gradient along a level surface (z = constant) or an isobaric surface (p = constant) is known.

Since a change $\delta\alpha$ of a function α in the (x-z) plane is given by

$$\delta\alpha = (\partial\alpha/\partial x)_z \, dx + (\partial\alpha/\partial z)_x dz \tag{11.6.1}$$

where the first term on the right-hand side gives the x-component of the gradient of α along the z = constant surface, the variation of α in the x-direction along the c-surface (see Fig. 11.8) is given by:

$$(\partial\alpha/\partial x)_c = (\partial\alpha/\partial x)_z + (\partial z/\partial x)_c (\partial\alpha/\partial z) \tag{11.6.2}$$

A similar expression can be written for the y-component of the gradient.

The transformation of the gradient of any arbitrary function from z to c co-ordinate is then given by the vector operator

$$\nabla_c = \nabla_z + (\nabla_c z)\partial/\partial z$$

i.e.,
$$\nabla_z = \nabla_c - (\nabla_c z)\partial/\partial z \tag{11.6.3}$$

Equation (11.6.3) is quite general and may be used for transformation to any vertical co-ordinate 'c' which is a function of z or p, such as an isobaric surface, an isentropic surface or a sigma (σ) surface (where $\sigma = p/p_0$).

As examples of the use of (11.6.3), let us consider the following transformations:

(a) From height (z=constant) to isobaric (p=constant) surface

In this case, we put $c = p$, and since we wish to compute the gradient of z along the isobaric (p = constant) surface, we apply the operator (11.6.3) to write

$$\nabla_p p = \nabla_z p + (\nabla_p z)\partial p/\partial z$$

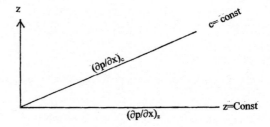

Fig. 11.8 Transformation of pressure gradient force from 'z' to 'c' co-ordinate

where the left-hand side disappears, and, since $\partial p/\partial z = -\rho g$ (the hydrostatic relation), the right-hand side reduces to the required relation

$$(1/\rho)\nabla_z p = g\,\nabla_p z = \nabla_P \Phi \tag{11.6.4}$$

where Φ is the geopotential ($= g\,z$).

Thus, on the isobaric surface, density does not appear explicitly and the horizontal pressure gradient is given by the gradient of the geopotential at constant pressure surface.

(b) From height (z=constant) to an isentropic (θ=constant) surface

In this case, we put $c = \theta$, and apply the operator (11.6.3) to p which varies along the $z =$ constant surface and write

$$\nabla_\theta p = \nabla_z p + (\nabla_\theta z)\partial p/\partial z \tag{11.6.5}$$

With the aid of the hydrostatic relation and re-arranging, (11.6.5) may be written

$$(1/\rho)\nabla_z p = (1/\rho)\nabla_\theta p + g\,(\nabla_\theta z) \tag{11.6.6}$$

Since $\theta = T\,(1000/p)^\kappa$, $\partial p/\partial z = -\rho g$, and $p = \rho RT$ (equation of state), it can be shown that the right-hand side of (11.6.6) is equal to $\nabla_\theta(c_p T + gz)$. Thus, along with (11.6.4), we have the relationships

$$(1/\rho)\nabla_z p = \nabla_P \Phi = \nabla_\theta(c_p T + gz) \tag{11.6.7}$$

where $(c_p T + gz)$ is called the dry static energy. It is also called the Montgomery potential or isentropic stream function.

(c) From isobaric (p=constant) to sigma (σ-constant) surface

Here we put $c = \sigma$, and, since $\sigma = p/p_s$, we obtain, using (11.6.3), the transformation equation

$$\nabla_P() = \nabla_\sigma() - (\sigma/p_s)\nabla p_s \partial()/\partial\sigma \tag{11.6.8}$$

Where ( ) is any variable.

11.7 The Equations of Motion in Spherical Co-ordinate System

It is well-known that the surface of the earth may be treated as spherical for most meteorological purposes. So a curvilinear co-ordinate system in which the horizontal co-ordinate axes follow the curvature of the earth is more appropriate for

11.7 The Equations of Motion in Spherical Co-ordinate System

Fig. 11.9 Spherical co-ordinate system on the earth's surface

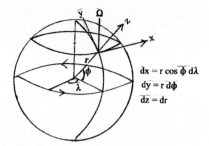

resolving the relative motion into components than any other co-ordinate system. In the spherical co-ordinate system (shown in Fig. 11.9), the horizontal co-ordinates are replaced by longitude λ (to the east) and latitude ϕ (to the north) and the earth's radius vector r is taken as the vertical co-ordinate. The relationships with the corresponding Cartesian co-ordinates in measurement of distances and velocity components are as follows:

$$\delta x = a \cos \phi \, \delta\lambda; \quad \delta y = a \, \delta\phi; \quad \delta z = \delta r$$
$$u = (a \cos \phi) d\lambda/dt; \quad v = a \, d\phi/dt; \quad w = dr/dt \quad (11.7.1)$$

In the spherical co-ordinate system, the velocity vector changes direction as a parcel moves along the earth's surface. So, we compute the velocity and acceleration in this co-ordinate system as follows: Let **i, j, k** be the unit vectors in the eastward, northward and vertical direction respectively. Then, we write for the velocity vector

$$\mathbf{V} = u\,\mathbf{i} + v\,\mathbf{j} + w\,\mathbf{k},$$

and, for the acceleration following motion,

$$d\mathbf{V}/dt = u \, d\mathbf{i}/dt + v \, d\mathbf{j}/dt + w \, d\mathbf{k}/dt$$
$$+ \mathbf{i} \, du/dt + \mathbf{j} \, dv/dt + \mathbf{k} \, dw/dt \quad (11.7.2)$$

In (11.7.2), the rate of change of the unit vectors may be evaluated as follows:

For d **i**/dt, we note that the unit vector **i** varies only along the x-direction, so we may write,

$$d\mathbf{i}/dt = u \, d\mathbf{i}/dx \quad (11.7.3)$$

Since the vector d**i** lies in a plane perpendicular to the axis of the earth and always directed towards the same axis, it has components in the northward and the upward directions, as shown in Fig. 11.10.

With the aid of Fig. 11.10, we, therefore, write

$$d\mathbf{i}/dx = (\mathbf{j} \sin \phi - \mathbf{k} \cos \phi)/(a \cos \phi) \quad (11.7.4)$$

The unit vector **j** varies with longitude and latitude but not with **k**. So we write

$$d\mathbf{j}/dt = u \, d\mathbf{j}/dx + v \, d\mathbf{j}/dy$$

Fig. 11.10 The components of the vector di along **j** and **k** axes

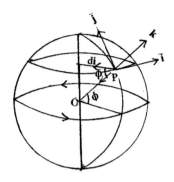

The components of the variations in these directions are

$$d\mathbf{j}/dx = -\mathbf{i}\tan\phi/a, \quad d\mathbf{j}/dy = -\mathbf{k}/a \qquad (11.7.5)$$

Fig. 11.11 shows how the unit vector **j** varies with (a) longitude, and (b) latitude. Similarly, we see that the unit vector **k** does not vary in the vertical but has variations along (a) longitude and (b) latitude. So we write

$$d\mathbf{k}/dt = u\, d\mathbf{k}/dx + v\, d\mathbf{k}/dy \qquad (11.7.6)$$

where $d\mathbf{k}/dx = \mathbf{i}/a$; and $d\mathbf{k}/dy = \mathbf{j}/a$

The variations of k along longitude and latitude are shown in Fig. 11.12 (a, b) respectively:

Substituting from (11.7.3) to (11.7.6) in (11.7.2), we get for the acceleration of motion in spherical co-ordinates the expression

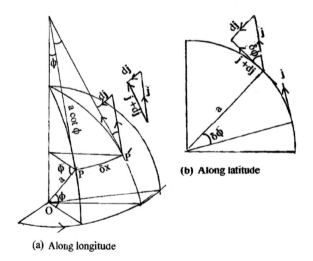

(a) Along longitude

Fig. 11.11 Components of the d**j** vector along (a) longitude and (b) latitude

11.8 The Equation of Continuity

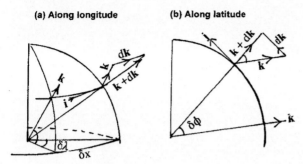

Fig. 11.12 Components of the **dk** vector along (**a**) longitude and (**b**) latitude

$$d\mathbf{V}/dt = [du/dt - uv\ \tan\phi/a + uw/a]\mathbf{i} + [dv/dt + (u^2\ \tan\phi)/a + vw/a]\mathbf{j}$$
$$+ [dw/dt - (u^2 + v^2)/a]\mathbf{k} \quad (11.7.7)$$

Finally, substituting (11.7.7) in (11.5.1), we get the components of the equations of motion in spherical co-ordinates

$$du/dt - uv\ \tan\phi/a + uw/a = -\alpha\partial p/\partial x + 2\Omega v \sin\phi - 2\Omega w \cos\phi + v\ \partial^2 u/\partial z^2 \quad (11.7.8)$$

$$dv/dt + u^2\ \tan\phi/a + vw/a = -\alpha\partial p/\partial y - 2\Omega u \sin\phi + v\ \partial^2 v/\partial z^2 \quad (11.7.9)$$

$$dw/dt - (u^2 + v^2)/a = -\alpha\partial p/\partial z + 2\Omega u \cos\phi - g \quad (11.7.10)$$

The momentum equations (11.7.8–11.7.10) are non-linear, since they contain terms which are quadratic in the dependent variables. Even the time derivatives of the dependent variables u, v, w, following motion are non-linear partial differential equations, as given by the relation for du/dt

$$du/dt = \partial u/\partial t + u\ \partial u/\partial x + v\partial u/\partial y + w\ \partial w/\partial z$$

and similar ones for dv/dt and dw/dt. The terms involving 'a' on the left-side of the momentum equations (11.7.8–11.7.10) are called the curvature terms, since they are due to the curvature of the earth.

11.8 The Equation of Continuity

The law of conservation of mass requires that the rate of change of mass following motion should be zero. That is,

$$d(\rho\ \delta V)/dt = 0$$

Or,
$$-(1/\rho)d\rho/dt = (1/\delta V)\ d(\delta V)/dt$$

In the above expression, the right-hand side gives the rate of expansion of the volume element δV per unit time, which is equivalent to the three-dimensional divergence of the wind, $\nabla \cdot \mathbf{V}$.

So the above equation can be written in the Lagrangian form

$$d\rho/dt = -\rho \nabla \cdot \mathbf{V} = -\rho(\partial u/\partial x + \partial v/\partial y + \partial w/\partial z) \tag{11.8.1}$$

This is the continuity equation for change of density and volume of an individual parcel of the fluid following motion. It states that the density of the parcel will decrease if there is divergence of air from the volume and increase if there is convergence into it.

Since, $\qquad d\rho/dt = \partial \rho/\partial t + \mathbf{V} \cdot \nabla \rho = -\rho \, \nabla \cdot \mathbf{V}$

Equation (11.8.1) may also be written in the Eulerian form

$$\partial \rho/\partial t = -\nabla \cdot \rho \mathbf{V} \tag{11.8.2}$$

The Eq. (11.8.2) states that the rate of change of mass in a unit volume of a fluid is equal to the convergence or divergence of mass through the boundaries of the volume. Equations (11.8.1) and (11.8.2) are only the different expressions of the principle of conservation of mass.

If we assume that the air is incompressible, the left-hand side of (11.8.2) vanishes and we write

$$\nabla \cdot \mathbf{V} = \partial u/dx + \partial v/\partial y + \partial w/\partial z = 0$$
$$\text{Or,} \quad \partial w/\partial z = -(\partial u/\partial x + \partial v/\partial y) = -\nabla_H \cdot \mathbf{V} \tag{11.8.3}$$

where $\nabla_H \cdot \mathbf{V}$ denotes horizontal divergence.

11.9 The Thermodynamic Energy Equation

The thermodynamic energy equation is virtually a restatement of the First law of thermodynamics (3.2) in the form

$$\delta Q/dt = c_v \, dT/dt + p \, d\alpha/dt$$

Or, using the equation of state for an ideal gas, and the relation, $c_p - c_v = R$, the above equation may be written as

$$\delta Q/dt = c_p \, dT/dt - \alpha \, dp/dt \tag{11.9.1}$$

where the left-side denotes the rate of diabatic heating, and the right-side constitutes the adiabatic response of the atmosphere in the form of an increase of its internal energy and doing work against external pressure, and the symbols have their usual meanings (see Chap. 3).

11.10 Scale Analysis and Simplification of the Equations of Motion

Motions in the atmosphere occur on different space and time scales. From molecular vibrations to very large planetary-scale flow, there is a vast range of motions and each type of motion has its own characteristic scale of length, velocity and time. Although the equations of motion (11.7.8–11.7.10) are quite general, it may be possible for any particular type of motion to simplify them by eliminating terms which may be irrelevant or vanishingly small, and retaining those which more closely represent the motion being considered. This process of simplification and approximation is called scale analysis. It is like weighing the terms with an atmospheric scale and finding out which is heavy and which is light. We present the average length-scale of a few familiar types of motion systems in Table 11.1.

The scale is based on commonly-observed values of the basic variables and their observed variations in time and space for the particular type of motion system. For example, for large-scale tropical or midlatitude motion systems, we adopt the following typical values of the scaling parameters.

Length, $L \sim 10^6$ m
Horizontal wind velocity, $V \sim 10$ m s^{-1}
Time, $t\ (L/V) \sim 10^5$ s
Depth, $D \sim 10^4$ m
Vertical velocity, $W \sim 10^{-2}$ m s^{-1}
Radius of the earth, $L \sim 10^6$ m
Horizontal pressure fluctuation Δp
Normalized by density ρ, $\Delta p/\rho \sim 10^3$ m^2 s^{-2}

It should be pointed out that the time scale 't' refers to the advective time scale for the motion system and is usually the time taken for the system to move through one wavelength at approximately the velocity of the wind. The vertical velocity 'w' is usually not directly measured but can be computed in principle from the

Table 11.1 Length scale of some types of atmospheric motions

Type of motion	Horizontal scale (m)
Molecular vibration (Mean free path)	10^{-7}
Small-scale eddy	10^{-2}
Dust devil	10
Tornado	10^2
Cumulus cloud	10^3
Cloud clusters	$(1-10) \times 10^4$
Cyclone/Hurricane/Typhoon	$(1-10) \times 10^5$
Synoptic-scale wave	10^6
Planetary-scale wave	10^7

divergence of the horizontal wind. Computations show that w is about two orders of magnitude smaller than the horizontal velocity. The frictional terms in Eqs. (11.7.8) and (11.7.9) are not considered, since it is held that except in the narrow boundary layer close to the earth's surface, friction does not play an important role.

We now consider a synoptic-scale wave disturbance centered at latitude, say $\phi = 30°$, where each of the components of the earth's angular velocity vector, viz., the vertical, $2\Omega \sin\phi$, and the horizontal, $2\Omega \cos\phi$, has magnitude of the order of $10^{-4}\,\text{s}^{-1}$. The vertical component is usually denoted by 'f' and called the Coriolis parameter in meteorology. The scale analysis of the various terms of the horizontal momentum equations (11.7.8) and (11.7.9) for this wave system is shown in Table 11.2 which shows that the magnitudes of the terms involving the curvature of the earth as well as the horizontal component of earth's angular velocity vector are relatively unimportant for the type of motion under consideration. The most important terms which appear to be in approximate geostrophic balance are the pressure gradient force and the Coriolis force, since they are approximately of the same order of magnitude (10^{-3}). This balance was first noted by Buys Ballot on weather charts in 1857.

11.10.1 The Geostrophic Approximation and the Geostrophic Wind

The approximate balance between the pressure gradient force and the Coriolis force gives us the geostrophic relationships:

$$u_g = -(1/\rho f)\,\partial p/\partial y$$
$$v_g = (1/\rho f)\,\partial p/\partial x$$

Or, in vector notation, $\mathbf{V}_g = -(1/\rho f)\,\mathbf{k} \times \nabla p$ \hfill (11.10.1)

where $\mathbf{V}_g (= u_g\,\mathbf{i} + v_g\,\mathbf{j})$ is called the geostrophic wind.

Table 11.2 Scale analysis of the horizontal momentum equations

	Terms					
	1	2	3	4	5	6
(x-comp)	du/dt	$-2\Omega v \sin\phi$	$+2\Omega w \cos\phi$	$+u w/a$	$-(u v \tan\phi)/a$	$= -\alpha\partial p/\partial x$
(y-comp)	dv/dt	$+2\Omega u \sin\phi$		$+v w/a$	$+(u^2 \tan\phi)/a$	$= -\alpha\partial p/\partial x$
Scales	V^2/L	$V \times 10^{-4}$	$W \times 10^{-4}$	V W/a	V^2/a	$((\Delta p)/\rho)/L$
Order of Magnitudes of the terms (ms^{-2})						
	10^{-4}	10^{-3}	10^{-6}	10^{-8}	10^{-5}	10^{-3}

11.10.2 Scale Analysis of the Vertical Momentum Equation

The scale analysis of the vertical component of the momentum equation(11.7.10) carried out on the same lines as in Table 11.2 shows that the order of magnitude of the various terms are as given in Table 11.3.

It follows from Table 11.3 that the two terms which are in close balance with each other are the 3rd and the 5th terms. They yield the well-known hydrostatic relation (2.3.4)

$$(1/\rho)\, \partial p/\partial z \sim -g \qquad (11.10.2)$$

Table 11.3 Scale analysis of the vertical momentum equation

Terms				
1	2	3	4	5
dw/dt	$-(u^2+v^2)/a$	$-\alpha\, \partial p/\partial z$	$-2\Omega u \cos \phi$	$-g$
Scale WV/L	V^2/L	$(\Delta p/\rho)/D$	$V \times 10^{-4}$	$-g$
Order of magnitude of the terms (ms^{-2})				
10^{-7}	10^{-5}	10	10^{-3}	10

Chapter 12
Simplified Equations of Motion – Quasi-Balanced Winds

12.1 Introduction

The scale analysis in Chap. 11 has shown that in the equations of motion (11.7.8–11.7.10), all terms are not equally important and that it is possible to eliminate some of them in order to retain the more important ones which may be in approximate balance. The sorting of this kind permits us to obtain useful relationships between the distributions of pressure, temperature and wind in the atmosphere. It will be shown in the present chapter that the simplified equations offer relationships of great practical importance in dealing with the real atmosphere, when used in isobaric co-ordinate and in natural co-ordinate systems.

12.2 The Basic Equations in Isobaric Co-ordinates

12.2.1 Horizontal Momentum Equations

On considerations of scale analysis, the horizontal momentum equation (11.4.7) may be written in the simplified vector form

$$d\mathbf{V}/dt + 2\mathbf{\Omega} \times \mathbf{V} = -(1/\rho)\, \nabla_z p \qquad (12.2.1)$$

where $\mathbf{V} = u\mathbf{i} + v\mathbf{j}$ is the horizontal velocity vector, and ∇_z is horizontal Del operator.

We can transform (12.2.1) to the isobaric co-ordinate system by noting that in this system d/dt is to be written in the form

$$d/dt \equiv \partial/\partial t + (\mathbf{V}\cdot\nabla_p) + \omega\, \partial/\partial p \qquad (12.2.2)$$

where $\omega(= dp/dt)$ is called 'omega' or vertical p-velocity and where $\omega\, \partial/\partial p$ has replaced $w\, \partial/\partial z$. Since pressure decreases with height in the atmosphere, a negative value of ω signifies upward motion and a positive value downward motion.

Further, as shown in (11.6.7), $-(1/\rho)\nabla_z p = -\nabla_p \Phi$

The simplified horizontal momentum equations in isobaric co-ordinates may, therefore, be written as

$$\partial u/\partial t + (\mathbf{V}\cdot\nabla_p)\, u + \omega \partial u/\partial p = -\partial \Phi/\partial x + f v \qquad (12.2.3)$$

$$\partial v/\partial t + (\mathbf{V}\cdot\nabla_p)\, v + \omega \partial v/\partial p = -\partial \Phi/\partial y - f u \qquad (12.2.4)$$

where $f = 2\Omega \sin\phi$.

In Eqs. (12.2.3) and (12.2.4), the density does not appear explicitly and this constitutes a great advantage of the isobaric co-ordinate system over the Cartesian co-ordinate system of the equations of motion.

12.2.2 The Continuity Equation

The equation of continuity in isobaric co-ordinates may be derived as follows:

In Sect. 11.8, we showed that the rate of change of mass following motion in Cartesian co-ordinates is given by (11.8.1). Since $\partial w/\partial z = \partial \omega/\partial p$, we can write the continuity equation following motion in isobaric co-ordinates as

$$d\rho/dt = -\rho\,(\partial u/\partial x + \partial v/\partial y + \partial \omega/\partial p) \qquad (12.2.5)$$

However, if we transform the total derivative on the left-side of (12.2.5) to the isobaric co-ordinate system using (12.2.2), we obtain, after re-arranging, the flux form of the continuity equation

$$\partial \rho/\partial t = -\{\nabla_p\,(\rho \mathbf{V}) + \partial(\rho\omega)/\partial p\} \qquad (12.2.6)$$

12.2.3 The Thermodynamic Energy Equation

In the isobaric co-ordinate system, the thermodynamic energy equation (11.9.1) may be written in the form

$$dT/dt = (\alpha\,\omega + \delta Q/dt)/c_p \qquad (12.2.7)$$

Expanding the left side of (12.2.7) with the aid of (12.2.2) and re-arranging, we obtain

$$\partial T/\partial t = -\mathbf{V}\cdot\nabla T + \sigma\,\omega + (1/c_p)\,\delta Q/dt \qquad (12.2.8)$$

where $\sigma\,(= \kappa T/p - \partial T/\partial p)$ is the static stability parameter, $\omega\,(= dp/dt)$ is the vertical p-velocity, and $\kappa = R/c_p$.

Since $T = (-p/R)\partial \Phi/\partial p$, the Eq. (12.2.8) may also be expressed in terms of $\partial \Phi/\partial p$.

12.3 Balanced Flow in Natural Co-ordinates

As already mentioned in Chap. 10, diabatic heating or cooling occurs due to convergence of net radiative as well as sensible and latent heat fluxes, and it is possible to get an approximate measure of it in a steady state ($\partial T/\partial t = 0$) by evaluating the adiabatic effects it produces in the form of horizontal and vertical thermal advections of heat as given by the first and the second terms on the right-hand side of equation (12.2.8). Usually, in the tropics the second term approximately balances the third term, but in midlatitudes the first and second terms together balance the third term.

12.3 Balanced Flow in Natural Co-ordinates

It is convenient to deal with horizontal motion depicted on weather maps in Natural co-ordinates. Depending upon the terms in balance, we may define some idealized balanced flows. This was realized more than a century ago.

12.3.1 Velocity and Acceleration in Natural Co-ordinate System

The natural co-ordinate system is characterized by a system of unit vectors, **t, n, k**, in which **t** is in the direction of the wind, **n** is in a direction normal to the wind (positive to the left of the wind), and **k** is in the vertical direction. This means that $\mathbf{k} = \mathbf{t} \times \mathbf{n}$.

Let the horizontal velocity along a curve $s = s(t)$ in this co-ordinate system be denoted by $\mathbf{V} = V\mathbf{t}$, where $V = ds/dt$, and **t** is a unit vector along the curve.

Then, the acceleration **a** is given by

$$\mathbf{a} = d\mathbf{V}/dt = d(V\mathbf{t})/dt = (dV/dt)\mathbf{t} + V\, d\mathbf{t}/dt \qquad (12.3.1)$$

[Note that **t** is a unit vector, while t denotes time]

To evaluate d**t**/dt, we assume that the wind changes direction and that after time δt it turns through an angle δψ in an anticlockwise direction (Fig. 12.1).

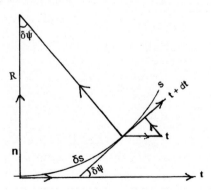

Fig. 12.1 Evaluation of d**t**/dt

Let R be the radius of curvature of the path of the parcel. Then it can be shown from Fig. 12.1 that

$$d\mathbf{t}/dt = d\psi/dt\ \mathbf{n} = V/R\ \mathbf{n} \qquad (12.3.2)$$

since $d\psi/dt$, the angular velocity, is equal to V/R.

Substituting for $d\mathbf{t}/dt$ from (12.3.2) in (12.3.1), we obtain

$$\mathbf{a} = (dV/dt)\ \mathbf{t} + (V^2/R)\ \mathbf{n} \qquad (12.3.3)$$

12.3.2 The Gradient Wind

In (12.3.3), dV/dt denotes the tangential acceleration, while V^2/R gives the radial acceleration of the moving parcel towards the center of the curve. The latter is also called the centripetal acceleration, since for a unit mass of the parcel it signifies the force with which it is continuously attracted towards the center while it moves along the curve. The curvature effect produces a centrifugal acceleration along the outward normal. Thus, the centripetal and the centrifugal accelerations are equal and opposite of each other.

Since the direction in which the Coriolis force acts is at right angle to the velocity vector, we may write it in the natural co-ordinates, as follows:

$$f\mathbf{k}\times\mathbf{V} = fV(\mathbf{k}\times\mathbf{t}) = fV\ \mathbf{n} \qquad (12.3.4)$$

Also, the horizontal pressure gradient may be written

$$\nabla_p = (\partial p/\partial s)\ \mathbf{t} + (\partial p/\partial n)\ \mathbf{n} \qquad (12.3.5)$$

Using (12.3.4) and (12.3.5), the horizontal momentum equations (12.2.1) may be resolved into the tangential and radial components as follows:

$$dV/dt = -(1/\rho)\ \partial p/\partial s \qquad (12.3.6)$$

$$V^2/R + fV = -(1/\rho)\ \partial p/\partial n \qquad (12.3.7)$$

It follows from (12.3.6) that if the pressure does not vary along the curve, i.e., $\partial p/\partial s = 0$, the tangential wind speed V remains constant along an isobar. The wind blowing along the curved path is called the gradient wind and (12.3.7) is the famous gradient wind equation.

The approximate balance wind that appears to closely conform to the observed wind in the field of a large-scale motion system on a synoptic weather map is the gradient wind given by (12.3.7) which represents a three-way balance between the pressure gradient force, Coriolis force and the centrifugal force in a circular p-field with radius R and is obtained by solving (12.3.7) for V.

Since (12.3.7) is a quadratic equation, its two roots are given by

12.3 Balanced Flow in Natural Co-ordinates

$$V = -fR/2 \pm [(f^2 R^2/4) - (R/\rho)(\partial p/\partial n)]^{1/2} \qquad (12.3.8)$$

Depending upon the values and signs of R and $\partial p/\partial n$, we may have several possible values of V in (12.3.8), but some of them are clearly not admissible because of the requirement that V be real and non-negative. No solution of (12.3.8) is admissible in which the quantity under the radical is negative. The admissible values of V, however, are found to be better estimates of the observed winds than the geostrophic winds. Fig. 12.2 shows the balance between the forces for the two types of gradient flow commonly observed in the northern hemisphere: (a) a regular low; (b) a regular high.

A point to note here is that the force that deviates the down-the-pressure-gradient movement to a movement along the isobar is the Coriolis force.

In the event of a lack of balance amongst the forces, it is not impossible to have an anomalous flow, such as an anticyclonic flow around a low pressure area, but such anomalies are rare and may occur as a passing phase only when dV/dt is large, i.e., does not obey (12.3.7).

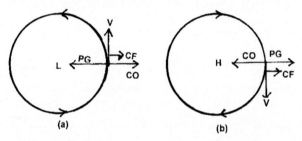

Fig. 12.2 Balance of forces in two types of gradient flow commonly observed in the northern hemisphere: (a) a regular low (L); (b) a regular high (H). PG denotes pressure gradient force, CO Coriolis force, and CF centrifugal force. V denotes the balanced wind

12.3.3 The Geostrophic Wind

In the very restrictive case in which an isobar is straight, i.e., has no curvature, R tends to be infinity and the curvature term V^2/R in (12.3.7) vanishes. In that case, (12.3.7) reduces to the balance relation

$$fV = -(1/\rho) \, \partial p/\partial n$$
$$\text{Or,} \qquad V_g = -(1/\rho f) \, \partial p/\partial n \qquad (12.3.9)$$

Equation (12.3.9) represents a balance between the Coriolis force and the pressure gradient force and is called geostrophic balance. The wind speed V_g is called the geostrophic wind speed and denoted by V_g. The balance is shown schematically in Fig. 12.3.

Fig. 12.3 Geostrophic balance between the pressure gradient force (denoted by PG) and the Coriolis force (denoted by CO) when the isobar has no curvature L denotes Low pressure, H High

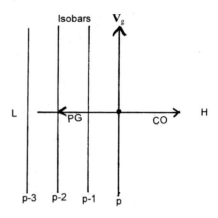

12.3.4 Relationship Between the Geostrophic Wind and the Gradient Wind

We have seen from (12.3.9) that in natural co-ordinates, the geostrophic wind is given by the relation

$$V_g = -(1/\rho f) \, \partial p/\partial n$$

Substituting this value in the expression for the gradient wind (12.3.7), we obtain

$$V_g/V = 1 + (V/f\,R) \qquad (12.3.10)$$

The ratio V_g/V may be evaluated for different types of atmospheric flow. In general, in the northern hemisphere cyclonic flow in which f R is positive, V_g is larger than V, whereas in anticyclonic flow in which f R is negative, V_g is smaller than V. In middle and high latitudes, the difference between V_g and V seldom exceeds 10–20 %. In low latitudes, however, the term V/fR, which is defined as the Rossby number and usually denoted by Ro, is of the order of unity and the difference between V and V_g is large enough to justify use of the gradient wind instead of the geostrophic wind.

12.3.5 Inertial Motion

When the horizontal pressure field is flat, i.e., there is no pressure gradient in any direction, the balance relation (12.3.7) reduces to the simple form

$$V^2/R + fV = 0$$
$$\text{or,} \quad R = -V/f \qquad (12.3.11)$$

According to (12.3.6), V is constant in such a case. Further, if we consider a flow at a fixed latitude ϕ where f is constant, R becomes constant. This means that the flow

12.3 Balanced Flow in Natural Co-ordinates

will be along a circle (in the clockwise sense) with constant radius R, which we call an inertia circle.

The period of the inertial oscillation, P, is given by

$$P = -2\pi R/V = -2\pi/f = \{(1/2)\text{day}\}/|\sin\phi| \qquad (12.3.12)$$

The period P is the time taken by a Foucault pendulum to turn through 180°.

Pure inertial oscillation does not appear to be of any consequence in the atmosphere except for land-sea breeze and the nocturnal jet (see Chap. 14). However, in the ocean where the flow velocities are much slower than in the atmosphere, the radius of the inertial circle is much smaller and significant amount of energy has been detected in currents which oscillate with the inertial period.

12.3.6 Cyclostrophic Motion

If the horizontal scale of the curved flow is very small, the curvature term, V^2/R, in (12.3.7) may become much more important than the Coriolis term, fV. In that case, the curvature term may alone balance the pressure gradient term. Thus,

$$(V^2/R) = -(1/\rho)\, \partial p/\partial n$$

or, $\qquad V = \{(-R/\rho)\, \partial p/\partial n\}^{1/2} \qquad (12.3.13)$

where V gives the speed of the cyclostrophic wind, and the radius R is positive in a cyclonic flow, negative in anticyclonic flow.

Figure 12.4 is a schematic showing the balance of forces in a cyclostrophic flow; (a) cyclonic, (b) anticyclonic.

The cyclostrophic flow can be cyclonic or anticyclonic. In either case, the pressure gradient force is always directed towards the center of curvature and the centrifugal force away from it. In (a), R is positive, but $\partial p/\partial n$ is negative, whereas in (b) R is negative, but $\partial p/\partial n$ is positive.

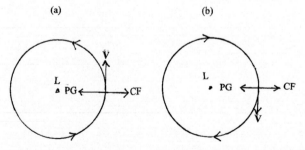

Fig. 12.4 Balance of forces in cyclostrophic flow: (**a**) cyclonic, (**b**) anticyclonic. PG denotes pressure gradient force, and CF the centrifugal force

Cyclostrophic flow usually occurs in fast-rotating small-scale atmospheric motion, such as a tornado or a midget cyclone. On account of very high speed of rotation, the ratio of the centrifugal force to the Coriolis force in such a flow, which is given by V/f R, is very large. We can have an idea of the order of magnitude of Ro in the case of a typical tornado from the following realistic values V, f and R:

$$V \sim 100\,\text{ms}^{-1}; \quad f \sim 10^{-5}\,\text{s}^{-1}; \quad R \sim 100\,\text{m}$$

These values give Ro $\sim 10^5$. A value of Ro of this magnitude stands in sharp contrast with that for large-scale flows in middle and high latitudes where Ro $\ll 1$. This provides a strong justification for neglecting the Coriolis force in comparison with the centrifugal force in cyclostrophic flow, especially in low latitudes. Also, an important point to note here is that when the scale of motion is reduced still further, as in a dust devil or water spout, the sense of rotation can be either cyclonic or anticyclonic.

The reason for this loss of control on the sense of rotation is not difficult to visualize. It is the Coriolis force which forces a parcel of air starting to move down the pressure gradient towards the center of low pressure to deviate from its movement till it attains a balance with the pressure gradient force and moves along the isobar. So when the deviating force is negligible, there is a tug of war between the centrifugal force and the pressure gradient force and the resulting sense of rotation can be somewhat arbitrary. In fact, it can be as often cyclonic as anticyclonic.

12.4 Trajectories and Streamlines

It is important to distinguish between two terms frequently used in meteorology in regard to air motion. These are: trajectories and streamlines. A trajectory may be defined as the path traced out by an individual moving parcel of air in a given period of time. In other words, if δs is a small segment of the path traced out by the parcel in time δt at wind speed V, the trajectory may be found by integrating the relation,

$$ds/dt = V(x,y,t) \qquad (12.4.1)$$

On the other hand, a streamline is defined as a straight or curved line which is everywhere tangential to the instantaneous wind vector **V**. The condition for this tangency is **V x dr** = 0, where **dr** is a length segment vector along the line.

Thus, a streamline is defined by the relation,

$$dy/dx = v(x, y, t_0)/u(x, y, t_0) \qquad (12.4.2)$$

where u, v are the components of the wind vector at location(x,y) at fixed time t_0. The streamline may be obtained by integrating (12.4.2).

12.4 Trajectories and Streamlines

It is important to note that the streamline refers to the position of a moving parcel of air at a particular instant of time and not to its successive positions in points of time which are given by a trajectory.

Trajectories and streamlines will coincide only in steady-state systems in which there is no local change of velocity with time. To find out a relation between the curvature of a trajectory and that of a streamline in a moving pressure system, we proceed as follows:

Let $\delta\beta$ be the angle between the directions of the wind while it moves over a length δs of the curve. Then, if R_t and R_s be the radii of the trajectory (t) and the streamline(s) respectively, we may write

$$d\beta/ds = 1/R_t; \quad \text{and,} \quad \partial\beta/\partial s = 1/R_s \qquad (12.4.3)$$

The rate of change of the wind direction following the motion may be written

$$d\beta/dt = V\, d\beta/ds = \partial\beta/\partial t + V\, \partial\beta/\partial s$$

Substituting from (12.4.3), we get for the local turning of the wind the relation

$$\partial\beta/\partial t = V\, (1/R_t - 1/R_s) \qquad (12.4.4)$$

Equation (12.4.4) states that the wind direction remains constant only if the curvatures of the trajectory and the streamline are the same.

It is interesting to see how the local turning of the wind occurs when a pressure system moves relative to some large-scale wind. For simplicity, let us assume that a cyclonic circular isobaric pattern is moving eastward without change of shape with a constant velocity **C**. Let us further assume that the wind circulating in the pressure system is given by the gradient wind equation (12.3.7). Since the isobars are streamlines, the local turning of the wind is entirely due to the motion of the isobars.

Thus, $\partial\beta/\partial t = -\mathbf{C}\cdot\nabla\beta = -C\, \partial\beta/\partial s \cos\zeta \qquad (12.4.5)$

where ζ is the angle between the streamlines and the direction of motion of the pressure system.

Substituting from (12.4.3) and (12.4.4) into (12.4.5), we get

$$R_s = R_t\, \{1 - (C/V) \cos\zeta\} \qquad (12.4.6)$$

We can use (12.4.6) to compute the curvature of the trajectory in any part of a moving pressure system. An example is shown in Fig. 12.5, after Holton (1979), in which a low pressure system with circular isobars moves eastward with a constant speed C relative to the wind speed V.

Figure 12.5 shows the type of trajectories that result for two speeds of the moving system relative to the wind: (a) when $C = V/2$ and (b) when $C = 2V$. The curvatures of the trajectories are shown for parcels initially located at the north, east, south and west of the center of the low for both the cases. For simplicity, the geostrophic wind

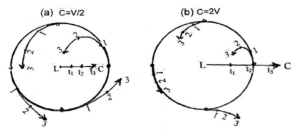

Fig. 12.5 Trajectories in the case of a moving low (L) pressure system with circular isobars. (**a**) C = V/2; (**b**) C = 2V. Numbers indicate successive stages of time t (After Holton, 1979)

is used for the computation instead of the gradient wind so that the wind blows along the isobars everywhere and it is assumed that there is no variation of the wind with distance from the center. The curvature of the trajectory is shown at successive times after leaving the initial location.

12.5 Streamline-Isotach Analysis

Streamline-isotach analysis is a common feature of weather maps in low latitudes, while isobaric analysis is preferred in midlatitudes. The distinction is a matter of the most information being in wind or pressure observations. A free-hand streamline-isotach analysis is hardly satisfactory, especially when wind observations on a weather map are few and far between. For this reason, it is recommended that the analysis be based on the use of isogons which are lines joining places which have the same wind direction and on which short line segments may be drawn across each isogon indicating the actual wind direction. The line segments so drawn are then connected by tangent curves. These curves are the desired streamlines. It is usually satisfactory to draw isogons at intervals of 30°, but in regions of small variations of wind direction they may be drawn at intervals of 15 or 10°. This method of drawing streamlines is due to Sandstrom (1910). Petterssen (1956) has described the Sandstrom's method of drawing streamlines and isotachs in great detail. Isotachs are lines drawn through places of equal wind speed. However, if wind observations are scanty, use may be made of certain pressure-wind relationships to obtain additional speed values between observations. For details of this method, the reader may refer to Petterssen's or any other standard book on weather analysis. In Fig. 12.6, we give an example of a streamline-isotach analysis which was prepared by the meteorological group of the International Indian Ocean Expedition (IIOE) (1962–1966) with the data then available.

After introduction of computer analysis of meteorological data as part of data assimilation schemes for numerical models, the traditional streamline-isotach analysis has gradually gone out of use in some of the advanced countries of the world.

Fig. 12.6 Streamline-isotach analysis prepared by the meteorological group of the International Indian Ocean Expedition (1962–1966) (Reproduced from Ramage and Raman, 1972, published by the National Science Foundation, Washington, D.C)

12.6 Variation of Wind with Height – The Thermal Wind

The variation of wind with height depends upon the distribution of density which, by the ideal gas law, is a function of temperature and pressure. When the air over a region is cold, the pressure falls off more rapidly with height than when it is warm. In other words, the thickness of an isobaric layer is smaller in cold air than in warm air. This follows from the hydrostatic equation and the ideal gas law and we may write the thickness equation in the form

$$\Phi_1 - \Phi_0 = R\,T \ln(p_0/p_1) \qquad (12.6.1)$$

where p_0, p_1 are the pressures at geopotential heights Φ_0, Φ_1 respectively, and T is the mean temperature of the layer (see Fig. 12.7).

Now, if there are two neighboring regions, one cold and the other warm, a horizontal temperature gradient exists between them, the effect of which upon the horizontal pressure gradient will lead to a vertical variation of the wind between the two regions. If we assume the wind to be largely geostrophic, we can compute the variation of the wind with height as follows:

$$\text{Let}\quad \mathbf{V}_g = (1/f)\mathbf{k}\times\nabla_p\Phi \qquad (12.6.2)$$

where $\mathbf{V}_g$ is the geostrophic wind vector, f is the Coriolis parameter, and $\nabla_p\Phi$ is the horizontal gradient of geopotential Φ along an isobaric surface.

We differentiate (12.6.2) with respect to pressure and obtain

$$\partial \mathbf{V}_g/\partial p = (1/f)\,\mathbf{k}\times\nabla_p(\partial\Phi/\partial p) \qquad (12.6.3)$$

Since, by the ideal gas law and the hydrostatic approximation, $\partial\Phi/\partial p = -RT/p$, where T is the mean temperature of the layer, we have, by substitution and rearrangement,

$$\partial\mathbf{V}_g/\partial p = -(R/fp)(\mathbf{k}\times\nabla_p T) \qquad (12.6.4)$$

where $\nabla_p T$ is the horizontal temperature gradient in the isobaric layer.

Integrating (12.6.4) between pressure surfaces p_0 and p_1, we get

$$(\mathbf{V}_g)_{p1} - (\mathbf{V}_g)_{p0} = (R/f)(\mathbf{k}\times\nabla_p T)\ln(p_0/p_1) \qquad (12.6.5)$$

If we denote the vector difference between the geostrophic winds at the two pressure surfaces by $\mathbf{V}_T$ and call it the thermal wind, then

$$\mathbf{V}_T = (R/f)(\mathbf{k}\times\nabla_p T)\,\ln(p_0/p_1) \qquad (12.6.6)$$

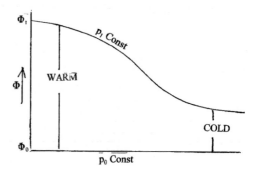

Fig. 12.7 Thickness of an isobaric layer in warm and cold air

12.6 Variation of Wind with Height – The Thermal Wind

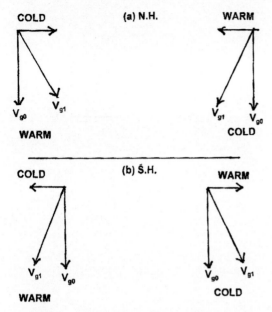

Fig. 12.8 Vertical variation of the geostrophic wind due to horizontal temperature gradient: C-cold, W-warm. (**a**) Northern hemisphere, (**b**) Southern hemisphere. $\mathbf{V}_T$ denotes thermal wind and $\mathbf{V}_g$ the geostrophic wind with suffixes 0 and 1 denoting lower and upper surface respectively

Equation (12.6.6) is the well-known thermal wind equation which controls the vertical variation of the wind with height in a region where there exists a horizontal gradient of temperature.

The components of the thermal wind vector along the x and the y axes may then be written as

$$u_T = -(R/f)(\partial T/\partial y)_p \ln(p_0/p_1) \quad (12.6.7)$$

$$v_T = (R/f)(\partial T/\partial x)_p \ln(p_0/p_1) \quad (12.6.8)$$

Equation (12.6.7) states that if, in the northern hemisphere, the temperature decreases with latitude, the westerly wind will strengthen with height, while the easterly wind will weaken. However, if the temperature increases with latitude, the westerly wind will weaken, while the easterly will strengthen with height. Examples of the former type are found in the formation of the westerly jetstream at about 250 mb in middle and high latitudes and of the latter type in the formation of the easterly jetstream at about 150 mb over the tropics. Similarly, (12.6.8) states that if the temperature increases (decreases) eastward, the southerly wind will strengthen (weaken) with height in the northern hemisphere. In other words, in the northern hemisphere, the thermal wind will blow so that it keeps cold air to the left and warm air to the right. The direction will reverse in the southern hemisphere. Fig. 12.8 illustrates how a northerly wind will turn with height in the two hemispheres.

Chapter 13
Circulation, Vorticity and Divergence

13.1 Definitions and Concepts – Circulation and Vorticity

Circulation and vorticity are two important parameters of a rotating motion. In the atmosphere, circulation consists of the physical movement of a parcel of air along the closed boundary of a surface area, while vorticity is the tendency of an infinitesimally small area of that surface to turn about an axis normal to it in the same sense as the circulation (see Fig. 13.1).

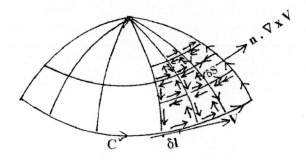

Fig. 13.1 Illustrating circulation and vorticity

If **V** denotes the velocity of a parcel of air and d**l** a line segment along the boundary of a closed surface, the circulation C along the boundary is measured by the line integral

$$C = \oint \mathbf{V} \cdot \mathbf{dl} \qquad (13.1.1)$$

By convention, C is treated as positive if the integration is carried out in the anti-clockwise direction, and negative if done in the clockwise direction.

Vorticity, on the other hand, is a vector field which gives a microscopic measure of the turning of the air over a unit area of a surface about an axis normal to it in the same sense as the circulation and is measured by the curl of the velocity vector, i.e., by $\nabla \times \mathbf{V}$. If we consider a surface area S the boundary of which is given by the length l, then, according to Stokes's theorem?

$$\oint \mathbf{V} \cdot d\mathbf{l} = \int\int_S (\nabla \times \mathbf{V}) \cdot \mathbf{n}\, dS \qquad (13.1.2)$$

where **n** is a unit vector normal to the unit area of the surface and the integral on the left-hand side is taken around the whole length l.

It may be noted in Fig. 13.1 that the flows along the common borders of all the unit areas cancel being in opposite directions except those along the outer borders which add up to the circulation along the outer boundary of the surface.

Thus, Stokes's theorem connects circulation and vorticity in solid-body rotation and the relationship is given by

$$\text{Circulation} = \text{vorticity} \times \text{area} \qquad (13.1.3)$$

The meaning of the term 'vorticity' can be further illustrated by taking the case of rotation of a solid circular disk with an angular velocity ω. A point at a distance r from the center of the disc has a linear velocity $r\omega$, so the circulation at this distance is $2\pi r^2 \omega$, where $2\pi r$ is the circumference of the circle of radius r. Dividing this by πr^2 which is the area of the circle, we obtain by (13.1.3) the vorticity of the disc as 2ω.

13.2 The Circulation Theorem

The problem of atmospheric circulation and vorticity was first studied by Helmholtz about the middle of the nineteenth century. He was followed by V. Bjerknes who in 1898 derived his famous circulation theorem. Here we give a brief review of his theorem. Let us first derive it in the absolute frame of reference by taking the line integral of the Newtonian equations of motion (11.3.2), by neglecting friction. Thus, we start with the equation

$$\oint (d\mathbf{V}_a/dt)_a \cdot d\mathbf{l} = -\oint \alpha \nabla p \cdot d\mathbf{l} - \oint \nabla \Phi \cdot d\mathbf{l} \qquad (13.2.1)$$

where the subscript 'a' denotes absolute motion in an absolute frame of reference.

The integrand on the left-hand side of (13.2.1) can be written

$$(d\mathbf{V}_a/dt)_a \cdot d\mathbf{l} = d(\mathbf{V}_a \cdot d\mathbf{l})_a/dt - \mathbf{V}_a \cdot (d(d\mathbf{l})/dt)_a \qquad (13.2.2)$$

Here, since **l** is a position vector, $(d\mathbf{l}/dt)_a = \mathbf{V}_a$, and the second term on the right-hand side of (13.2.2) reduces to $\mathbf{V}_a \cdot (d\mathbf{V}_a)$. But $\mathbf{V}_a \cdot d\mathbf{V}_a = (1/2) d(\mathbf{V}_a \cdot \mathbf{V}_a)$.

Hence, by substitution from (13.2.2), (13.2.1) may be written as

$$\oint d(\mathbf{V}_a \cdot d\mathbf{l})_a/dt = -\oint \alpha\, dp - \oint d\Phi + (1/2) \oint d(\mathbf{V}_a \cdot \mathbf{V}_a) \qquad (13.2.3)$$

13.2 The Circulation Theorem

The line integral of a perfect differential being zero, the second and the third terms on the right-hand side of (13.2.3) disappear, and we are left with

$$dC_a/dt = -\oint \alpha \, dp \qquad (13.2.4)$$

where we have written C_a for $\oint(\mathbf{V}_a \cdot d\mathbf{l})_a$ which constitutes absolute circulation, and the term $-\oint \alpha \, dp$ gives the number of unit isosteric-isobaric solenoids enclosed by the circulation curve.

In a barotropic atmosphere, where the density or specific volume is a function of pressure only, the isosteric (α = constant) surfaces coincide with the isobaric (p = constant) surfaces and the solenoidal term vanishes and the absolute circulation remains constant. This is the well-known Kelvin circulation theorem for the conservation of absolute circulation in a frictionless barotropic atmosphere. It corresponds to the law of conservation of absolute angular momentum in solid-body rotation in classical mechanics.

The real atmosphere, however, is almost always baroclinic and the solenoidal term plays an important role in atmospheric circulation. In Fig. 13.2(a, b) we show the distinction between the two types of atmospheres, so far as the distribution of the isosteric and isobaric surfaces are concerned. In a barotropic atmosphere, as there are no solenoids, there is no mechanism to change the circulation with time. On the other hand, in a baroclinic atmosphere, the surfaces of pressure (p) and specific volume (α) intersect producing isobaric-isosteric solenoids which bring about a baroclinic circulation, as shown in Fig. 13.2 (b). The direction of the circulation is found by turning the $\nabla\alpha$ vector through an angle ($< 180°$) so that it coincides with the $-\nabla p$ vector.

Examples of baroclinic circulation of the type shown in Fig. 13.2 (b) may be found in the generation of land and sea breezes and large-scale monsoons which are driven by differential heating between two parts of the earth's surface, usually between continents and neighbouring oceans.

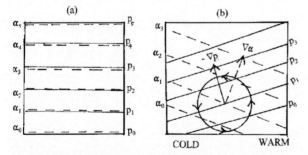

Fig. 13.2 (a) A barotropic atmosphere with no solenoids; (b) A baroclinic atmosphere with isosteric (α)-isobaric (p) solenoids. C - cold, W -warm. The *arrow* in circle shows the direction of circulation, turning the $\nabla\alpha$ vector towards the $-\nabla p$ vector

In meteorology, however, it is more convenient to work with the circulation produced by relative motion which is obtained by subtracting from the absolute circulation the circulation due to the rotation of the earth. Now, the circulation due to earth's rotation, denoted by C_e, may be expressed as $\oint \mathbf{V}_e d\mathbf{r}$, where $\mathbf{V}_e = \mathbf{\Omega} \times \mathbf{r}$, and dr is a line segment along the curved path of the circulation at the position vector $\mathbf{r}$. However, by Stokes's theorem,

$$C_e = \oint \mathbf{V}_e \cdot d\mathbf{r} = \oint (\nabla \times \mathbf{V}_e) \cdot \mathbf{n}\, dA \qquad (13.2.5)$$

where dA is a small area of the earth's surface at latitude ϕ, and the unit vector $\mathbf{n}$ is normal to the surface. If we choose the surface to be horizontal, then $\mathbf{n}$ is along the local vertical pointing outward. In that case, $(\nabla \times \mathbf{V}_e) \cdot \mathbf{n} = 2\Omega \sin \phi = f$, where f is the Coriolis parameter at latitude ϕ (see Fig. 13.3).

Equation (13.2.5) then yields for the rate of change of circulation due to earth's rotation, dC_e/dt, the expression

$$dC_e/dt = 2\Omega\, d(A \sin \phi)/dt = 2\Omega\, dA_e/dt \qquad (13.2.6)$$

where $A_e (= A \sin \phi)$ is the projection of the surface A onto the equatorial plane.

If we denote the relative circulation by C, its rate of change, dC/dt, is given by the difference between (13.2.4) and (13.2.6). Thus, we get

$$dC/dt = -\oint \alpha\, dp - 2\Omega\, dA_e/dt \qquad (13.2.7)$$

This is the well-known Bjerknes circulation theorem. It emphasizes the fact that on the rotating earth, the rate of change of a given circulation around an area is determined by the number of isobaric-isosteric solenoids enclosed by the area and the rate of expansion or contraction of the area with latitude.

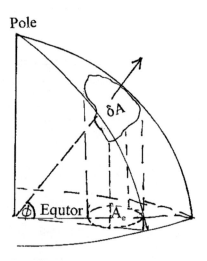

Fig. 13.3 Projection of the area δA of the earth's surface onto the equatorial plane

13.3 Absolute and Relative Vorticity

The absolute vorticity is defined as the curl of the absolute velocity, $\nabla \times V_a$, whereas the relative vorticity is given by the curl of the relative velocity, $\nabla \times V$. The velocity field V being three-dimensional, the vorticity field is also three-dimensional. However, in meteorology, we are largely concerned with horizontal motion and hence the vertical component of the vorticity. If η denotes the vertical component of the absolute vorticity and ζ that of the relative vorticity, then we may write

$$\eta = k \cdot \nabla \times V_a, \qquad \zeta = k \cdot \nabla \times V$$

The difference between the absolute and the relative vorticity is the vertical component of the vorticity due to the rotation of the earth, given by $k \cdot \nabla \times V_e$, which, as we have shown earlier, is equal to f, the Coriolis parameter.

Thus, for largely horizontal circulation, in Cartesian co-ordinates,

$$\zeta = \partial v/\partial x - \partial u/\partial y, \text{ and } \eta = \zeta + f \qquad (13.3.1)$$

13.4 Vorticity and Divergence in Natural Co-ordinates

The physical interpretation of vorticity in an atmospheric flow or circulation is often facilitated by expressing it in a natural co-ordinate system. The same remark holds for divergence which may be defined as the rate of increase of an area or volume per unit area or volume. For this, we take a streamline in a rectangular co-ordinate system and assume that at a point P along the streamline, the velocity is $V(u, v)$ and it makes an angle β with the x-axis (see Fig. 13.4).

Then, the components of the speed are: $u = V\cos\beta$ along the x-axis, and $v = V\sin\beta$ along an axis at right angles to it.

Differentiating u with respect to y and v with respect to x, we obtain

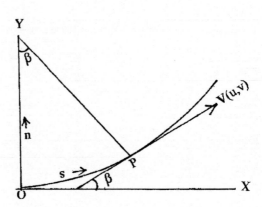

Fig. 13.4 Vorticity and divergence in natural co-ordinate system

$$\partial v/\partial x = (\partial V/\partial x)\sin\beta + (V\cos\beta)\partial\beta/\partial x$$
$$\partial u/\partial y = (\partial V/\partial y)\cos\beta - (V\sin\beta)\partial\beta/\partial y$$

If we now take a point O along the streamline where the x-axis is tangent to the streamline and measure the distance s along the streamline and n at right angles to it, and note that at O, $\beta = 0$, we can write vorticity(ζ) and divergence(D) in the form

$$\zeta = \partial v/\partial x - \partial u/\partial y = V\partial\beta/\partial s - \partial V/\partial n = VK_s - \partial V/\partial n \qquad (13.4.1)$$

$$D = \partial u/\partial x + \partial v/\partial y = \partial V/\partial s + V\partial\beta/\partial n \qquad (13.4.2)$$

The interpretation of (13.4.1) is simple. The vorticity ζ is positive if the streamline has a cyclonic curvature (anti-clockwise direction) and also if there is a cyclonic wind shear with speed increasing along the outward normal. On a streamline-isotach map, therefore, the maximum positive vorticity between a low and a high pressure area will be found where the streamline has both maximum cyclonic curvature and positive wind shear and the maximum negative vorticity where the streamline has both maximum anticyclonic curvature and negative wind shear. Similarly, it follows from (13.4.2) that $\partial V/\partial s$ represents a stretching of the flow downstream and V $\partial\beta/\partial n$ represents the effect of divergence or convergence of the streamlines. Thus, divergence is positive where the wind speed increases downstream and where air tends to stream in isobaric channels in which the speed varies inversely with the width of the channel. This is due to the circumstance that the pressure gradient force is very nearly balanced by the Coriolis force and divergence in the large-scale aircurrents is a very small quantity ($\sim 10^{-6} s^{-1}$).

An example, in the case of a W'ly jetstream (J) in the northern hemisphere, is shown in Fig. 13.5.

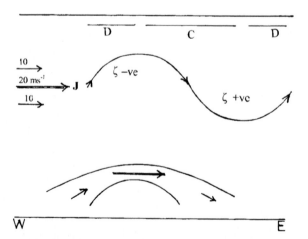

Fig. 13.5 Distribution of relative vorticity (ζ) and divergence (D) in a W'ly jetstream (J) in the northern hemisphere. Lower panel shows how an anticyclonic curvature with negative vorticity has divergence (D) upstream of the jet maximum and convergence (C) downstream

13.5 Potential Vorticity

The conservative property of potential temperature in adiabatic flow in a barotropic atmosphere has been used to derive a form of vorticity called potential vorticity which has been found to be extremely useful in tracking changes in atmospheric circulation. We derive an expression for it in this section. It was shown in sect. 13.4 that the absolute circulation remains constant in a barotropic atmosphere. This means that

$$dC_a/dt = 0 \tag{13.5.1}$$

From (13.1.3) and (13.3.1) it follows that

$$C_a = \eta\, A = (\zeta + f)A,$$

where A is the circulation area.

Substituting for C_a in (13.5.1), we obtain the relation

$$(\zeta + f)\, A = \text{Constant} \tag{13.5.2}$$

Equation (13.5.2) is an important relation, since it connects the area of the circulation with the absolute vorticity. It states that for absolute circulation in a barotropic atmosphere to remain constant, the area of the circulation varies inversely as the absolute vorticity.

It can be shown that as in a barotropic atmosphere, the solenoidal term vanishes in frictionless adiabatic flow in which a parcel of air is forced to move along a surface of constant potential temperature. On account of this conservative property, it can be shown that the absolute circulation as given by the relation (13.5.2) remains constant in adiabatic flow. However, in this case, for circulation to remain constant, we can find an expression for change of absolute vorticity as a function of the vertical separation of the potential temperature surfaces.

Since, mass is conserved, we can write

$$A = -g\, M/(\Delta p) = -\{gM/(\Delta\theta)\}(\Delta\theta/\Delta p) \tag{13.5.3}$$

where M is mass, $\Delta\theta$ is a small change in potential temperature, and Δp a small change in pressure.

For a constant potential temperature difference, $\Delta\theta$, i.e., for an isentropic layer, we may write (13.5.3) as

$$A = \text{Constant}/(\Delta p) \tag{13.5.4}$$

Substituting (13.5.4) in (13.5.2), we obtain

$$(\zeta + f)/(\Delta p) = \text{Constant} \tag{13.5.5}$$

The relation (13.5.5) assumes a particularly simple form in a homogeneous atmosphere in which the variation of density may be neglected. It then reduces to the form

$$(\zeta + f)/\Delta z = \text{constant} \qquad (13.5.6)$$

where Δz is the vertical distance between two constant potential temperature surfaces.

The relation (13.5.5) is a mathematical statement of the principle of conservation of potential vorticity in adiabatic, frictionless flow. According to it, the absolute vorticity is directly proportional to the pressure depth of the flow. This means that when a flow enters an area where its vertical depth changes, its absolute vorticity will change so as to conserve potential vorticity. However, this principle works in quite a different way for westerly and easterly flow when the flow approaches a high mountain, such as the north-south-oriented Western Ghats Mountain of peninsular India or the Andes of South America, or the northwest-southeast-oriented lofty Himalayas along the northern boundary of India.

Let us first illustrate the difference by taking the case of a zonal flow without any horizontal shear when Δp remains constant (see Fig. 13.6).

If the flow is westerly, it will develop cyclonic relative vorticity if it turns northward and anticyclonic relative vorticity if it turns southward. But, in turning northward, the westerly flow also acquires anticyclonic relative vorticity because of increase in the value of the Coriolis parameter, whereas a southward-turning flow develops cyclonic relative vorticity because of decrease in the value of the Coriolis parameter. Thus, to conserve absolute vorticity in this case, a westerly flow must remain predominantly zonal. The situation, however, is quite different with an easterly flow when Δp remains constant. It the easterly flow turns northward, it develops enhanced negative relative vorticity on account of both anticyclonic curvature and an increased Coriolis parameter, whereas a southward turn will lead to increased positive relative vorticity on account of both cyclonic curvature and decreased Coriolis parameter.

The situation, however, changes markedly when the flow approaches a mountain where the depth Δp varies. Here, also, the case of a westerly flow differs markedly from that of an easterly flow, as illustrated in Fig. 13.7 and Fig. 13.8 respectively.

The case of the westerly flow is depicted in Fig. 13.7 in two sections: (a) vertical, (b) horizontal. As potential vorticity is to be conserved, the flow will turn

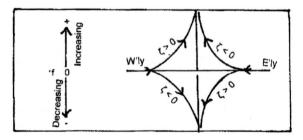

Fig. 13.6 The curvature and the Coriolis effects on a zonal flow for conservation of absolute vorticity

13.5 Potential Vorticity

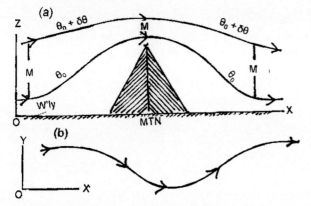

Fig. 13.7 Conservation of potential vorticity in the case of a westerly flow negotiating a mountain (MTN): (**a**) Vertical section; (**b**) horizontal section. M-Mass, θ potential temperature

anticyclonically southward on the windward side where its depth decreases (see lower panel) till it passes the top of the mountain, and then cyclonically on the lee-side where the depth increases.

In the case of an easterly flow as depicted in Fig. 13.8, it appears that an air parcel is able to sense the presence of the mountain from a distance and adjust its absolute vorticity by first turning southward in a cyclonic curvature before turning northward in an anticyclonic curvature, so as to partially offset the effect of an enhanced negative relative vorticity (on account of curvature as well as increased value of the Coriolis parameter) at the mountainside. In this way, it can cross the mountain and resume its easterly flow at the original latitude on the leeside.

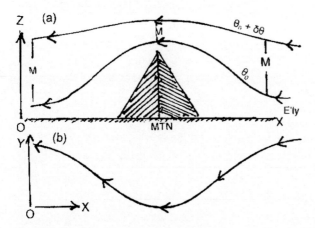

Fig. 13.8 Conservation of potential vorticity in the case of an easterly flow negotiating a mountain: (**a**) Vertical section, (**b**) horizontal section

13.6 The Vorticity Equation in Frictionless Adiabatic Flow

Bjerknes's circulation theorem (13.2.7) may be used to derive an expression for the rate of change of vorticity due to creation of solenoids as well as expansion or contraction of the circulation area over the earth's surface. If we denote the solenoidal term, $-\oint \alpha \, dp$ by N, we can write (13.2.7) in the form

$$d(\zeta+f)/dt = -(\zeta+f)(1/A)dA/dt + N/A \qquad (13.6.1)$$

where we have made use of the following relations:

$$C = A\zeta, \quad 2\Omega\sin\phi = f$$

Now, $(1/A) \, dA/dt$, which denotes the rate of change of the area A per unit area is called divergence, $\nabla \cdot \mathbf{V}$.

We may, therefore, write (13.6.1) in the form

$$d(\zeta+f)/dt = -(\zeta+f)\nabla \cdot \mathbf{V} + N/A \qquad (13.6.2)$$

13.7 The Vorticity Equation from the Equations of Motion

We now derive expressions for the rate of change of vorticity in the atmosphere without enforcing adiabatic conditions. For this purpose, we take the more general equations of frictionless motion which were derived in Chaps. 11 and 12 in Cartesian as well as isobaric co-ordinate systems.

13.7.1 Vorticity Equation in Cartesian Co-ordinates (x, y, z)

If we disregard friction, the horizontal momentum equations (11.5.1) may be written in the approximate form

$$\partial u/\partial t + u \, \partial u/\partial x + v \, \partial u/\partial y + w \, \partial u/\partial z = fv - (1/\rho)\partial p/\partial x \qquad (13.7.1)$$

$$\partial v/\partial t + u \, \partial v/\partial x + v \, \partial v/\partial y + w \, \partial v/\partial z = -fu - (1/\rho)\partial p/\partial y \qquad (13.7.2)$$

To get the vertical component of the vorticity, we differentiate (13.7.1) with respect to y and (13.7.2) with respect to x, and then obtain, by subtracting the former from the latter, the expression

$$\begin{aligned}
\partial \zeta/\partial t + u \, \partial \zeta/\partial x + v \, \partial \zeta/\partial y + w \, \partial \zeta/\partial z = &- (\zeta+f)(\partial u/\partial x + \partial v/\partial y) \\
&- (\partial w/\partial x \, \partial v/\partial z - \partial w/\partial y \, \partial u/\partial z) \\
&- v \, \partial f/\partial y + (1/\rho^2)[\partial p/\partial x \, \partial p/\partial y \\
&- \partial p/\partial y \, \partial p/\partial x] \qquad (13.7.3)
\end{aligned}$$

13.7 The Vorticity Equation from the Equations of Motion

where we have substituted ζ for the vertical component of the relative vorticity, $(\partial v/\partial x - \partial u/\partial y)$.

Since $df/dt = v\partial f/\partial y$, by re-arranging, we can write (13.7.3) in the form

$$d(\zeta+f)/dt = -(\zeta+f)(\partial u/\partial x + \partial v/\partial y) - \{(\partial w/\partial x\, \partial v/\partial z - \partial w/\partial y\, \partial u/\partial z)\} \\ + [(1/\rho^2)(\partial \rho/\partial x\, \partial p/\partial y - \partial \rho/\partial y\, \partial p/\partial x)] \quad (13.7.4)$$

The three terms on the right-hand side of (13.7.4) are called the divergence term, the tilting term and the solenoidal term respectively. In general, all the terms contribute to the rate of change of the vertical component of the absolute vorticity $(\zeta+f)$ following motion in the atmosphere.

The divergence or convergence of air is a powerful mechanism in controlling absolute vorticity. Divergence(convergence) decreases (increases) the absolute vorticity, The tilting term represents a contribution from the horizontally-oriented axis of rotation when it is vertically tilted by a non-uniform field of vertical motion, as illustrated in Fig. 13.9, in which the eastward or the u- component of the velocity varies in the vertical, producing a vorticity along the axis of y. Now, if the vertical motion field varies along the y-axis, it will tilt the y-component of the vorticity in the vertical to change the vertical component of the absolute vorticity. The tilting term may be said to be largely responsible for the helical movement of air parcels around the center of a fast-rotating mesoscale convective system, in which the strongest upward velocity is reached at some height above the surface, which separates the regime of low-level cyclonic circulation from that of the anticyclonic circulation above.

The solenoidal term in (13.7.4) represents the contribution of the solenoidal circulation per unit area to the rate of change of absolute vorticity, as already stated in (13.6.2). The following show this relationship:

Using Stokes's theorem, we may write (13.2.4) as

$$dC_a/dt = -\oint \alpha\, dp = -\oint \alpha \nabla p \cdot d\mathbf{l} = -\int\int_A \nabla \times (\alpha \nabla p) \cdot \mathbf{k}\, dA$$

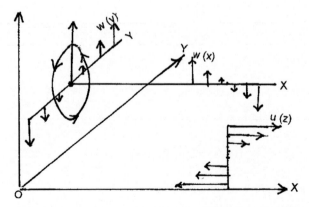

Fig. 13.9 Illustrating how the tilting term contributes to the generation of absolute vorticity

But, by vector identity, $\nabla \times (\alpha \nabla p) = \nabla \alpha \times \nabla p$,
Hence the above expression may be written as

$$dC_a/dt = -\int\int \mathbf{k} \cdot (\nabla \alpha \times \nabla p)\, dA$$

This may be compared with the solenoidal term in (13.7.4) which can be written

$$-(1/\rho^2)[\partial \rho/\partial x\, \partial p/\partial y - \partial \rho/\partial y\, \partial p/\partial x] = -\mathbf{k} \cdot (\nabla \alpha \times \nabla p), \text{ where } \alpha = 1/\rho.$$

It can, therefore, be seen that the solenoidal term truly represents a contribution to the rate of change of the absolute vorticity of the horizontal circulation per unit area.

The same result could be derived if in the solenoidal term we replaced density by temperature using the equation of state. When the circulation is predominantly horizontal, the contribution of this term is negligible.

It is easy to see that if all the three terms on the right-hand side of (13.7.4) are absent as in a barotropic atmosphere, the absolute vorticity following the motion remains conserved.

13.7.2 The Vorticity Equation in Isobaric Co-ordinates

A somewhat simpler form of the vorticity equation may be derived by using the equations of motion in isobaric co-ordinates.

Combining (12.2.3) and (12.2.4) and using vector notations, we can write the equations of motion in isobaric co-ordinates as

$$\partial \mathbf{V}/\partial t + (\mathbf{V} \cdot \nabla)_p \mathbf{V} + \omega \partial \mathbf{V}/\partial p = -\nabla \Phi - \mathbf{k} \times f \mathbf{V}$$

Or, $\quad \partial \mathbf{V}/\partial t = -\nabla [\Phi + (\mathbf{V} \cdot \mathbf{V})/2] - \mathbf{k} \times \nabla (\zeta + f) - \omega \partial \mathbf{V}/\partial p \quad (13.7.5)$

where we have used the vector identity, $(\mathbf{V} \cdot \nabla)\mathbf{V} = (1/2)(\mathbf{V} \cdot \mathbf{V}) + \mathbf{k} \times \zeta \mathbf{V}$, and put $\zeta = \mathbf{k} \cdot \nabla \times \mathbf{V}$.

We now operate on the vector equation (13.7.5) with the Del operator $\nabla \times$, and obtain, after some vector operations, the vorticity equation

$$\partial \zeta/\partial t = -\mathbf{V} \cdot \nabla_p(\zeta + f) - \omega \partial \zeta/\partial p - (\zeta + f)\nabla_p \cdot \mathbf{V} - \mathbf{k} \cdot \nabla \omega \times \partial \mathbf{V}/\partial p \quad (13.7.6)$$

Or, since $\partial f/\partial t$ and $\partial f/\partial p$ are both 0, we can write vorticity equation (13.7.6) as

$$d(\zeta + f)/dt = -(\zeta + f)\nabla_p \cdot \mathbf{V} - \mathbf{k} \cdot \nabla \omega \times \partial \mathbf{V}/\partial p \quad (13.7.7)$$

where the divergence of the velocity is along the isobaric surface.

13.8 Circulation and Vorticity in the Real Atmosphere (In Three Dimensions)

We have in the foregoing sections considered for simplicity mostly the vorticity of a circulation in the horizontal plane about a vertical axis. However, in the earth-atmosphere system, circulation depends upon the actual locations of heat sources and sinks, and, as such, may occur in any plane in space with vorticity in a direction normal to the plane of the circulation, according to Stokes's theorem. In a rectangular system of co-ordinates, a given circulation can, therefore, be resolved into three components so as to have the vorticity of each component along the co-ordinate axes, x (eastward), y (northward) and z (upward). The component circulation may be one or more of the following:

(a) Horizontal circulation about the vertical z-axis, involving wind components u and v along x and y axes respectively.

The Stokes's relation for this circulation may be written as

$$\oint (u \, dx + v \, dy) = \int\int (\partial v/\partial x - \partial u/\partial y) dx \, dy \qquad (13.8.1)$$

A purely horizontal circulation of this type is schematically shown in Fig. 13.10a.

(b) Zonal-vertical circulation about the y-axis, involving wind components u and w along the x and z axes respectively.

The Stokes's relation in this case is

$$\oint (u \, dx + w \, dz) = \int\int (\partial u/\partial z - \partial w/\partial x) dx \, dz \qquad (13.8.2)$$

The zonal-vertical circulation of this type, shown in Fig. 13.10b, is usually described as a Walker circulation.

(c) Meridional-vertical circulation about the x-axis, involving the v and w components of the wind along y and z axes respectively.

The Stokes's relation in this case is

$$\oint (v \, dy + w \, dz) = \int\int (\partial w/\partial y - \partial v/\partial z) \, dy \, dz \qquad (13.8.3)$$

The meridional-vertical circulation of this type, usually called a Hadley-type circulation, is shown in Fig. 13.10c.

The component circulations depicted in Fig. 13.10 (a, b, c) are, however, highly idealized. In nature, they always get superimposed on each other and what we actually observe is a resultant circulation which could be in any plane with vorticity about a direction normal to it.

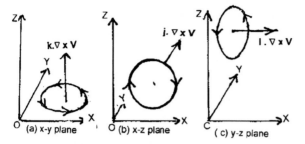

Fig. 13.10 Resolution of a circulation in space into component circulations with vorticity about co-ordinate axes: (**a**) Horizontal circulation in the x-y plane about a vertical z-axis, (**b**) Zonal-vertical Walker-type circulation in the x-z plane about the y-axis, and (**c**) Meridional-vertical Hadley-type circulation in the y-z plane about the x-axis

13.9 Vertical Motion in the Atmosphere

In the foregoing sections, we assumed air motion to be either along a horizontal or an isobaric, isentropic or a spherical surface. Over most parts of the globe, especially over high latitudes, this assumption is well borne out by observations and the wind is found to be largely horizontal. But there is also a vertical component of the air motion, the magnitude of which is ordinarily found to be about two orders of magnitude smaller than the horizontal component in large-scale motion systems. No matter how small this vertical motion may be, it is always significant. Measurements reveal that in some sub-synoptic and meso-scale circulation systems, such as cyclones and tornadoes, the vertical motion can be as large and important as the horizontal motion. Vertical motion plays a key role in the formation of such weather phenomena as cloud, precipitation, thunderstorms, etc. Because of its small magnitude, vertical velocity is difficult to measure directly. Several indirect methods have been devised to infer it from the observed winds and other atmospheric parameters. Three methods which are in common use for the purpose are: (1) the kinematic method which uses the equation of continuity in isobaric co-ordinates (12.2.6); (2) The adiabatic method which uses the thermodynamic energy equation (12.2.8); and (3) The vorticity method which uses different forms of the vorticity equation (13.7.6).

13.9.1 The kinematic Method

If we assume the atmosphere to be incompressible, the right-hand side of the continuity equation (12.2.6) may be put equal to zero. Thus, we get

$$\partial u/\partial x + \partial v/\partial y + \partial \omega/\partial p = 0$$

Or, $$\partial \omega/\partial p = -(\partial u/\partial x + \partial v/\partial y)_p \qquad (13.9.1)$$

13.9 Vertical Motion in the Atmosphere

where the right-hand side gives the wind divergence along the isobaric surface. Integrating (13.9.1) from p_0 at the earth's surface to any pressure surface p_1, we get

$$\omega(p_1) = \omega(p_0) - \int_{p_0}^{p_1} (\partial u/\partial x + \partial v/\partial y)_p \, dp \qquad (13.9.2)$$

Equation (13.9.2) can be used to compute the distribution of vertical velocity between the earth's surface and any desired pressure surface in the atmosphere. For this, the atmosphere between the two pressure surfaces are sub-divided into smaller pressure layers and the divergence is vertically averaged layerwise. These vertically-averaged divergence values are then used to integrate the vertical velocity generated by each layer starting from that at the lower boundary surface. The vertical velocity at the lower boundary is usually computed from the relation, $\omega(p_0) \simeq \mathbf{V}(p_0) \cdot \nabla p_0$

In practice, wind observations are available at a few standard pressure surfaces only, and the vertical velocity is computed on the basis of these data, usually between 1000 mb and 100 mb. However, on account of inaccuracies in wind observations, the computed divergence values are prone to large errors and the vertical velocity computed by the kinematic method are often unrealistic, especially in the upper troposphere and require large corrections or adjustments.

13.9.2 The Adiabatic Method

This method assumes absence of diabatic heating or cooling and uses the thermodynamic energy equation (12.2.8) in the following form:

$$\omega = (1/\sigma)(\partial T/\partial t + u \, \partial T/\partial x + v \partial T/\partial y) \qquad (13.9.3)$$

where σ is the static stability parameter.

The vertical velocity computed by this method is somewhat more reliable than the kinematic method in midlatitudes, but the method requires data at close intervals of time for calculation of the term $\partial T/\partial t$. Under steady state conditions, this term is zero and (13.9.3) may be written

$$\omega = (1/\sigma)(u \, \partial T/\partial x + v \partial T/\partial y) \qquad (13.9.4)$$

In low latitudes where diabatic heating or cooling contributes significantly to vertical motion, the adiabatic method fails to deliver goods.

13.9.3 The Vorticity Method

The vorticity method makes use of the vorticity equation (13.7.6) in pressure co-ordinates by substituting the divergence term $\nabla_p \cdot \mathbf{V}$ by $\partial \omega/\partial p$ and neglecting all terms involving ω on the right-hand side of the equation. This means that the tilting term and the vertical advection of vorticity term are both neglected. Further, if we assume a steady state for mean vertical motion over a period, say a week or month, the equation may be written in the simplified form.

$$\partial \omega / \partial p = \{1/(\zeta+f)\}[u\, \partial \zeta/\partial x + v\, \partial \zeta/\partial y + v\partial f/\partial y]_p \qquad (13.9.5)$$

where we have put $\partial f/\partial t$ and $\partial f/\partial x$ both equal to 0.

Vorticity is computed by using the relation: $\zeta = \partial v/\partial x - \partial u/\partial y + (u \tan \varphi)/a$, where u,v are the zonal and meridional components of the wind, φ is latitude and a is the mean radius of the earth. For computation of vorticity, one may use either the observed wind or the geostrophic wind. As in the kinematic method, Eq. (13.9.5) may be integrated with respect to pressure to secure values of ω at different pressure surfaces.

A comparative study (e.g., van den Dool, 1990; Sardeshmukh, 1993) of the ω-values computed by different methods suggests that vertical motion computed by the vorticity method by balancing divergence against vorticity advection may not be much better than those computed by the other methods and that the computed field of vorticity needs adjustment to be useful just as any of the other methods, but Sardeshmukh suggests from a scale argument that in most cases the adjustment to the vorticity field will be much smaller than that to the divergence field.

13.10 Differential Properties of a Wind Field

Vorticity, divergence and deformation are among the most important differential properties of a given wind field at a given instant. In fact, a given wind field can be kinematically analyzed to reveal the presence of these properties, as follows:

Let u(x, y) and v(x, y) be the horizontal velocity components of a wind field at a given instant in a Cartesian co-ordinate system(x, y). Then, choosing an arbitrary point as origin and expanding these components as functions of x and y in a Taylor series, we may write

$$u = u_0 + (\partial u/\partial x)_0\, x + (\partial u/\partial y)_0\, y + \text{terms of higher order}$$
$$v = v_0 + (\partial v/\partial x)_0\, x + (\partial v/\partial y)_0\, y + \text{terms of higher order} \qquad (13.10.1)$$

where the subscript '$_0$' denotes values at the chosen origin.

Since we are here concerned with the first order differentials only, we may neglect terms of the higher order. Then we regroup the first order terms so as to write (13.10.1) in the form

$$\begin{aligned} u &= u_0 + (1/2)(D+F)\, x + (1/2)(r-\zeta)y \\ v &= v_0 + (1/2)(r+\zeta)\, x + (1/2)(D-F)y \end{aligned} \qquad (13.10.2)$$

13.10 Differential Properties of a Wind Field

where we have written

$$D = [(\partial u/\partial x) + (\partial v/\partial y)]_0, \qquad \zeta = [(\partial v/\partial x) - (\partial u/\partial y)]_0$$
$$F = [(\partial u/\partial x) - (\partial v/\partial y)]_0, \qquad r = [(\partial v/\partial x) + (\partial u/\partial y)]_0 \quad (13.10.3)$$

We now rotate the system of co-ordinates by a certain angle ψ_0, such that $r = 0$, i.e., $\partial v/\partial x = -\partial u/\partial y$.

Equation (13.10.2) may then be written as

$$u = u_0 + (1/2) \, F \, x + (1/2) \, D \, x - (1/2) \, \zeta \, y$$
$$v = v_0 - (1/2) \, F \, y + (1/2) \, D \, y + (1/2) \, \zeta \, x \qquad (13.10.4)$$

If, instead of rotating the system of co-ordinates by ψ_0, we had rotated it by an angle $(\psi_0 + \pi/2)$ or $(\psi_0 - \pi/2)$, r would again be zero. So, we may choose between these rotations such that F is positive, while D and ζ may have either sign. The derivations leading to (13.10.4) can easily be verified by performing the rotation.

The terms in (13.10.4) represent four types of operations with distinctly different differential properties. These are:

$$u_0, v_0 = \text{translation},$$
$$\partial u/\partial x - \partial v/\partial y = F = \text{Deformation},$$
$$\partial u/\partial x + \partial v/\partial y = D = \text{Divergence, or expansion}$$
$$\partial v/\partial x - \partial u/\partial y = \zeta = \text{rotation or vorticity} \qquad (13.10.5)$$

Some remarks on these differential properties are as follows:

13.10.1 Translation, (u_0, v_0)

This means a uniform physical movement of a chain or surface of particles constituting the motion without involving any change of shape. It involves a movement with the same speed and in the same direction.

13.10.2 Divergence, Expansion (D)

This term signifies an expansion or contraction of the area (or volume, in three dimensions) of a closed physical curve. Let us consider here an infinitesimal area, $\delta A = \delta x \, \delta y$. Differentiating (following the motion), we get

$$d(\delta x \, \delta y)/dt = \delta y \, d(\delta x)/dt + \delta x \, d(\delta y)/dt \qquad (13.10.6)$$

Since δx and δy represent distances between particles along the co-ordinate axes, $\delta x \delta y$ represents an infinitesimal area, δA, in the xy-plane and the differentiation represents simply an expansion or contraction of this area. Similarly, differentiation of the line elements δx and δy may be regarded as stretching (or shrinking) of these elements.

Thus,

$$d/(\delta x)/dt = \delta u = (\partial u/\partial x)\delta x, \qquad d(\delta y)/dt = \delta v = (\partial v/\partial y)\delta y$$

Now, from (13.10.4), we obtain

$$\partial u/\partial x = (1/2)(D+F), \qquad \partial v/\partial y = (1/2)(D-F)$$

and hence, and also from (13.10.6), we obtain

$$D = (1/\delta A) \, d \, (\delta A)/dt = (\partial u/\partial x + \partial v/\partial y) \qquad (13.10.7)$$

It, therefore, follows that the divergence of a two-dimensional flow is the areal expansion per unit area per unit time, and this expansion is independent of translation, deformation, and rotation. The opposite process, i.e., areal contraction per unit area per unit time is called Convergence ($D < 0$).

13.10.3 Deformation

A deformation means a change of shape of the area within a closed physical curve, such as $\delta A = \delta x \, \delta y$. The shape of this area may be defined by the ratio $\delta x/\delta y$. Differentiating this ratio (following the motion), we get

$$d(\delta x/\delta y)/dt = [\delta y\{d(\delta x)/dt\} - \delta x\{d(\delta y)/dt\}]/(\delta y)^2$$

Using the same principle as in 13.10.2, and re-arranging the terms, we find

$$\{1/(\delta x/\delta y)\} \{d(\delta x/\delta y)/dt\} = (\partial u/\partial x - \partial v/\partial y) = F$$

Thus, F, which is called deformation, is independent of translation, divergence, and rotation. Perhaps, we may explain the deformation process as follows: If the original area δx δy were a unit square, and the deformation F represents the rate at which this square is transformed into a rectangle of the same size, so that the area would be stretched in the direction of the x-axis (called the axis of dilatation) and compressed (by an equal amount) in the direction of the y-axis (called the axis of contraction). Similarly, if the original area were circular, F would give the rate at which the circular area would be converted into an elliptical area.

In a pure deformation field in which the x-axis serves as the dilatation axis and the y axis as the axis of contraction, or vice versa, the streamlines are hyperbolas with the aforementioned axes as asymptotes.

13.10.4 Rotation

It is easy to show that the last terms on the right-hand side of (13.10.4) represent the horizontal components of a circulation as given by the Stokes's relation (13.8.1), in which ζ represents the vertical component of vorticity and, therefore, a rotation about a vertical axis. Thus,

$$\zeta = \mathbf{k} \cdot \nabla \mathbf{x} \mathbf{V} = \partial v/\partial x - \partial u/\partial y$$

13.11 Types of Wind Fields – Graphical Representation

The component motions of a linear wind field which exhibit the above-mentioned differential properties of air flow individually are:

A, uniform translation;
B, x-component of deformation, or the dilatation;
C, the y-component of deformation, or the contraction;

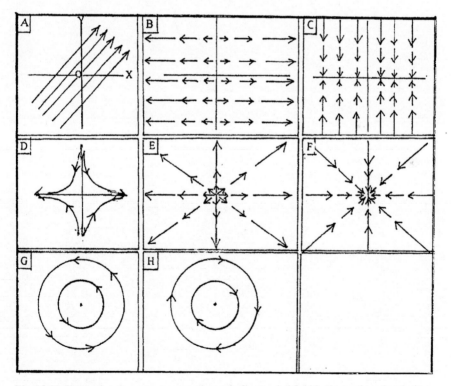

Fig. 13.11 Illustrating the component motions of a linear wind field in the northern hemisphere

D, the total deformation;
E, divergence, or areal expansion;
F, convergence, or areal contraction;
G, positive rotation; and
H, negative rotation

These component motions are shown schematically in Fig. 13.11

In D, the point where the axes of dilatation and contraction intersect is called a 'Col'.

In the real atmosphere, however, wind fields are much more complicated and the simple linear wind fields of the type shown in Fig. 13.11 seldom occur by themselves. More often than not, the differential properties are superimposed on one another in a given wind field. Further, the flow patterns can be central, i.e., symmetrical with reference to a center, or may not be related to a center. Fig. 13.12 shows some central circulation patterns corresponding to pressure systems with closed isobars and how the patterns are transformed when deformation and divergence are superimposed on rotation.

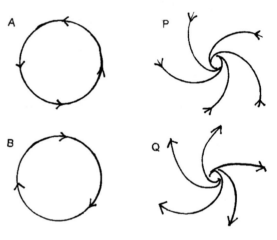

Fig. 13.12 Central patterns without straight streamlines, corresponding to pressure patterns with closed isobars: **A**, **B** are cases of pure rotation. **P** shows a case in which deformation and convergence have been superimposed on rotational field, A. In **Q**, deformation and divergence have been superimposed on rotational field at B

Chapter 14
The Boundary Layers of the Atmosphere and the Ocean

14.1 Introduction

In the atmosphere, the boundary layer may be defined as the thin layer extending from the earth's surface upward in which the airflow strongly experiences the effect of the earth's surface friction. Since the days of Osborne Reynolds (1842–1912) who experimented with the motion of viscous fluids in pipes, it has been known that when the velocity of a viscous fluid exceeds a certain critical limit, the initial laminar flow breaks down into irregular turbulent eddies, resulting in a rapid mixing of the fluid elements. The transition from laminar to turbulent flow appears to occur when the ratio of the inertial to frictional accelerations, which is called the Reynolds number, exceeds a certain critical limit.

Atmospheric motion, especially in the boundary layer, is almost always turbulent, indicating that conditions here are not quite the same as in a fluid pipe. Yet, when the atmosphere is thermally stable, several aspects of turbulent flow in which the horizontal momentum is transported vertically by the movement of small-scale eddies may be studied by applying the laws of large-scale motion. Prandtl's mixing-length hypothesis postulates that the frequent fluctuations of velocity at a point in a turbulent atmosphere near the ground are caused by the random movement of eddies, which are nothing but air in bulk of different momentum, from one level to another. In the process, the eddies carry the momentum of their original level and deliver it to the destination level and thereby produce a fluctuation of momentum at the new level. In this respect, the mixing-length may be regarded as somewhat analogous to the mean free path in molecular motion. This hypothesis enables us to parameterize the fluctuations in terms of the mean motion and work out the additional stresses, known as Reynolds stresses, which are brought about by the moving eddies. It enables us to derive an expression for the co-efficient of eddy viscosity on the model of that of molecular viscosity. Near the ground where the characteristic scale of the eddy at any level is assumed to be proportional to its distance from the ground, a logarithmic profile is found to closely represent the variation of the wind with height. This is called the surface layer in which the shearing stress varies little with height. Above this lies a layer of about a kilometer or so in height in which there is a three-way balance between the pressure gradient force, the Coriolis force and the eddy shearing stress. This is the well-known Ekman layer through which the wind speed and direction vary spirally so as to become nearly geostrophic at the top of the layer. Above the Ekman layer is the free atmosphere where the influence of the earth's surface friction is assumed to be minimal.

However, as it is known from observations, turbulence in the earth's boundary layer is greatly influenced by the thermal stability of the lower atmosphere. According to L.F. Richardson (1920), the flow can remain turbulent only so long as the rate of supply of energy by the Reynolds stresses is at least as great as the work required to be done to maintain the turbulence against gravity. Thus, it is the ratio of the gravitational stability arising from the vertical gradient of potential temperature and Reynolds's stresses, which determines the growth or decay of turbulence in the atmosphere. The importance of this ratio, which is called the Richardson number, in studies of atmospheric turbulence, can hardly be over-emphasized. Subsequent studies have shown that Richardson's work marked an important advance in our understanding of the phenomenon of atmospheric turbulence in the boundary layer. In this chapter, while we trace the early developments of the study of frictional effects in the boundary layer, we also review some of the later studies involving the effects of both friction and thermal stability.

The chapter also gives a brief introduction to the boundary layer of the ocean, created largely by the effect of windstress at the ocean surface, which is known as the Ekman layer of the ocean. A brief treatment of this problem is given, following Gill (1982).

14.2 The Equations of Turbulent Motion in the Atmosphere

One familiar with meteorological observations is aware of the presence of noise in the observations in the form of fluctuations of different frequency and amplitude. An example of this noise in the record of a sensitive anemometer exposed at a height of about 1 m or so above the ground at a fixed location on a hot summer afternoon is shown schematically in Fig. 14.1

We may interpret Fig. 14.1 as follows: If V (t) is the value of the observed windspeed at time t, the values at successive times can be averaged over a reasonably short time interval, say T, so that the observed value at any instant may be expressed

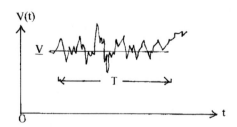

Fig. 14.1 An example of fluctuations of the windspeed, V (t), with time t at a height of about 1 m above ground on a hot summer afternoon with T is the sampling period

14.2 The Equations of Turbulent Motion in the Atmosphere

as the sum of the averaged value $\underline{V}$ and a small deviation therefrom, $V'(t)$. That is, $V(t) = \underline{V} + V'(t)$. Here we have used the underlining to denote the time-average and the prime to denote the deviation therefrom. The interval T is so chosen as to be long enough to average out the eddy fluctuations but short enough to reveal the trends in the large-scale flow. Therefore, if u, v, w are the components of the instantaneous velocity vector along the rectangular co-ordinate axes x, y, and z respectively, they may be expressed in terms of the time-mean and the deviations as follows:

$$u = \underline{u} + u'; \qquad v = \underline{v} + v'; \qquad w = \underline{w} + w' \tag{14.2.1}$$

In 11.2, we derived an expression for the force of friction per unit mass using the co-efficient of kinematic viscosity, v. With the inclusion of this force in the equations of motion in large-scale viscous flow, the horizontal momentum equations may be written as

$$\partial u/\partial t + u\, \partial u/\partial x + v\, \partial u/\partial y + w\, \partial u/\partial z = -(1/\rho)\partial p/\partial x + fv + v\, \partial^2 u/\partial z^2 \tag{14.2.2}$$

$$\partial v/\partial t + u\, \partial v/\partial x + v\, \partial v/\partial y + w\, \partial v/\partial z = -(1/\rho)\, \partial p/\partial y - fu + v\, \partial^2 v/\partial z^2 \tag{14.2.3}$$

where the last terms on the right-hand side of both the equations express the effect of molecular viscosity with v denoting the co-efficient of kinematic viscosity.

With the aid of the continuity equation (11.8.2) which may be written in the form

$$\partial \rho/\partial t + \partial(\rho u)/\partial x + \partial(\rho v)/\partial y + \partial(\rho w)/\partial z = 0 \tag{14.2.4}$$

we put the momentum equations in the flux form. We do this for the u-component by multiplying (14.2.2) by ρ and (14.2.4) by u and adding the two to obtain the following:

$$\partial(\rho u)/\partial t + \partial(\rho uu)/\partial x + \partial(\rho uv)/\partial y + \partial(\rho uw)/\partial z$$
$$= -\partial p/\partial x + \rho f\, v + \mu \partial^2 u/\partial z^2 \tag{14.2.5}$$

where μ denotes the co-efficient of molecular viscosity ($= \rho v$).

The flux form of the equation for the v-component is obtained similarly by multiplying (14.2.3) by ρ and (14.2.4) by v and adding the two to give

$$\partial(\rho v)/\partial t + \partial(\rho uv)/\partial x + \partial(\rho vv)/\partial y + \partial(\rho vw)/\partial z = -\partial p/\partial y - \rho fu + \mu \partial^2 v/\partial z^2 \tag{14.2.6}$$

We now partition the flow between the mean motion and turbulent fields by substituting for u, v, and w from (14.2.1) and average (14.2.5) and (14.2.6) in time by neglecting the small fluctuations in pressure and density. While averaging, we neglect the average of the product of a mean value and a fluctuation, so that the average of terms like $\underline{u}v'$ and $\underline{v}w'$ disappear, since $\underline{u'} = \underline{v'} = \underline{w'} = 0$, and the term like ρuv becomes

$$\{\rho uv\} = \rho\{(\underline{u}+u')(\underline{v}+v')\} = \rho\{\underline{uv}+\underline{u'v'}\}$$

After averaging, the Eqs. (14.2.5) and (14.2.6), with the aid of the averaged continuity equation, become respectively

$$\partial \underline{u}/\partial t + \underline{u}\,\partial \underline{u}/\partial x + \underline{v}\,\partial \underline{u}/\partial y + \underline{w}\,\partial \underline{u}/\partial z = -(1/\rho)\,\partial \underline{p}/\partial x + f\,\underline{v}$$
$$+ v\,\partial^2 \underline{u}/\partial z^2 - (1/\rho)\{\partial(\rho\,\underline{u'u'})/\partial x + \partial(\rho\,\underline{u'v'})/\partial y + \partial(\rho\,\underline{u'w'})/\partial z)\} \quad (14.2.7)$$

$$\partial \underline{v}/\partial t + \underline{u}\,\partial \underline{v}/\partial x + \underline{v}\,\partial \underline{v}/\partial y + \underline{w}\,\partial \underline{v}/\partial z = -(1/\rho)\,\partial \underline{p}/\partial y - f\,\underline{u}$$
$$+ v\,\partial^2 \underline{v}/\partial z^2 - (1/\rho)\{\partial(\rho\,\underline{u'v'})/\partial x + \partial(\rho\,\underline{v'v'})/\partial y$$
$$+ \partial(\rho\,\underline{v'w'})/\partial z)\} \quad (14.2.8)$$

The Eqs. (14.2.7) and (14.2.8) are revealing, in that they bring out not only the shearing stresses due to molecular motion (the third terms on the right-hand side of the equations) but also the shearing stresses generated by the eddies (the last terms within the second bracket on the right-hand side of the equations). The latter, i.e., the shearing stresses due to eddies are called the Reynolds stresses in honour of Osborne Reynolds. In a fully turbulent atmosphere, the components of the Reynolds stresses are of about the same order of magnitude along the co-ordinate axes, but in practice the vertical variation is much larger than the horizontal variations. So, if we retain the vertical components of the stresses only, (14.2.7) and (14.2.8) may be written in the simpler and more concise form

$$d\underline{u}/dt = -(1/\rho)\partial\underline{p}/\partial x + f\,\underline{v} + (1/\rho)\,\partial(\tau_{zx} + \tau'_{zx})/\partial z \quad (14.2.9)$$

$$d\underline{v}/dt = -(1/\rho)\,\partial\underline{p}/\partial y - f\,\underline{u} + (1/\rho)\,\partial(\tau_{zy} + \tau'_{zy})/\partial z \quad (14.2.10)$$

where we have expressed the horizontal accelerations following mean motion by making the following substitutions:

$$d/dt = \partial/\partial t + u\,\partial u/\partial x + v\,\partial v/\partial y + w\,\partial/\partial z;$$

$\tau_{zx}(=\mu\partial\,\underline{u}/\partial z)$ and $\tau_{zy}(=\mu\partial\,\underline{v}/\partial z)$ for the components of the shearing stresses due to molecular viscosity;
and
$\tau'_{zx}(=-\rho\,\underline{u'\,w'})$ and $\tau'_{zy}(=-\rho\,\underline{v'\,w'})$ for the components of the shearing stresses due to eddy viscosity.

In a fully turbulent flow, the shearing stresses due to eddy viscosity are much larger than those due to molecular viscosity. So, by neglecting the small effect of the molecular viscosity, the Eqs. (14.2.9) and (14.2.10) can be further simplified to the final form

$$d\,\underline{u}/dt = -(1/\rho)\,\partial\,\underline{p}/\partial x + f\,\underline{v} + (1/\rho)\,\partial\tau'_{zx}/\partial z \quad (14.2.11)$$

$$d\,\underline{v}/dt = -(1/\rho)\,\partial\,\underline{p}/\partial y - f\,\underline{u} + (1/\rho)\,\partial\,\tau'_{zy}/\partial z \quad (14.2.12)$$

14.3 The Mixing-Length Hypothesis – Exchange Co-efficients

For unaccelerated balanced motion, Eqs. (14.2.11) and (14.2.12) may be looked upon as the horizontal components of the momentum equation in vector form

$$(1/\rho)\, \partial \tau'/\partial z + f\, \mathbf{k} \times (\mathbf{V}_g - \underline{\mathbf{V}}) = 0$$

or,

$$(1/\rho)\, \partial \tau'/\partial z + f\, \mathbf{k} \times \mathbf{V}' = 0 \qquad (14.2.13)$$

where we have put $\mathbf{V}'$ for the ageostrophic mean wind $(\mathbf{V}_g - \underline{\mathbf{V}})$ and used the relations,

$f\, \mathbf{k} \times \mathbf{V}_g = (1/\rho) \nabla_H p$; and $\tau' = \tau'_{zx}\, \mathbf{i} + \tau'_{zy}\, \mathbf{j}$, where $\mathbf{i}$ and $\mathbf{j}$ are unit vectors.

14.3 The Mixing-Length Hypothesis – Exchange Co-efficients

Prandtl hypothesized that turbulent fluctuations can be parameterized in terms of the mean field variables if we postulate a characteristic mixing length l' for the eddy motion somewhat like the mean free path in molecular motion. According to this hypothesis a parcel of air at level z which is displaced vertically through a height interval l' carries the mean momentum of the original level to the new level where it mixes with the air at the new level to cause a fluctuation in the mean momentum of the new level. This means that there is no mixing, whatsoever, in between the original level and the new level. The extent of the fluctuation will depend upon the magnitude of the mixing length and the vertical gradient of the mean velocity. Thus, for the u-component, $u' = -l'\, \partial \underline{u}/\partial z$

Since $\partial \underline{u}/\partial z$ is usually positive in the boundary layer, l' is positive for downward displacement $(w' < 0)$ and negative for upward displacement $(w' > 0)$.

We may, therefore, write

$$\tau'_{zx} = -\rho\, \underline{u'\, w'} = \rho\, \underline{w'\, l'}\, \partial \underline{u}/\partial z = A_x\, \partial \underline{u}/\partial z \qquad (14.3.1)$$

Schmidt called A_x, which has been put for $\rho \underline{w'\, l'}$, the Austausch or exchange co-efficient for the x-momentum.

Observations of velocity fluctuations along the three co-ordinate axes suggest that they are of about the same order of magnitude, and, to a reasonable approximation, of about the same magnitude. So we may write, $w' = l'\, \partial \underline{u}/\partial z$, since w' and l' are of opposite sign by the above-mentioned convention of sign. Thus,

$$\tau'_{zx} = -\rho \underline{u'\, w'} = \rho \overline{l'^2}\, (\partial \underline{u}/\partial z)^2 = \rho\, l_x^2\, (\partial \underline{u}/\partial z)^2$$
$$= \rho\, K_x\, (\partial \underline{u}/\partial z) = A_x (\partial \underline{u}/\partial z) \qquad (14.3.2)$$

where l_x is the mean mixing-length and $K_x (= A_x/\rho)$, which is equal to $l_x^2 (\partial \underline{u}/\partial z)$, is called the co-efficient of eddy viscosity for the vertical transport of x-momentum. Following an analogous procedure, we can obtain an expression for the vertical eddy transport of the y-momentum, τ'_{zy}. By comparing (14.3.2) with (11.2.1), we conclude that the exchange co-efficient A has the same significance in turbulent flow as the co-efficient of kinematic viscosity μ in molecular motion.

14.4 The Vertical Structure of the Frictionally-Controlled Boundary Layer

Depending upon the rate at which the shearing stress varies with height, the frictional layer has been divided into two sub-layers, as shown in Fig. 14.2

They are: (i) the Prandtl or surface layer and (ii) the Ekman or transition layer. Above the frictionally-controlled layer lies the free atmosphere in which the eddy stresses are regarded as negligible.

14.4.1 The Surface Layer

This is the bottom layer of the atmosphere in intimate contact with the earth's surface, hence dominated by friction. According to observations in a neutrally stable atmosphere, the distribution of wind with height within the first 20 m or so of the surface suggests that the shearing stress remains more or less constant within the layer. This means that the shearing stress at a height z within this layer is about the same as that at the surface, τ_0. Hence, we may write (14.3.2) in the form

$$\tau_0/\rho = K \, \partial u/\partial z \qquad (14.4.1)$$

where we have written u for the time-averaged wind $\underline{u}$.

Since density varies only slightly within the surface layer, it may be treated as independent of height. According to Prandtl, τ_0/ρ has the dimension of velocity-squared and is written as u^{*2}, where u^* is called the friction velocity. Also, the

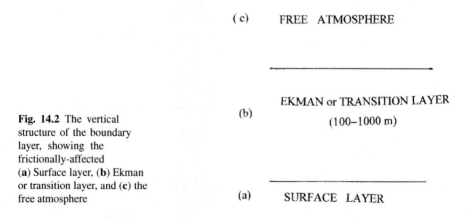

Fig. 14.2 The vertical structure of the boundary layer, showing the frictionally-affected (**a**) Surface layer, (**b**) Ekman or transition layer, and (**c**) the free atmosphere

14.4 The Vertical Structure of the Frictionally-Controlled Boundary Layer

dimension of K is velocity times a length. Now, the question arises as to what velocity and what length should be used for K in the case. The logical choice is the friction velocity u^* for the velocity and a length proportional to height z above the surface. This means that we put $K = \kappa z u^*$, where κ, the constant of proportionality, is von Karman constant ($\kappa = 0.4$). Making the substitutions in (14.4.1) and simplifying, we obtain

$$\partial u/\partial z = u^*/\kappa z \qquad (14.4.2)$$

Integrating (14.4.2) with respect to z, we get the logarithmic velocity profile,

$$u = (u^*/\kappa)\ln(z/z_0) \qquad (14.4.3)$$

where z_0, which is called the roughness length, is the constant of integration chosen so that $u = 0$ at $z = z_0$. The roughness length varies widely with the physical characteristics of the surface and the average height of the obstacles to airflow. Values of z_0 found over some natural surfaces are: 0.5 cm over smooth lawn and snow surfaces; 3.2 cm over low grass; 3.9 cm over high grass; 4.5 cm over a wheat field. When z_0 is very large as over dense vegetation, a modified logarithmic wind profile as given below has been found to represent the observations somewhat better.

$$u = (u^*/\kappa)\ln\{(z-d)/z_0\} \qquad (14.4.4)$$

where d is called the datum-level displacement.

Experience shows that inspite of several assumptions involved in the derivation, (14.4.3) gives a fairly satisfactory representation of the vertical wind profile in the earth's surface layer. Also, in this layer, since the wind direction remains more or less constant with height, it follows that the shearing stress vector is parallel to the wind vector. Hence with arbitrary orientation of the co-ordinate axes, we put, $v/u = (\partial v/\partial z)/(\partial u/\partial z)$. This relation between the wind shear and the wind may be used as the lower boundary condition for the Ekman layer.

It should be noted, however, that the logarithmic wind profile (14.4.3) holds only in a neutrally stable surface layer. The profile changes when the atmosphere becomes thermally stable or unstable. After carefully examining the vertical wind and temperature values between 0.5 m and 13 m above ground, under different stability conditions, Deacon (1949) proposed a modified wind profile

$$\partial u/\partial z = (u^*/\kappa z_0)(z/z_0)^{-\beta} \qquad (14.4.5)$$

where β is a decreasing function of the Richardson criterion Ri, defined by him for the purpose, by the expression, $Ri = (\theta_5 - \theta_{0.2})/u_1^2$, where the subscripts indicate the levels (m) at which the potential temperature θ and the wind velocity u were measured.

Since the stability of the layer is a function of Ri which depends upon the vertical variation of θ, it follows that β is less than, equal to, or greater than 1 in stable, neutral, or unstable atmosphere respectively. If (14.4.5) is integrated from z_0 (where $u = 0$) to z, we obtain

$$u = u^*[(z/z_0)^{1-\beta} - 1]/\{\kappa(1-\beta)\} \qquad (14.4.6)$$

Deacon showed that when β is very nearly equal to 1, the observed wind profile approaches the logarithmic wind profile (14.4.3).

The effect of thermal stability on the wind profile can be seen readily if one examines the diurnal variation of the low-level winds at a place under different thermal stability conditions. In the early morning hours when there is an inversion of temperature with height near the surface and the atmosphere is thermally stable, the wind is light or variable at the surface but strong at some height above the ground. In the afternoon, with rapid warming of the surface, the inversion is replaced by a lapse rate of temperature and the atmosphere becomes thermally unstable. In such condition, turbulent mixing occurs which transfers momentum downward resulting in a decrease of the vertical wind shear except, perhaps, in a very shallow layer very close to the ground. The characteristic diurnal variations brought about in the vertical wind profile of the lower boundary layer by the effects of friction and thermal stability are discussed further under 'nocturnal jet' later in this section.

14.4.2 The Ekman or Transition Layer

Above the surface layer, the structure of the boundary layer is determined by the vector relation (14.2.13) which may be written in the form

$$K \, \partial^2 \underline{V}/\partial z^2 + f \, \mathbf{k} \times (\underline{V}_g - \underline{V}) = 0 \qquad (14.4.7)$$

where we have assumed that K is invariant with height in this layer and made use of the relation (14.4.1).

If u,v are the horizontal components of the mean velocity vector $\underline{V}$, (14.4.7) yields the following equations:

$$K \partial^2 u/\partial z^2 + f(v - v_g) = 0 \qquad (14.4.8)$$

$$K \partial^2 v/\partial z^2 - f(u - u_g) = 0 \qquad (14.4.9)$$

Equations (14.4.8) and (14.4.9) can be solved to determine the departure of the observed wind from geostrophic balance in the Ekman layer. We do this in two steps, following a treatment given by Holton (1979). In the first step we ignore the presence of the surface layer and assume that the Ekman layer starts from the ground ($z = 0$), instead of from the top of the surface layer. The boundary conditions on u, v, then require that the velocity components disappear at the ground and assume geostrophic values at great distances from the ground. That is, the boundary conditions are:

$$u = 0, v = 0, \quad \text{at } z = 0; \quad \text{and } u = u_g, v = v_g, \text{ as } z \to \infty \qquad (14.4.10)$$

14.4 The Vertical Structure of the Frictionally-Controlled Boundary Layer

To solve (14.4.8) and (14.4.9), we multiply (14.4.9) by $i \equiv \sqrt{(-1)}$ and then add it to (14.4.8). The result is the complex equation

$$K\partial^2(u+vi)/\partial z^2 - i\,f(u+iv) = -i\,f(u_g + v_g) \tag{14.4.11}$$

We can arrive at a simple solution of (14.4.11) if we assume that the geostrophic wind does not vary with height and that we choose the x-axis along the geostrophic wind so that $v_g = 0$. Equation (14.4.11) may then be written as

$$\partial^2(u+iv-u_g)/\partial z^2 - (1+i)^2 m^2(u+iv-u_g) = 0 \tag{14.4.12}$$

where we have put m for $\sqrt{(f/2K)}$.

The general solution of (14.4.12) may be written

$$u + vi = u_g + A\exp[(1+i)mz] + B\exp[(-(1+i)mz] \tag{14.4.13}$$

Applying the boundary conditions (14.4.10), we note that in the northern hemisphere ($f > 0$), $A = 0$, and $B = -u_g$.

So, using the Euler formula, $\exp(i\psi) = \cos\psi + i\sin\psi$, and equating the real and imaginary parts, we may write the permissible solution of (14.4.13) as

$$u = u_g[1 - \exp(-mz)\cos mz] \tag{14.4.14}$$

$$v = u_g[\exp(-mz)\sin mz] \tag{14.4.15}$$

This is the well-known Ekman spiral solution, so called in honour of the famous Swedish oceanographer, V.W. Ekman, who in 1902 first developed the theory for the boundary layer of the ocean (Ekman, 1905). The structure of the spiral is best shown by a hodograph presented in Fig. 14.3 in which the components of the wind are plotted as a function of height. Thus the points on the curve represent the values of u and v for different values of mz as one moves away from the origin. The wind becomes parallel to the geostrophic wind at a height $Z_g = \pi/m$, though the magnitude of the wind at this height was slightly greater than the geostrophic wind. Z_g is then the depth of the Ekman layer.

Observations show that the wind attains its geostrophic value at a height of about 1 km above ground. Using this value for Z_g at a place where $f = 10^{-4}\mathrm{s}^{-1}$, we get a value of about $5\,\mathrm{m}^2\,\mathrm{s}^{-1}$ for K. In Sect. 14.3, we put $K = l'^2 \partial u/\partial z$, where l' is the mixing-length for an eddy.

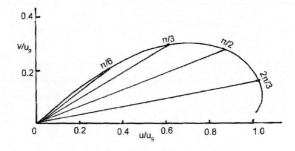

Fig. 14.3 Hodograph of the Ekman spiral solution with points marked on the graph showing the values of mz, which is a non-dimensional measure of height z

Fig. 14.4 Balance of forces in the Ekman layer

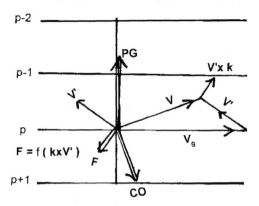

So, if we take a value of $5\,\text{ms}^{-1}\,\text{km}^{-1}$ for the wind shear $\partial \underline{u}/\partial z$, we get a value of about 30 m for l'. This value of the mixing length in the transition layer of depth about 1 km must be regarded as quite reasonable, if the mixing-length concept is to be useful.

Figure 14.4 shows a three-way balance of the frictional force **F** with the pressure gradient force **PG** and the Coriolis force **CO** in the Ekman layer, where **V** is the observed wind, $\mathbf{V}_g$ is the geostrophic wind, $\mathbf{V}'$ is the vector difference $(\mathbf{V}-\mathbf{V}_g)$ (called the ageostrophic wind), f is Coriolis parameter, and **k** is a unit vector pointing upward from the plane of **V** and $\mathbf{V}_g$.

However, the ideal Ekman layer solution discussed above is seldom realized in practice because of the observed large increase of the eddy exchange co-efficient with height near the surface of the earth. For this reason, the given solution is applicable to the region only above the surface layer. A more satisfactory solution can, therefore, be obtained by combining the above solution with the logarithmic wind profile solution for the surface layer. We again treat the eddy viscosity co-efficient K as constant but apply it to the region above the surface layer. Thus, instead of $u + vi = 0$, at $z = 0$, we let at the boundary between the two layers,

$$u + v\,i = C_0 \exp(i\alpha) \qquad (14.4.16)$$

where C_0 is the magnitude of the wind at the top of the surface layer and α is the angle between the wind and the isobars in the surface layer.

Observations show that in the surface layer the wind shear is more or less parallel to the direction of the observed wind. Applying this condition now to the wind in the Ekman layer, we write

$$u + vi = C\,\partial(u + vi)/\partial z \qquad (14.4.17)$$

where C is a constant.

For convenience, we let $z = 0$ at the bottom of the Ekman layer. Then using (14.4.16) and the condition that $u \to u_g$ as $z \to \infty$, the solution of (14.4.11) may be written

$$u + vi = [C_0 \exp(i\alpha) - u_g] \exp\{-(1+i)mz\} + u_g \qquad (14.4.18)$$

Using (14.4.18) in the boundary condition (14.4.17) and equating the real and imaginary parts, we obtain

$$C_0 \cos\alpha = mC[C_0(\sin\alpha - \cos\alpha) + u_g]$$
$$C_0 \sin\alpha = mC[-C_0(\sin\alpha + \cos\alpha) + u_g]$$

Elimination of C from these equations gives

$$C_0 = u_g(\cos\alpha - \sin\alpha) \qquad (14.4.19)$$

Substituting this value of C_0 in (14.4.18) and equating the real and imaginary parts, we obtain

$$u = u_g[1 - \sqrt{2}\sin\alpha \exp(-mz)\cos(mz - \alpha + \pi/4) \qquad (14.4.20)$$
$$v = u_g \sqrt{2}\sin\alpha \exp(-mz)\sin(mz - \alpha + \pi/4) \qquad (14.4.21)$$

One can readily see that for $\alpha = \pi/4$, the modified expressions (14.4.20–14.4.21) for the spiralling wind reduce to the classical version (14.4.14–14.4.15).

Experience shows that although the modified version is a slight improvement on the classical one, neither of them represents the real conditions in the atmosphere in a satisfactory manner due to effects of several other factors on the boundary layer besides friction. Transience and baroclinic effects are prominent amongst these other factors. Convective currents generated by horizontal temperature gradient in an unstable baroclinic layer create additional turbulence which modifies the turbulence due to friction. But, even in steady barotropic conditions, the ideal Ekman pattern is seldom realized.

It is further observed that even in a neutrally buoyant atmosphere, the frictional inflow into a lower pressure area generates a secondary circulation the horizontal and vertical scales of which are comparable with the depth of the boundary layer. Thus, it is not possible to parameterize this circulation in terms of the mixing-length theory. However, the circulation transfers momentum vertically and thereby reduces the angle between the boundary-layer wind and the geostrophic wind. The secondary circulation set up by friction in the boundary layer is of great importance in meteorology.

14.5 The Secondary Circulation – The Spin-Down Effect

The vertical motion in the secondary circulation referred to at the end of the previous section can be computed from the mass flux into the low-pressure area by the v-component of the boundary-layer wind given by (14.4.15). Thus,

$$M = \int_0^{z_g} \rho\, v\, dz = \int_0^{z_g} \rho\, u_g \exp(-\pi z/z_g)(\sin \pi z/z_g)\, dz \qquad (14.5.1)$$

If it is assumed that the density remains invariant in the boundary layer, we may write the mass continuity equation in the form

$$\partial(\rho w)/\partial z = -\partial(\rho u)/\partial x - \partial(\rho v)/\partial y \qquad (14.5.2)$$

Integrating (14.5.2) from 0 to z_g, we obtain for the vertical velocity at the top of the boundary layer the expression

$$(\rho w)_{z_g} = -\int_0^{z_g} [\partial(u)/\partial x + \partial(v)/\partial y]\, dz$$

Here, we have assumed that $w = 0$ at $z = 0$. Substituting for u and v from (14.4.14) and (14.4.15) and again assuming that u_g is invariant with x, we can re-write the expression for the vertical mass flux at the top of the boundary layer where $v_g = 0$, in the form

$$(\rho w)_{z_g} = -(\partial/\partial y) \int_0^{z_g} \rho u_g \exp(-\pi z/z_g) \sin(\pi z/z_g)\, dz \qquad (14.5.3)$$

Comparing (14.5.3) with (14.5.1), we note that the the vertical mass flux at the top of the boundary layer is equal to the horizontal convergence of mass in the boundary layer. Since $-\partial u_g/\partial y = \zeta_g$, where ζ_g is the geostrophic vorticity, it follows from (14.5.3) that the vertical velocity at the top of the boundary layer is given by

$$w_{zg} = \zeta_g \sqrt{(K/2f)} \qquad (14.5.4)$$

The relation (14.5.4) which states that the vertical motion at the top of the boundary layer is directly proportional to the geostrophic vorticity of the layer is important, since it tells us how friction communicates mass from the layer to the free atmosphere directly through a secondary vertical circulation rather than through the slow process of molecular diffusion. For example, in a typical synoptic-scale circulation vortex in midlatitudes in which $\zeta_g \simeq 1.0 \times 10^{-5} s^{-1}$, $K \simeq 5\ m^2 s^{-1}$ and $f \simeq 1.0 \times 10^{-4}\ s^{-1}$, the vertical velocity given by (14.5.4) works out to be about $1.58\ mm\ s^{-1}$.

Figure 14.5 shows schematically the direction of the frictionally-generated secondary circulation in a cyclonic vortex in a barotropic atmosphere.

An important effect of the secondary circulation in the atmosphere is what is known as the spin-down effect. This means that in the absence of any other disturbing factor, the vorticity of the azimuthal circulation in the outflow region will continually decrease with time. This can be readily shown by taking the vorticity equation for a barotropic atmosphere and integrating it over the height interval from the top of the boundary layer z_g to the tropopause level H. Since the atmosphere is

14.5 The Secondary Circulation – The Spin-Down Effect

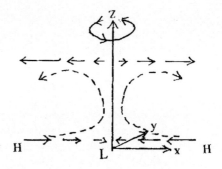

Fig. 14.5 A schematic of the secondary circulation generated by frictional convergence in the boundary layer of a cyclonic vortex in a barotropic atmosphere

treated as barotropic, the solenoidal term is zero and we write the vorticity equation (13.6.2) in the form

$$d(\zeta+f)/dt = -(\zeta+f)(\partial u/\partial x + \partial v/\partial y) = f\partial w/\partial z \tag{14.5.5}$$

where we have used the continuity equation and neglected ζ compared to f on the right-hand side. Neglecting the latitudinal variation of f and integrating (14.5.5) from z_g to H, we obtain

$$\int_{z_g}^{H} (d\zeta/dt)\,dz = f \int_{w(z_g)}^{w(H)} dw \tag{14.5.6}$$

If we now assume that the vertical velocity vanishes at the tropopause and that the vorticity in the layer $(H - z_g)$ remains at its geostrophic value at the level z_g, then substituting the value of $w(z_g)$ from (14.5.4), we obtain the differential equation

$$d\zeta_g/dt = -fw(z_g) = -\sqrt{(fK/2H^2)}\zeta_g \tag{14.5.7}$$

where we have neglected z_g in comparison with H.

Integration of (14.5.7) with respect to time yields

$$\zeta_g = \zeta_g(0)\exp\{-\sqrt{(fK/2H^2)}t\} \tag{14.5.8}$$

where $\zeta_g(0)$ is geostrophic vorticity at time 0.

It, therefore, follows from (14.5.8) that in a northern hemisphere barotropic vortex the geostrophic vorticity will decay with time to e^{-1} of its original value in time $t \equiv H(2/fK)^{-1/2}$. This e-folding time is actually what is called the spin-down time. For a value of H = 10 km, K = 10 m^{2}s^{-1}, and f = 1.0×10^{-4}s^{-1}, the spin-down time is about 4 days. This time-frame should be compared with the time that would be required for spin-down to the same extent by eddy diffusion through the same height. In this latter case, using the relation, $t = H^2/2K$, and using a value of 5 m^{2}s^{-1} for K, we obtain $t \approx 100$ days.

We can understand the spin-down effect in the atmosphere in another way. Here, as the fluid elements flow outward above the boundary layer, the outflow experiences a Coriolis force to the right in the clockwise direction, which opposes the anticlockwise azimuthal velocity and thereby slows it down. Holton (1979) cites the example of spin-down in a tea-cup where a similar secondary circulation is set up by the effect of friction at the bottom of the cup. Here, as part of the vertical circulation, the outflow at the top transports fluid from low-momentum to high-momentum region and thereby, according to the principle of conservation of angular momentum, spins down the azimuthal circulation.

In nature, examples of secondary circulation set up by flow over rough terrain are so numerous that the phenomenon may be said to be almost ubiquitous. There is indication of its occurrence over deserts where long, parallel rows of sand-dunes are piled up by high winds (Hanna, 1969). According to Woodcock (1942), the flight pattern of soaring gulls indicates the presence of line updrafts under certain conditions. Cases of glider pilots experiencing long lifting bands at heights up to 2–3 km have been discussed by Kuettner (1959). Kuettner (1967) has also reported many observations of parallel cloud lines which he attributed to helical motion in the boundary layer. Elongated cloud streets and roll clouds frequently seen in satellite cloud pictures are believed to be indications of occurrence of secondary circulation in the earth's boundary layer on a large scale.

However, it is important to recognize that in the creation of most of these phenomena, friction and buoyancy are both involved. In fact, over several areas, strong surface heating and vertical lapse rates of temperature cause low-level convection and horizontal temperature gradients cause thermal winds which may augment or counteract the effect of friction. Joseph and Raman (1966) reported the occurrence of a low-level westerly wind maximum of 20–30 m s^{-1} at a height of about 1.5 km a.s.l. over the southern Indian peninsula during July. Above the low-level maximum, the westerly wind continuously decreased in speed with height and reversed direction to attain an easterly wind maximum of about 40–50 m s^{-1} at an altitude of about 14–15 km a.s.l. The low-level westerly jetstream could be explained only as the combined effect of the meridional temperature gradient and the boundary-layer friction. Under the effect of friction alone, the westerly surface wind would increase in speed to become geostrophic at the top of the boundary layer, but the effect of a horizontal temperature gradient with warm air to the north and cold air to the south produces an easterly thermal wind which counteracts and arrests the further increase in the speed of the westerly wind with height. The result is a low-level westerly jetstream of the kind observed, besides the easterly jet stream at high levels which appears to be wholly thermally generated and controlled (Saha, 1968).

14.5.1 The Nocturnal Jet

An interesting feature of the wind in the boundary layer is that at many inland stations, a jet-like strong wind appears during night at a fairly low level in the atmosphere. It is called the nocturnal jet. The basic reason for its occurrence is given by Blackadar (1957).

14.5 The Secondary Circulation – The Spin-Down Effect

During the heat of the day, the boundary layer is deep and momentum is well-mixed by convective turbulence up to a height of about 1 km or more above the ground. At night, the ground cools by radiation and a stable layer develops within about 200 m of the ground. The frictional effects are then confined near the ground and a wind jet appears at the top of this stable layer where momentum is now concentrated. However, the region above (200–1000 m), which in day-time was a part of the turbulent boundary layer, is released from frictional influence, i.e., the stress term in that region suddenly drops to zero after sundown. In response, the wind begins an inertial oscillation that proceeds until mixing resumes next day. The daily cycle brings about a cyclonic rotation of the wind relative to the geostrophic wind during daytime and an anticyclonic rotation during the night. These opposite turnings of the wind between day and night have actually been observed in field experiments as well as reproduced in numerical modeling (Thorpe and Guymer, 1977).

14.5.2 Turbulent Diffusion and Dispersion in the Atmosphere

The atmosphere being turbulent, any gaseous or particulate matter released or ejected into it, such as gases, vapours, smoke, dust particles, etc., is dispersed by the eddies that may be present at the time. However, the speed and extent of dispersion in terms of height and distance traveled depends upon not only the state of turbulence but also the thermal stability of the atmosphere.

Normally, there is a large diurnal variation in the level of turbulent diffusion in the atmosphere Under thermally stable conditions and light winds as during a clear night when there is little convection and turbulence in the air, vertical mixing is restricted to the surface layers only and the released material may spread out horizontally and slowly settle down to the earth as it travels downstream. In an unstable atmosphere with strong winds as during a hot summer afternoon, there is rapid mixing of the ejected material with the upper layers of the atmosphere by large-scale convection and turbulent eddies. However, as turbulence and convection die down during the night, some of the material in the upper layer may come back to the lower layer and may even settle down to the earth. Gases, vapours and light particulate matter, such as fine dust and smoke, etc., however, may rise to higher levels and travel longer distances unless and until they are washed down by precipitation.

Atmospheric diffusion of the kind which brings about pollution in the atmosphere with harmful effects on humans, animals and plants is of great concern. In this category, one may include effluents from chemical and industrial chimneys, exhaust gases from road vehicles, aeroplanes, etc. which burn fossil fuels, radioactive leakage from atomic power plants, bomb explosions, volcanic eruptions, to name only a few. Once ejected into the atmosphere, the concentration of the pollutants in the air needs to be known in order to assess risks of adverse effects on life and property on earth. The risk is definitely greater when the pollutants are released in the atmosphere at the ground or lower levels than from chimneys or towers at sufficiently high levels where winds are usually stronger to disperse them quickly.

14.6 The Boundary Layer of the Ocean – Ekman Drift and Mass Transport

When a wind blows over a calm ocean, it forces the water at the surface to move along with it thereby producing an ocean current. Owing to viscosity of ocean water, the influence of the surface stress in dragging water is transmitted downward to a depth varying from 10 m to 100 m. In this thin layer, the ocean current decreases in strength and its direction veers with depth producing what is known as the Ekman drift in the shape of a spiral, as shown in Fig. 14.6.

In the ocean, the windstress at the surface accelerates the current in the boundary layer. Thus, along with the pressure gradient force, the forces that act in the oceanic boundary layer are practically the same as in the atmospheric boundary layer. We may, therefore, write the linearized horizontal momentum equations for a fluid on a uniformly-rotating earth in the simplified form

$$\partial u/\partial t - fv = -(1/\rho)\partial p'/\partial x + (1/\rho)\partial X/\partial z \qquad (14.6.1)$$

$$\partial v/\partial t + fu = -(1/\rho)\partial p'/\partial y + (1/\rho)\partial Y/\partial z \qquad (14.6.2)$$

where (X,Y) are the components of the stress vector, (u, v) are the components of the current velocity vector, p' is the perturbation pressure at the surface and the other terms have their usual meanings.

In the above equations, we have considered the vertical variation of the stress only and not the horizontal variation, because of the extraordinarily large scale of the horizontal variation compared with the scale of the vertical variation. Also, we have treated the density as constant within the shallow boundary layer, since water is much more incompressible than air.

From (14.6.1) and (14.6.2), it may be seen that the two forces that tend to accelerate the current are the pressure gradient force and the force due to the variation of the Ekman stress. It is convenient to consider these two effects separately. The part (u_p, v_p) of the current velocity driven by the pressure gradient force is determined by the equations

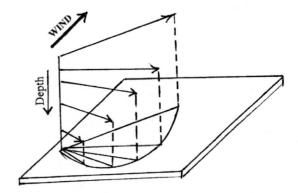

Fig. 14.6 Schematic representation of a wind-driven ocean current in deep water, showing the decreases in velocity and change of direction at intervals of depth (the Ekman spiral).

14.6 The Boundary Layer of the Ocean – Ekman Drift and Mass Transport

$$\partial u_p/\partial t - f\,v_p = -(1/\rho)\partial p'/\partial x; \quad \partial v_p/\partial t + f u_p = -(1/\rho)\partial p'/\partial y \quad (14.6.3)$$

and assume the geostrophic value in a steady-state flow ($\partial u_p/\partial t = \partial v_p/\partial t = 0$). The other part of the current velocity which is driven by the stress vector and which we denote by (u_E, v_E) satisfies the equations

$$\partial u_E/\partial t - f v_E = -(1/\rho)\partial X/\partial z; \quad \partial v_E/\partial t + f u_E = -(1/\rho)\partial Y/\partial z \quad (14.6.4)$$

Thus, the total velocity that appears in (14.6.1) and (14.6.2) may be considered as the sum

$$u = u_p + u_E; v = v_p + v_E$$

Now, the windstress (X, Y) is zero outside the boundary layer. So, the integration of (14.6.4) with respect to z across the layer yields the relation (when boundary below)

$$\rho(\partial U_E/\partial t - f V_E) = -X_s; \quad \rho(\partial V_E/\partial t + f U_E) = -Y_s \quad (14.6.5)$$

where (X_s, Y_s) is the value of the stress vector at the surface, and (U_E, V_E) is the Ekman volume transport, relative to the pressure-driven flow, as given by the relations, $U_E = \int u_E\,dz$, and $V_E = \int v_E\,dz$.

Since the density is regarded as constant, the Ekman mass transports are given by ($\rho U_E, \rho V_E$). The sign of the stress term depends on whether the boundary surface is below or above the layer. The Eq. (14.6.5) applies to the case of the atmospheric boundary layer or to the ocean's bottom boundary layer in which the boundary is below the layer. In the case of the wind-driven ocean current, the boundary surface is above the layer. Hence the signs of the stress term are reversed in the oceanic case and the integral of (14.6.4) across the layer gives (when boundary above)

$$\rho(\partial U_E/\partial t - f V_E) = X_s; \quad \rho(\partial V_E/\partial t + f U_E) = Y_s \quad (14.6.6)$$

In the steady state, (14.6.6) gives

$$\rho U_E = Y_s/f, \text{ and } \rho V_E = -X_s/f \quad (14.6.7)$$

In steady-state flow, the Ekman transport is directed at right angles relative to the surface stress in the northern hemisphere. In the atmosphere, the transport direction is to the left of the surface stress, whereas in the ocean it is to the right, as shown schematically in Fig. 14.7, which shows the magnitude and direction of the Ekman transport above and below the ocean surface.

It follows from (14.6.5) and (14.6.7) that the sum of the Ekman mass transports above and below the ocean surface is zero.

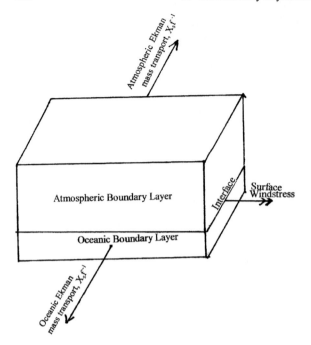

Fig. 14.7 A schematic showing the directions and magnitudes of the Ekman mass transports in relation to the surface windstress in the atmospheric and oceanic boundary layers in the northern hemisphere. The directions are reversed in the southern hemisphere

The same argument cannot, however, be applied to the Ekman volume transports because of large differences in density between the atmosphere and the ocean. The direction of the Ekman transports in the cases of both atmosphere and ocean reverses in the southern hemisphere.

14.7 Ekman Pumping and Coastal Upwelling in the Ocean

The Ekman mass transport between neighbouring areas in the boundary layer often causes convergence or divergence of mass and thereby vertical motion which can be computed with the aid of the continuity equation, as was done for the atmospheric boundary layer in Sect. 14.5.

In the oceanic case, we again neglect the density variations and integrate the continuity equation in the form, $\partial u/\partial x + \partial v/\partial y + \partial w/\partial z = 0$, with respect to z using the boundary condition $w = 0$ at the ocean surface (boundary above) and obtain

$$\partial(\rho U_E)/\partial x + \partial(\rho V_E)/\partial y - \rho w_E = 0 \qquad (14.7.1)$$

Substituting from (14.6.7) and neglecting latitudinal variation of the Coriolis parameter f, we obtain

14.7 Ekman Pumping and Coastal Upwelling in the Ocean

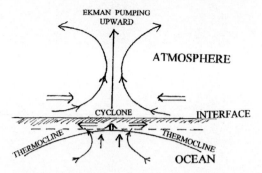

Fig. 14.8 A vertical section through the center of a cyclone over the ocean in the northern hemisphere illustrating the formation of Ekman pumping in the atmosphere above and the ocean below the interface. Note the directions of Ekman transports (indicated by double-shaft arrows) in the atmospheric and the oceanic boundary layers and the raising of the thermocline in the ocean below the cyclone center where Ekman pumping is upward

$$\rho w_E = \partial(Y_s/f)/\partial x - \partial(X_s/f)/\partial y = (1/f)\mathbf{k} \cdot \nabla \times (\tau_s) \qquad (14.7.2)$$

where τ_s is the surface stress vector with components (X_s, Y_s).

Thus, the vertical velocity w_E, which is called the Ekman pumping velocity, is, according to (14.7.2), $(\rho f)^{-1}$ times the vertical component of the curl of the surface wind stress vector. Note that it has the same sign in the ocean as in the atmosphere (14.5.4).

When a wind blows over a coastal region so as to have a strong horizontal component parallel to the coastline, the Ekman transport in the oceanic boundary layer is at right angles to the direction of the windstress and can be either inward towards the coast, or outward away from it, depending upon the direction of the windstress and the hemisphere in which the coast is located. If the Ekman transport is outward and there is divergence of water from the top, mass continuity requires cold water from deep layers to upwell to replace the water being removed. The upwelled water is rich in nutrients for plant and marine life. That is why the worlds's important fisheries are found in coastal regions where there is intense upwelling for most of the year. Famous upwelling regions are the coasts of California, northwestern Africa and Somalia in the northern hemisphere and Chile-Peru, Southwestern Africa and Southwestern Australia in the southern hemisphere. According to an estimate (Gill, 1982), coastal upwelling is 30–100 times stronger than open ocean upwelling.

When a cyclone moves over an ocean surface, the low pressure in its central area causes convergence of air and strong upward motion in the atmospheric boundary layer above the interface, while there is upwelling and divergence of water below the center of the cyclonic circulation, which results in cooling of the ocean surface. This is illustrated by a schematic in Fig. 14.8.

Chapter 15
Waves and Oscillations in the Atmosphere and the Ocean

15.1 Introduction

In Table 11.1, we referred to a few types of atmospheric waves which concern meteorologists most of the time because of their apparent relations with the formations of weather and climate. However, in the atmosphere as well as the ocean, depending upon the fluctuations in pressure, temperature and wind, several types of waves and oscillations may be excited.

A fluid parcel when displaced from its position of equilibrium in a stable atmosphere by an external force will tend to return to its original state by a series of movements which usually take the form of oscillations or wave motions. For example, a perturbation in the pressure field will generate sound waves which vibrate longitudinally and propagate at normal temperature and pressure with a velocity of $\simeq 330\,\mathrm{m\,s^{-1}}$ in the direction of vibration. Here the restoring force is pressure as per the compressibility of the fluid. A displacement in the density field produces gravity waves which are transverse waves and oscillate in a direction at right angles to the direction of propagation. They also move fast at an average speed of $\simeq 200\,\mathrm{m\,s^{-1}}$. In the case of the gravity waves, the restoring force is the pull of the Earth's gravity. On the other hand, in a barotropic atmosphere certain types of slow-moving planetary waves are excited by the variation of the Coriolis parameter with latitude. These are called the Rossby waves. Pure Rossby waves oscillate in the horizontal plane and usually move westward relative to the mean wind with a phase velocity which is generally much smaller($\simeq 5\text{--}10\,\mathrm{m\,s^{-1}}$) than that of the sound or the gravity waves. Of these, the Rossby waves are the ones which are of the greatest meteorological interest because of their association with synoptic-scale weather.

The above-mentioned waves and oscillations which are perturbations in the mean motion of the atmosphere can be derived from the general equations of motion only after making some simplifying assumptions, since the equations are nonlinear and cannot be solved by any known method. A perturbation technique has been devised to linearize them.

However, before we describe this technique and apply it to the case of atmospheric waves, we describe some of the general properties of harmonic motion, such as amplitude, frequency, period, wave-length, phase velocity, etc., starting with the working of a simple pendulum which is an example of a non-propagating simple harmonic motion.

15.2 The Simple Pendulum

In a simple pendulum, a round metallic bob of mass m is suspended from a support with a weightless string of length about 1m. In its rest position, the string hangs vertically under the force of gravity, mg (see Fig. 15.1).

The bob is then displaced to one side through a small angle θ to a position P where it is let go to oscillate freely. Passing through the vertical at O, the bob will move to the other side to a position P′ and then move back to continue the oscillation. At P, the equilibrium is reached when the force of displacement is balanced by the opposing force of gravity and the relevant equation is

$$m\, l\, d^2\theta/dt^2 = -m\, g \sin\theta \qquad (15.2.1)$$

where l denotes the length of the pendulum. Since θ is small, $\sin\theta$ differs little from θ. So, (15.2.1) may be written

$$d^2\theta/dt^2 = -(g/l)\theta = -\mu^2\,\theta \qquad (15.2.2)$$

where $\mu \cong (g/l)^{1/2}$.

The general solution of the differential equation (15.2.2) may be written

$$\theta = A\cos\mu t + B\sin\mu t = \theta_0\cos(\mu t - \alpha) \qquad (15.2.3)$$

where A, B, θ_0 and α are all constants which can be determined from initial conditions.

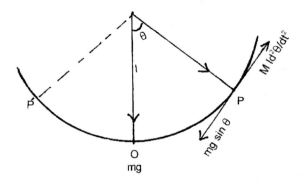

Fig. 15.1 The oscillations of a simple pendulum

Thus the solution is a periodic oscillation with an amplitude θ_0 (the angle of maximum displacement) and a frequency μ (or a period $2\pi/\mu$). The angle $(\mu t - \alpha)$ is called the phase angle which varies linearly in time by a factor of 2π radians for every period of oscillation, and α is the phase difference between two oscillations.

In the case of an oscillation or wave propagating in space, say along the x-axis with a velocity c, another term is to be taken into consideration to determine the phase, and that is the wave number k which represents the number of waves of a particular wavelength λ around a latitude circle. Thus, k being equal to $2\pi/\lambda$, the phase of the travelling wave is $kx - (\mu t - \alpha)$, or, $(kx - \mu t + \alpha)$. For an observer moving with the wave, the phase speed c is constant and given by the relation, $c = \mu/k$.

15.3 Representation of Waves by Fourier Series

Though atmospheric waves are never purely sinusoidal, a wave or perturbation can be represented as a function of longitude by a Fourier series in terms of a zonal mean plus a series of sine and cosine terms. Thus, a function f(x) may be written as

$$f(x) = \sum_{m=1}^{\infty} (A_m \sin k_m x + B_m \cos k_m x) \qquad (15.3.1)$$

where $k_m = 2\pi m/L$ is the wave number of the mth wave (m being the number of waves and L the length of a latitude circle) and A_m and B_m are the amplitudes of the sine and the cosine parts respectively of the mth wave. Here, the value of the zonal mean of the function is assumed to be zero.

The co-efficients A_m and B_m are calculated by first multiplying both sides of (15.3.1) by $\sin k_n x$ and then integrating them around a latitude circle and applying the orthogonality relationships

$$\int_0^L \sin(2\pi mx/L) \sin(2\pi nx/L) \, dx = 0 \text{ or } L/2, \text{ according as } m \neq n \text{ or } m = n,$$

and

$$\int_0^L \cos(2\pi mx/L) \sin(2\pi nx/L) \, dx = 0, \text{ for any m,n combination}$$

(15.3.2)

The calculation yields, $\quad A_m = (2/L) \int_0^L f(x) \sin(2\pi mx/L) \, dx \qquad (15.3.3)$

A similar calculation resulting from multiplication of both sides of (15.3.1) by $\cos k_n x$ and integration around a latitude circle and application of orthogonality relationships gives

$$B_m = (2/L) \int_0^L f(x) \cos(2\pi m x/L)\, dx \tag{15.3.4}$$

Thus,
$$f_m(x) = A_m \sin mx + B_m \cos mx \tag{15.3.5}$$

where $f_m(x)$ is called the mth harmonic of the function $f(x)$ and A_m and B_m are called the Fourier co-efficients of that harmonic.

A Fourier harmonic can also be expressed more conveniently and in a more compact form by using complex exponential notation as in the Euler formula

$$e^{ix} = \cos x + i \sin x, \text{ where } i = \sqrt{(-1)}$$

Thus, we can write,
$$f_m(x) = \text{Re}\{C_m \exp(i\, k_m x)\} \tag{15.3.6}$$

where Re denotes the real part of { }, and C_m is a complex co-efficient.

Comparison of (15.3.6) with (15.3.5) shows that

$$A_m = -\text{Im}\{C_m\}, \text{ and } B_m = \text{Re}\{C_m\}$$

The computation of Fourier co-efficients enables us to identify the wave components which contribute to an observed perturbation of a meteorological variable, such as the height of an isobaric surface along a latitudinal circle. Sometimes one or more dominant wave components may be identified which account for most of the observed variation. If only a qualitative representation is desired, then in most cases a limited number of components may suffice.

15.4 Dispersion of Waves and Group Velocity

In the case of a harmonic oscillator, the frequency of the oscillation depends only on the physical characteristics of the oscillator and not on the motion itself. For a propagating wave, however, the frequency depends on the wave number as well as the physical characteristics of the medium. Thus, the phase velocity depends on the wave number, since $c = \mu/k$, unless, of course, μ is a function of k. Waves in which the phase speed varies with the wave number are called dispersive waves and the formula relating the frequency with the wave number is called the dispersion relation.

However, not all waves in the atmosphere are dispersive. For example, acoustic waves are non-dispersive. Their phase speed is constant, regardless of the wave number. When two waves of the same amplitude but differing slightly in frequency and wave number are superposed on each other in the course of their propagation in the same direction, the amplitude of the resulting wave fluctuates as the interacting waves get in and out of phase. Amplification occurs where they get in phase and energy gets concentrated, whereas the amplitude dies down where they get out of phase. The occurrence of such periodic ups and downs of amplitude in acoustics is known as 'beats'. The resulting wave thus moves as a wave group with its amplitude fluctuating in time (see Fig. 15.2).

15.5 The Perturbation Technique

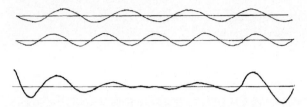

Fig. 15.2 Superposition of two waves of equal amplitude but different wave number and frequency The resulting wave group is shown in the lower part

In what follows, we derive an expression for the propagation of a wave group. Let two waves of equal amplitude but differing in frequency by $2\Delta\mu$ and wave number $2\Delta k$ be superimposed on each other.

Then, the amplitude of the resulting wave may be written as

$$A(x,t) = \exp[i\{(k+\Delta k)x - (\mu+\Delta\mu)t\}] + \exp[i\{(k-\Delta k)x - (\mu-\Delta\mu)t\}]$$

Re-arranging, we get

$$A(x,t) = [\exp\{i(\Delta k\, x - \Delta\mu\, t)\} + \exp\{-i(\Delta k\, x - \Delta\mu\, t)\}]\exp\{i(kx-\mu t)\}$$

Or,
$$A(x,t) = 2\cos(\Delta k\, x - \Delta\mu\, t)\exp\{i(kx-\mu t)\} \qquad (15.4.1)$$

The meaning of (15.4.1) is as follows: It represents a high-frequency carrier wave of wavelength $2\pi/k$ and phase velocity μ/k, whose amplitude is modulated by a wave-like periodic variation, or a wave-group, of wavelength $2\pi/\Delta k$ and phase speed $\Delta\mu/\Delta k$. Thus, the amplitude of the wave-group which is indicated by an envelope in the lower part of Fig. 15.2 travels at a speed which is different from the speed of the carrier wave. If C_g denotes the group velocity, its relationship with the phase speed c of the carrier wave may be found as follows:

At the limit when $\Delta k \to 0$, $C_g = d\mu/dk$. Since, $c = \mu/k$ and $k = 2\pi/\lambda$, where λ is the wavelength, we find

$$C_g = d(ck)/dk = c - \lambda(dc/d\lambda) \qquad (15.4.2)$$

It may be noted that when the wave-group is non-dispersive (i.e., $dc/d\lambda = 0$), it travels at the same speed as the carrier wave.

15.5 The Perturbation Technique

The basic assumptions of this technique are the following:

Assumption 1. A meteorological variable, such as pressure, temperature, wind or density, can be split into two states: a basic state that is assumed to remain invariant with time (t) and the co-ordinate axis along which it is measured (say, x), and a perturbed state that changes with these variables. For example, if we apply the technique to the zonal component of the wind u, it can be written as

$$u(x,t) = \underline{u} + u'(x,t), \quad (15.5.1)$$

where $\underline{u}$ (u underlined) denotes the mean basic state of u, and u' (primed) the perturbation.

Assumption 2. The basic state variables must on their own satisfy the governing equations when there is no perturbation. This obviously requires that the perturbation field must remain small enough for the product of the perturbation variables to be neglected. This requires that $|u'/\underline{u}| \ll 1$, and, for the nonlinear term $u\partial u/\partial x$ of the governing equations, for example, we can write:

$$u\partial u/\partial x = \underline{u}\partial u'/\partial x + u'\partial u'/\partial x,$$

But, since $\underline{u}\,\partial u'/\partial x \gg u'\partial u'/\partial x$, the assumption leads to the result

$$u\partial u/\partial x = \underline{u}\,\partial u'/\partial x \quad (15.5.2)$$

The technique, therefore, enables us to reduce the nonlinear differential equations to linear differential equations in perturbation variables in which the basic state variables appear as constant co-efficients to the perturbation terms. The linearized equations can then be solved by known techniques to yield information regarding the properties of the perturbation. Since, in most cases, a perturbation is assumed to have a sinusoidal form, the solution enables us to find the various characteristics of the wave form such as its frequency, wavelength, phase velocity, etc.

15.6 Simple Wave Types

As examples of pure waves, we consider three types of wave motion in this section:

(a) Acoustic or sound waves in which the fluid movements are longitudinal, i.e., they move in the direction of wave propagation and the restoring force is compressibility of the fluid;
(b) Gravity waves in which the fluid movements are transverse, i.e., perpendicular to the direction of wave propagation and the restoring force is the earth's gravity; and
(c) Rossby waves which are horizontal transverse waves perpendicular to the direction of wave propagation and in which the restoring force is the variation of the Coriolis parameter with latitude.

15.6 Simple Wave Types

In what follows, we discuss some aspects of their formation and propagation:

(a) Acoustic or sound waves

In sect. 3.4.5, we showed how the laws of thermodynamics could be applied to deduce the Laplace's equation for the velocity of sound in air. The basic idea there was that the process of compression and rarefaction produced by the sound waves occurred so rapidly that the changes could only be termed as adiabatic. Here, we use the perturbation method to derive an expression for the velocity of sound in air. For simplicity, we assume that the oscillation is entirely longitudinal and occurs along the x-axis with no transverse component along the y or z axis. This means that in the momentum equations, $v = w = 0$ and there is no dependence of the perturbation variables along the y and the z axes. With these restrictions, we write the momentum, continuity and thermodynamic energy equations as

$$du/dt + (1/\rho)\partial p/\partial x = 0 \qquad (15.6.1)$$

$$d\rho/dt + \rho\partial u/\partial x = 0 \qquad (15.6.2)$$

$$d\ln\Theta/dt = 0 \qquad (15.6.3)$$

where u is the zonal component of the wind, p is pressure, ρ is density and Θ is potential temperature, and $d/dt = \partial/\partial t + u\partial u/\partial x$.

Since $\Theta = (p/\rho R)(1000/p)^{R/C_p}$, we can eliminate Θ from (15.6.3) and write

$$(1/\gamma)d\ln p/dt - d\ln\rho/dt = 0 \qquad (15.6.4)$$

where $\gamma = c_p/c_v$. Eliminating ρ between (15.6.4) and (15.6.2), we get

$$(1/\gamma)d\ln p/dt + \partial u/\partial x = 0 \qquad (15.6.5)$$

We now divide the dependent variables, u, p and ρ into their basic state portions (underlined) and the perturbed portions (primed) and write

$$u(x,t) = \underline{u} + u'(x,t)$$

$$p(x,t) = \underline{p} + p'(x,t)$$

$$\rho(x,t) = \underline{\rho} + \rho'(x,t) \qquad (15.6.6)$$

Substituting (15.6.6) into (15.6.1) and (15.6.5), we get

$$\partial(\underline{u}+u')/\partial t + (\underline{u}+u')\partial(\underline{u}+u')/\partial x + \{1/(\underline{\rho}+\rho')\}\partial(\underline{p}+p')/\partial x = 0$$

$$\partial(\underline{p}+p')/\partial t + (\underline{u}+u')\partial(\underline{p}+p')/\partial x + \gamma(\underline{p}+p')\partial(\underline{u}+u')/\partial x = 0$$

Since, by the perturbation theory, $|\rho'/\underline{\rho}| <<\ll 1$, we can simplify the density term $\{1/(\underline{\rho}+\rho')\}$ to $(1/\underline{\rho})$. Neglecting the products of the perturbation quantities and noting the basic state parameters are constant, we can write the linearized perturbation equations

$$(\partial/\partial t + \underline{u}\partial/\partial x)u' + (1/\underline{\rho})\partial p'/\partial x = 0 \tag{15.6.7}$$

$$(\partial/\partial t + \underline{u}\partial/\partial x)p' + \gamma\underline{p}\partial u'/\partial x = 0 \tag{15.6.8}$$

Differentiating (15.6.7) with respect to x and eliminating $\partial u'/\partial x$ from (15.6.8), we obtain

$$(\partial/\partial t + \underline{u}\partial/\partial x)^2 p' - (\gamma\underline{p}/\underline{\rho})\partial^2 p'/\partial x^2 = 0 \tag{15.6.9}$$

The differential equation (15.6.9) is a wave equation. We assume a solution of the form

$$p' = A\exp\{ik(x - ct)\} \tag{15.6.10}$$

Substituting in (15.6.10), we find that the phase velocity c must satisfy the relation

$$(-ikc + i\underline{u}k)^2 - (\gamma\underline{p}/\underline{\rho})(ik)^2 = 0 \tag{15.6.11}$$

Solving (15.6.11) for c, we get

$$c = \underline{u} \pm \sqrt{(\gamma\underline{p}/\underline{\rho})} = \underline{u} \pm \sqrt{(\gamma R\underline{T})} \tag{15.6.12}$$

Equation (15.6.10) is, therefore, a solution of (15.6.9) provided that the phase velocity is given by the relation (15.6.12) which states that the adiabatic speed of the sound wave relative to the zonal wind is $\pm\sqrt{(\gamma\underline{p}/\underline{\rho})}$ or $\pm\sqrt{(\gamma R\underline{T})}$. The mean zonal wind in (15.6.12) plays the role of Doppler shifting the sound wave, so that frequency of the wave is given by the relation

$$\mu = kc = k\underline{u} \pm k\sqrt{(\gamma R\underline{T})} \tag{15.6.13}$$

According to (15.6.13), the frequency of the sound wave will appear to be higher when an observer is downstream of a source than when he is upstream.

(b) The gravity wave

Most of us are familiar with the type of waves that are produced on the surface of water in a pond when, for example, a stone is thrown into it. Where the stone drops, the energy of the impact creates a set of waves at the water surface which propagate outward in all directions in ever-widening circles till the energy is dissipated by friction. This is clearly a case of gravity waves the formation of which can be understood in the following way (see a schematic in Fig. 15.3).

Let us take a vertical section along the x-axis through a pond and assume that a stone falls at a point O on the water surface where it creates a depression of the surface to a certain depth. The removal or divergence of water from O lowers the hydrostatic pressure at O but increases it at the side point A where water is diverted

15.6 Simple Wave Types

Fig. 15.3 Vertical section illustrating the formation of water waves and their propagation

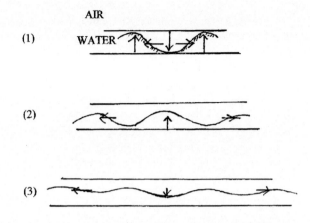

to and converges. The gradient of pressure then drives the water back to O thereby raising the pressure at O due to convergence but lowering it at A due to divergence from there. As a result of this movement, the water level rises at O but falls at A. The fall of pressure at A then in turn leads to convergence at A but divergence at the side point B. In this way, the pattern of divergence and convergence set in motion by the impact of the stone propagates outward in the form of a pressure or gravity wave. It should be noted that pressure at individual points increases (decreases) as the water level rises (falls) as the wave passes.

In what follows, we use a two-layer model to derive an expression for the phase velocity of a gravity wave moving along the interface between two layers of fluids of different densities. We assume that the layers are homogeneous incompressible fluids of constant densities ρ_2 in the upper layer and ρ_1 in the lower, with $\rho_1 > \rho_2$. The stratification is, therefore, vertically stable. In a geophysical fluid, however, the assumption of a uniform density in a medium, such as the ocean or the atmosphere, is seldom valid, though across their interface, sharp differences in their density do occur. So, the treatment given here would apply to such an interface between two media with sharp differences in densities. The working of the model is shown schematically in Fig. 15.4, which is a vertical section in the x-z plane through the interface.

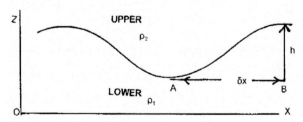

Fig. 15.4 Gravity wave in a two-layer model

For simplicity, we assume that there is no horizontal pressure gradient in the upper layer. We can get the horizontal pressure gradient in the lower layer by vertical integration of the hydrostatic equation. Thus, between the points B and A in Fig. 15.4 we have the pressure difference

$$\delta p = p_B - p_A = g(\rho_1 - \rho_2)(\partial h/\partial x)\delta x$$

where $\partial h/\partial x$ is the slope of the interface and δx is the distance between the points.

In the limit $\delta x \to 0$, the pressure gradient $= g(\Delta\rho)\partial h/\partial x$, where $\Delta\rho = \rho_1 - \rho_2$.

We use this value of the pressure gradient in the x- momentum equation and obtain

$$\partial u/\partial t + u\partial u/\partial x + w\partial u/\partial z = -g(\Delta\rho/\rho_1)\partial h/\partial x \tag{15.6.14}$$

The continuity equation in this case is

$$\partial u/\partial x + \partial w/\partial z = 0 \tag{15.6.15}$$

Since the pressure gradient is independent of z, u is also independent of z, provided that $u \neq u(z)$ initially. This makes $\partial u/\partial z = 0$ in (15.6.14). On vertical integration of (15.6.15) from $z = 0$ to $z = h$, therefore, we get

$$w(h) - w(0) = -\int_0^h \partial u/\partial x \delta z = -h\partial u/\partial x$$

If we assume the lower boundary to be a level surface, $w(0) = 0$. Further, since $w(h) = dh/dt = \partial h/\partial t + u\partial h/\partial x$, we can write the continuity equation (15.6.15) as

$$\partial h/\partial t + \partial(hu)/\partial x = 0 \tag{15.6.16}$$

The Eqs. (15.6.14) and (15.6.16) are a closed set in variables u and h and can be solved by the perturbation method.

We let

$$u = \underline{u} + u'; \text{ and } h = H + h'$$

where $\underline{u}$ is the constant basic state zonal velocity and H is the mean depth of the lower layer with $H \gg h'$. The perturbation forms of (15.6.14) and (15.6.16), after neglecting the product of the perturbation variables, are then

$$\partial h'/\partial t + \underline{u}\partial h'/\partial x + g(\Delta\rho/\rho_1)\partial h'/\partial x = 0 \tag{15.6.17}$$

$$\partial u'/\partial t + \underline{u}\partial u'/\partial x + H\partial u'/\partial x = 0 \tag{15.6.18}$$

Eliminating u' between (15.6.17) and (15.6.18), we obtain

$$(\partial/\partial t + \underline{u}\partial/\partial x)^2 h' - gH(\Delta\rho/\rho_1)\partial^2 h'/\partial x^2 = 0 \tag{15.6.19}$$

15.6 Simple Wave Types

Equation (15.6.19) is a wave equation in h'. We, therefore, seek a wave-type solution

$$h' = A \exp\{ik(x - ct)\}$$

Substituting in (15.6.19), we find that the assumed solution satisfies the equation, only if

$$c = \underline{u} \pm (g\, H\Delta\rho/\rho_1)^{1/2} \qquad (15.6.20)$$

If the two layers are air and water, $\Delta\rho/\rho_1 \simeq 1$, and (15.6.20) simplifies to

$$c = \underline{u} \pm (gH)^{1/2}$$

The expression $(gH)^{1/2}$ is called the shallow water wave speed for the simple reason that it can only be valid for a fluid in which the depth H is much smaller than the wave length, so that the vertical velocity remains small enough for the hydrostatic approximation to be valid. For an average ocean depth of 4 Km, the phase velocity of the surface gravity waves works out to be $200 \,\text{m s}^{-1}$.

However, according to (15.6.20), the phase velocity of the gravity waves along the interface between two layers very much depends upon the value of the ratio, $\Delta\rho/\rho_1$, besides the depth of the lower layer. The ratio can be approximated to 1 only as long as ρ_2 is negligible compared to ρ_1, as in the case between air and water. Gravity waves can also form inside the ocean where there is a slight density difference between the upper mixed layer and the lower and denser deep ocean along an interface called the thermocline. If we assume a value of 0.001 for the ratio $\Delta\rho/\rho_1$, it follows from (15.6.20) that the phase speed of the internal gravity waves that will occur in this case and travel along the thermocline will be only one-tenth of that of the surface gravity waves. The internal gravity waves in the ocean can travel both horizontally and vertically, but those travelling vertically are reflected from the upper and the lower boundaries to remain trapped inside the ocean as stationary waves.

(c) Rossby waves

The wave type that is most important in connection with the large-scale meteorological processes and directly related to observed weather and climate is the Rossby or planetary wave. Under barotropic conditions, the Rossby wave is an absolute vorticity conserving motion and caused by the variation of the Coriolis parameter with latitude, or what is called the β-effect.

We can qualitatively understand its propagation in the atmosphere in the following way (see Fig. 15.5). Here, we assume the atmosphere to be barotropic.

Let a chain of fluid parcels be initially located along a latitude circle where the Coriolis parameter is f_0 at time t_0. A fluid parcel is then displaced meridionally so as to reach latitude with Coriolis parameter f_1 after time t_1 after covering a distance δy.

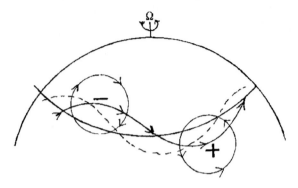

Fig. 15.5 Vorticity generation and westward propagation of Rossby waves

Since the absolute vorticity η is conserved following motion in a barotropic atmosphere(13.7.4) and it is given by the relation, $\eta = \zeta + f$, where ζ is the relative vorticity of the parcel, we can write down the relation between the vorticities before and after displacement as follows:

$(\zeta + f)_{t1} - f_{t0} = 0$, since the parcel has no vorticity at time t_0.

$$\text{Or, } \zeta_{t1} = f_{t0} - f_{t1} = -\beta \, \delta y \tag{15.6.21}$$

where $\beta = \partial f/\partial y$ is assumed to be constant.

The meaning of (15.6.21) is clear. The latitudinal displacement of the parcel generates perturbation relative vorticity, negative for northward displacement and positive for southward displacement from the initial latitude, as shown in Fig. 15.5.

The westward gradient of perturbation relative vorticity thus generated by the parcel oscillation forces the Rossby wave to move westward.

The dispersion relationship and the phase velocity of the barotropic Rossby waves can be derived formally by finding wave solutions of the linearized vorticity equation as follows: Since the absolute vorticity is conserved following the motion, we write its vertical component as

$$(\partial/\partial t + u\partial/\partial x + v\partial/\partial y)\zeta + v\partial f/\partial y = 0 \tag{15.6.22}$$

For simplicity, we assume that the motion consists of a basic state zonal velocity $\underline{u}$ and a horizontal perturbation (u', v') with perturbation relative vorticity $\zeta'(= \partial v'/\partial x - \partial u'/\partial y)$. Further, we define a perturbation stream function ψ so that $u' = -\partial \psi/\partial y$, and $v' = \partial \psi/\partial x$, and $\zeta' = \nabla^2 \psi$.

The perturbation form of (15.6.22) is then

$$(\partial/\partial t + \underline{u}\, \partial/\partial x)\nabla^2 \psi + \beta \partial \psi/\partial x = 0 \tag{15.6.23}$$

where $\beta \equiv \partial f/\partial y$ and we have neglected the products of the perturbation terms.

We assume a solution of (15.6.23) of the form

$$\psi = \text{Re}\{A \exp(i\phi)\} \tag{15.6.24}$$

where $\phi = kx + ly - \mu t$, k and l being the wave numbers in the zonal and the meridional directions respectively and μ the frequency.

Substitution of (15.6.24) into (15.6.23) yields

$$-(-\mu + k\underline{u})(k^2 + l^2) + k\beta = 0$$

which we solve for μ and obtain

$$\mu = \underline{u}\, k - \beta k/(k^2 + l^2) \qquad (15.6.25)$$

Since, $c_x = \mu/k$, we have finally

$$c_x - \underline{u} = -\beta/(k^2 + l^2) \qquad (15.6.26)$$

Thus, the Rossby waves propagate westward relative to the background zonal wind with a phase velocity which depends upon the wave numbers and hence increases with wavelengths. They are, therefore, dispersive waves. For very long waves, their westward phase velocity may at times be large enough to equal the basic state mean zonal wind velocity with the result that the waves become stationary with respect to the ground. If the wavelength exceeds the critical value for stationariness, the Rossby waves may actually retrogress. For a typical midlatitude disturbance with a zonal wavelength of 6,000 km and latitudinal width of 3,000 km, the Rossby wave speed relative to the zonal wind, computed from (15.6.26) is approximately $-6\,\mathrm{m\,s^{-1}}$.

Pure Rossby waves, however, are rare occurrences in the real atmosphere where other types of waves such as the gravity and sound waves are also excited. It is possible to use the linearized versions of the full primitive equations to study atmospheric waves following a procedure similar to that for the barotropic vorticity equation, though the procedure is rather involved. It is found that the free oscillations occurring in a hydrostatic gravitationally stable atmosphere consist of both westward and eastward moving gravity waves somewhat modified by the rotation of the earth and westward-propagating Rossby waves which are slightly modified by gravitational stability. These free oscillations are the normal modes of the atmosphere and they are continually excited by the various forces acting in the atmosphere.

15.7 Internal Gravity (or Buoyancy) Waves in the Atmosphere

Unlike the ocean which has a top boundary, the atmosphere has no real upper boundary. So, the case for the occurrence of internal gravity waves in the atmosphere is different from that in the ocean. In fact, the internal gravity waves in the atmosphere are nothing but buoyancy oscillations that form only in a stably stratified atmosphere and can propagate both horizontally and vertically. They are known to form on the leeside of mountains, which are called lee or mountain waves. They are believed to be an important mechanism for transporting momentum and energy to higher levels of the atmosphere. Clear air turbulence experienced by aircraft at high altitudes is also believed to be caused by vertically propagating internal gravity waves. In the equatorial stratosphere, they are believed to be responsible for eastward-propagating Kelvin waves and westward-propagating inertia-gravity waves, as well as the well-known quasi-biennial oscillation.

15.7.1 Internal Gravity (Buoyancy) Waves – General Considerations

Let us limit our discussion of internal gravity waves to an x-z plane in which the phase lines $\delta s(\phi = \text{const})$ of the parcel oscillations in the propagating wave are tilted at an angle α to the vertical (see Fig. 15.6).

It was shown in (3.4.12) that in the case of buoyancy oscillations in a stably stratified atmosphere, the buoyancy force acting on the parcel is given by the expression, $-N^2 \delta z$, where $N = \{(g/\theta)(\partial \theta/\partial z)\}^{1/2}$ is the frequency of the oscillations and z is the vertical co-ordinate. In the present case, the parcel oscillations occur in a direction which is not vertical but inclined to the vertical at an angle α. So, since $\delta z = \delta s \cos \alpha$, and its projection along δs is $\delta z \cos \alpha$, the buoyancy force on the oscillating particle is $-N^2(\delta s \cos \alpha) \cos \alpha$. The momentum equation of the parcel oscillation along δs may, therefore, be written

$$d^2(\delta s)/dt^2 = -(N \cos \alpha)^2 \delta s \tag{15.7.1}$$

The general solution of (15.7.1) is

$$\delta s = A \, \exp\{\pm i(N \cos \alpha) t\}$$

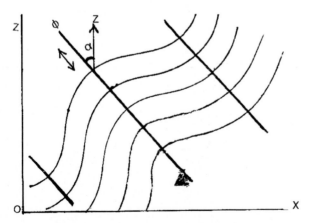

Fig. 15.6 Tilt of the path of parcel oscillations (constant phase φ - heavy lines) to the vertical by angle α. The *blunt arrow* shows the direction of phase propagation

15.7 Internal Gravity (or Buoyancy) Waves in the Atmosphere

Thus when the phase lines are inclined to the vertical by an angle α, the frequency of the buoyancy oscillations is given by $N \cos \alpha$. This heuristic result can be verified by considering the linearized equations for internal gravity waves in the x-z plane in a stably stratified incompressible atmosphere. We simplify the equations by making the Bossiness approximation in which the density is treated as constant except where it is coupled to gravity and the vertical scale of the motions is less than the scale height H(8 km).

We write the basic equations, neglecting rotation, as follows:
The momentum equations:

$$\partial u/\partial t + u\partial u/\partial x + W\partial u/\partial z + (1/\rho)\, \partial p/\partial x = 0 \quad (15.7.2)$$

$$\partial w/\partial t + u\partial w/\partial x + w\partial w/\partial z + (1/\rho)\, \partial p/\partial z + g = 0 \quad (15.7.3)$$

The continuity equation:

$$\partial u/\partial x + \partial w/\partial z = 0 \quad (15.7.4)$$

The thermodynamic energy equation:

$$\partial \theta/\partial t + u\partial \theta/\partial x + w\partial \theta/\partial z = 0 \quad (15.7.5)$$

where the potential temperature θ is given by

$$\theta = (p/\rho R)(p_s/p)^k, \quad (15.7.6)$$

where p_s is surface pressure and $k = R/c_p$.

We now linearize the above equations by assuming a motionless basic state with constant density ρ_0 and putting

$$p = \underline{p}(z) + p'; \ \rho = \rho_0 + \rho'; \ \theta = \underline{\theta}(z) + \theta'; \ u = u'; \ w = w' \quad (15.7.7)$$

The basic state pressure $\underline{p}(z)$ must then satisfy the relation

$$\partial \underline{p}/\partial z = -\rho_0 g \quad (15.7.8)$$

and the basic state temperature may be expressed as

$$\ln \underline{\theta} = \gamma^{-1} \ln \underline{p} - \ln \rho_0 + \text{const} \quad (15.7.9)$$

The linearized equations are then obtained by the perturbation technique, i.e., by substituting from (15.7.7) in Eqs. (15.7.2–15.7.5) and neglecting products of the perturbation variables. For example, we simplify the last two terms in (15.7.3) as

$$(1/\rho)\partial p/\partial z + g = \{1/(\rho_0 + \rho')\partial(\underline{p} + p')/\partial z\} + g$$
$$\simeq (1/\rho_0)(1 - \rho'/\rho_0)\partial \underline{p}/\partial z + (1/\rho_0)\partial p'/\partial z + g$$
$$= (1/\rho_0)\partial p'/\partial z + (\rho'/\rho_0)g \quad (15.7.10)$$

where (15.7.8) has been used to eliminate $\underline{p}$.

Similarly, the perturbation form of (15.7.9) may be obtained by noting that

$$\ln\{\underline{\theta}(1+\theta'/\underline{\theta})\} = \gamma^{-1}\ln\{\underline{p}(1+p'/\underline{p})\} - \ln\{\rho_0(1+\rho'/\rho_0)\} + \text{const} \quad (15.7.11)$$

which, with the aid of (15.7.9), can be simplified to

$$\theta'/\underline{\theta} \simeq \gamma^{-1}p'/\underline{p} - \rho'/\rho_0$$

Or,
$$\rho' \simeq -\rho_0\theta'/\underline{\theta} + p'/c_s^2 \quad (15.7.12)$$

where $c_s = \sqrt{(\gamma \underline{p}/\rho_0)}$ is the velocity of sound waves.

For buoyancy wave motions, the density fluctuations due to pressure change are much smaller than those due to temperature change, i.e., $|p'/c_s^2| \ll |\rho_0\theta'/\underline{\theta}|$

Hence, (15.7.12) may be approximated to

$$\theta'/\underline{\theta} = -\rho'/\rho_0 \quad (15.7.13)$$

Using (15.7.10) and (15.7.13), the linearized equations may now be written as

$$\partial u'/\partial t + (1/\rho_0)\partial p'/\partial x = 0 \quad (15.7.14)$$
$$\partial w'/\partial t + (1/\rho_0)\partial p'/\partial z - (\theta'/\underline{\theta})g = 0 \quad (15.7.15)$$
$$\partial u'/\partial x + \partial w'/\partial z = 0 \quad (15.7.16)$$
$$\partial \theta'/\partial t + w'\partial \underline{\theta}/\partial z = 0 \quad (15.7.17)$$

We eliminate p' from (15.7.14) and (15.7.15) by subtracting $\partial(15.7.14)/\partial z$ from $\partial(15.7.15)/\partial x$ and get the relation

$$\partial(\partial w'/\partial x - \partial u'/\partial z)/\partial t - (g/\underline{\theta})\partial \theta'/\partial x = 0 \quad (15.7.18)$$

We now use (15.7.16) and (15.7.17) to eliminate u' and θ' and obtain

$$\partial^2(\partial^2 w'/\partial x^2 + \partial^2 w'/\partial z^2)/\partial t^2 + N^2\partial^2 w'/\partial x^2 = 0 \quad (15.7.19)$$

where $N^2 \equiv g\, d(\ln\underline{\theta})/dz$ is the square of the buoyancy frequency and is assumed to be constant.

Equation (15.7.19) is a wave equation in w'. We seek its wave solution of the form

$$w' = \text{Re}\{A\exp(i\phi)\} \quad (15.7.20)$$

where $\phi = kx + mz - \mu t$ is the phase in which k and m may be regarded as the components of a vector(k,m) along the x- and the z-axis respectively and μ is the frequency of the wave. In terms of wavelengths, $k = 2\pi/\lambda_x$ and $m = 2\pi/\lambda_z$, where λ_x and λ_z are the wavelengths along the respective axis.

15.7 Internal Gravity (or Buoyancy) Waves in the Atmosphere

By substituting (15.7.20) in (15.7.19) we obtain the dispersion relatioship

$$\mu = \pm Nk/(k^2+m^2)^{1/2} \qquad (15.7.21)$$

In terms of wavelengths,

$$\mu = \pm N\lambda_z/(\lambda_x^2+\lambda_z^2)^{1/2}$$
$$= \pm N\cos\alpha \qquad (15.7.22)$$

where α is the angle between the phase line $\phi = $ const and the vertical (Fig. 15.6).

Obviously, (15.7.22) is in agreement with the parcel oscillation frequency heuristically derived earlier (15.7.1). Thus, the tilt of the phase lines of the internal gravity waves depends only on the ratio of the wave frequency to the buoyancy frequency (μ/N) and is independent of the wavelength.

If we let $k > 0$ and $m < 0$, it follows that for the phase $\phi = kx + mz$ to remain constant, an increase of x must be accompanied by an increase of z. This means that in this case the phase lines tilt eastward with height, as shown in Fig. 15.7.

The choice of μ positive in (15.7.21), therefore, implies that the phase propagation is eastward and downward in this case with the horizontal and the vertical components of the phase velocity being given respectively by: $c_x = \mu/k$ and $c_z = \mu/m$. The components of the group velocity C_g are:

$$C_{gx} = \partial\mu/\partial k = Nm^2/(k^2+m^2)^{3/2} > 0$$
$$C_{gz} = \partial\mu/\partial m = -Nkm/(k^2+m^2)^{3/2} > 0 \qquad (15.7.23)$$

Since the energy propagates with the group velocity, it follows from (15.7.23) that for internal gravity waves downward phase propagation is accompanied by upward energy propagation.

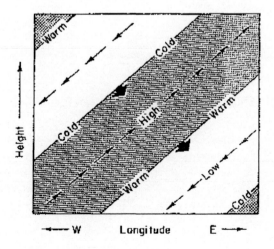

Fig. 15.7 Idealized cross-section in the x-z plane showing the phases of the pressure, temperature and velocity perturbations for an internal gravity wave. *Thin arrows* show the perturbation velocity field and and the *blunt arrows* the phase velocity (Reproduced from Wallace and Kousky, 1968, with permission of American Meteorological Society).

15.7.2 Mountain Lee Waves

A well-known example of internal gravity waves in the atmosphere are lee waves which form on the leeside of mountains when air parcels are forced to rise against a north-south oriented mountain range in a thermally stable atmosphere. If the vertical motions associated with these waves are strong enough and the air is sufficiently moist, condensation may occur in the rising parts of the waves which become visible in the form of rows of lee clouds. Such cloud formations are common occurrences on the leeside of mountains such as the French Alps (Gerbier and Berenger, 1961) and several other mountain ranges of the world.

Since lee waves are observed to remain stationary with respect to the ground under prevailing wind conditions, it follows that to an observer moving with the westerly wind, the constant phase lines of the lee waves will appear to be gradually moving westward. This means that the constant phase lines will tilt westward with height, allowing phase to move downward and wave energy to be transported upward. This westward tilt of the phase lines is shown in Fig. 15.6. To an observer moving with the mean zonal wind $\underline{u}$, the frequency of the lee waves generated by a sinusoidal-shaped mountain is given by: $\mu = -\underline{u}/k$.

Thus, from (15.7.21),

$$\underline{u} = N/(k^2 + m^2)^{1/2} \qquad (15.7.24)$$

We can calculate the value of m from (15.7.24) for given values of $\underline{u}$, k and N to determine the tilt of the phase lines with height. Since waves propagate vertically, we must have $m^2 > 0$ (that is, m is positive). With a positive m, it follows from (15.7.24) that $\underline{u}$ must be less than N/k.

From this, we conclude that conditions which are favourable for the formation of lee waves are generally: a wide mountain range (small k), strong thermal stability (large N), and light mean zonal wind (small $\underline{u}$) in the vertical. At times, when the atmosphere above the mountain height is stably stratified under a strong thermal inversion, the amplitudes of the lee waves can be quite powerful to cause strong downslope winds and intense clear air turbulence on the ground behind the mountain.

15.8 Dynamics of Shallow Water Gravity Waves

A problem of fluid dynamics of great geophysical interest is how mass and velocity fields adjust to each other in a rotating fluid the depth of which is much smaller than the horizontal scale of the perturbation of the flow. In a series of papers, Rossby (1937, 1938a, b; 1940) addressed this problem and threw light on several of the flow characteristics. In this section, we give a brief review of some of his findings, following a treatment by Gill (1982).

15.8.1 The Adjustment Problem – Shallow Water Equations in a Rotating Frame

Let us consider a rotating fluid that is initially at rest relative to a frame of reference that is rotating with a uniform angular velocity f/2 (where f is the Coriolis parameter) about a vertical axis. The fluid motion is considered relative to this frame and is supposed to be a small perturbation of height h from the state of relative rest at all times. The horizontal scale of the perturbation is assumed to be large compared to the depth, so that the hydrostatic approximation (p' = −ρgh, where p' is perturbation pressure and ρ is fluid density) can be applied

The momentum equations, after making the hydrostatic approximation, are:

$$\partial u/\partial t - fv = -g\partial h/\partial x \qquad (15.8.1)$$

$$\partial v/\partial t + fu = -g\partial h/\partial y \qquad (15.8.2)$$

where (u, v) are the velocity components along the x, y axes respectively and are independent of depth.

The continuity equation is:

$$\partial h/\partial t = -H(\partial u/\partial x + \partial v/\partial y) \qquad (15.8.3)$$

where H is the undisturbed depth of the fluid.

Taking the divergence of the momentum equations (i.e., taking $\partial/\partial x$ of (15.8.1) and $\partial/\partial y$ of (15.8.2) and adding) and substituting from (15.8.3), we obtain

$$\partial^2 h/\partial t^2 - c^2(\partial^2 h/\partial x^2 + \partial^2 h/\partial y^2) + f H \zeta = 0 \qquad (15.8.4)$$

where we have put $c^2 = gH$, and $\zeta = (\partial v/\partial x - \partial u/\partial y)$.

Since ζ is the relative vorticity of the fluid, we need to know how it changes with time in order to solve (15.8.4). In this regard, the principle of conservation of potential vorticity (13.5.6) is of fundamental importance. By taking the curl of the momentum equations (15.8.1–15.8.2) and eliminating divergence with the aid of the continuity equation (15.8.3), we obtain the relation

$$\partial(\zeta/f - h/H)/\partial t = 0 \qquad (15.8.5)$$

which states that potential vorticity is invariant with time.

We now define a quantity Q', given by

$$Q' = (\zeta/H - fh/H^2), \qquad (15.8.6)$$

and call it the perturbation potential vorticity. Since (15.8.5) expresses the fact that Q' retains its initial value at each point at all times, we have the relation

$$Q'(x, y, t) = Q'(x, y, 0) \qquad (15.8.7)$$

Equation (15.8.7) signifies an infinite memory of an inviscid rotating fluid to retain its initial perturbation potential vorticity and can be exploited to find an equilibrium solution for a particular initial state without considering details of the transient motion at finite times in between.

Following Rossby (1938a), we consider a case for which Q' is nonzero. The particular initial conditions that we consider are: $u = v = 0$ and the surface elevation h is given by the expression,

$$h = -h_0 \, \text{sgn}(x), \qquad (15.8.8)$$

where sgn(x) is the sign function (sign of x) defined by
sgn(x) = 1 or -1, according as $x < 0$ or $x > 0$
The integral of (15.8.5) in this case is

$$\zeta/f - h/H = (h_0/H)\text{sgn}(x) \qquad (15.8.9)$$

Substitution of (15.8.9) in (15.8.4) gives an equation in h alone, viz.,

$$\partial^2 h/\partial t^2 - c^2(\partial^2 h/\partial x^2 + \partial^2 h/\partial y^2) + f^2 h = -f\,H^2 Q'(x,y,0) = -f^2 h_0 \, \text{sgn}(x) \qquad (15.8.10)$$

15.8.2 The Steady-State Solution: Geostrophic Adjustment

We now assume that the gravitational adjustment process leads ultimately to a steady state which can be given by the time-independent solution of (15.8.10). Further, we assume that in the steady state, a geostrophic balance is reached in which the pressure gradient is balanced by the Coriolis acceleration. Since the initial condition is independent of y, we assume that the solution at all subsequent times will be independent of y. Thus, from (15.8.1) and (15.8.2), we obtain the geostrophic balance

$$\begin{aligned} f\,u &= -g\partial h/\partial y, \\ f\,v &= g\partial h/\partial x \end{aligned} \qquad (15.8.11)$$

Equation (15.8.11) has the property that the flow is along the isobars.

Using (15.8.6) and (15.8.7) and substituting from (15.8.9), we get the steady-state version of (15.8.10) as

$$-c^2(\partial^2 h/\partial x^2 + \partial^2 h/\partial y^2) + f^2 h = -f\,H^2 Q'(x,y,0) = -f^2 h_0 \, \text{sgn}(x) \qquad (15.8.12)$$

For the present case, we ignore the y-dependence in (15.8.12) and get

$$-c^2 \partial^2 h/\partial x^2 + f^2 h = -f^2 h_0 \text{sgn}(x) \qquad (15.8.13)$$

The solution of (15.8.13) that is continuous and antisymmetric about $x = 0$ is given by

$$h/h_0 = \begin{bmatrix} -1+\exp(-x/a) & \text{for } x > 0 \\ 1-\exp(x/a) & \text{for } x < 0 \end{bmatrix} \qquad (15.8.14)$$

15.8 Dynamics of Shallow Water Gravity Waves

where a is a length scale of great importance for the behaviour of rotating fluids subject to gravitational restoring forces and is given by

$$a = c/|f| = (gH)^{1/2}/|f| \tag{15.8.15}$$

The length scale 'a' is called the Rossby radius of deformation, in honour of Rossby. The modulus sign is used in (15.8.15) to ensure that 'a' is always positive, since f can have either sign.

The velocity field associated with the solution (15.8.15) may be derived from the geostrophic relations (15.8.11), which gives

$$v = -(gh_0/fa)\exp(-|x|/a) \tag{15.8.16}$$

Equation (15.8.16) shows that the flow is not in the direction of the pressure gradient but at right angles to it along the contours of the surface elevation that are parallel to the line of initial discontinuity.

Thus, the geostrophic equilibrium solution corresponding to adjustment from an initial state that is one of rest but has uniform infinitesimal surface elevation of $-h_0$ for $x > 0$ and elevation $+h_0$ at $x < 0$ has the following characteristics:

(a) the equilibrium surface level h tends towards the initial level $-h_0$ as x tends to plus infinity, $+h_0$ as x tends to minus infinity.
(b) The corresponding equilibrium velocity distribution has a speed maximum directed along the initial discontinuity in level with maximum velocity equal to $h_0(g/H)^{1/2}$.

15.8.3 Energy Transformations

Energy transformations in a rotating fluid differ from those in a nonrotating fluid. To show this, we first consider the nonrotating case.

(a) Energetics of shallow water motion in a nonrotating frame (f = 0)

The momentum equations (15.8.1 and 15.8.2) in this case are reduced to:

$$\partial u/\partial t = -g\partial h/\partial x \tag{15.8.17}$$

$$\partial v/\partial t = -g\partial h/\partial y \tag{15.8.18}$$

and the continuity equation (15.8.3) is:

$$\partial h/\partial t = -H(\partial u/\partial x + \partial v/\partial y) \tag{15.8.19}$$

The rate of change of kinetic energy of the flow per unit area, obtained by multiplying (15.8.17) by ρHu and (15.8.18) by ρHv and adding, is:

$$\partial\{(1/2)\rho H(u^2+v^2)\}/\partial t = -\rho g\{Hu(\partial h/\partial x)+Hv(\partial h/\partial y)\} \qquad (15.8.20)$$

The potential energy per unit area is given by

$$\int_{-H}^{h} \rho g z \, dz = (1/2)\rho g(h^2-H^2) \qquad (15.8.21)$$

From (15.8.21), the perturbation potential energy per unit area is $(1/2)\rho g h^2$.

Thus, the rate of change of perturbation potential energy per unit area is obtained by multiplying (15.8.19) by ρgh to give

$$\partial\{(1/2)\rho g h^2\}/\partial t = -\rho g\{h\partial(uH)/\partial x + h\partial(vH)/\partial y\} \qquad (15.8.22)$$

The equation for total (kinetic + potential) perturbation energy is obtained by adding (15.8.20) and (15.8.22) to give

$$\partial\{(1/2)\rho H(u^2+v^2)+(1/2)\rho g h^2\}/\partial t + \partial(\rho g Huh)/\partial x + \partial(\rho g Hvh)/\partial y = 0 \qquad (15.8.23)$$

In the special case in which there is no variation with y, the integral of (15.8.23) with respect to x in the region $|x|<X$ yield

$$\partial E/\partial t + F(X,t) - F(-X,t) = 0 \qquad (15.8.24)$$

where

$$E = \int_{-X}^{X} \{(1/2)\rho H(u^2+v^2)+(1/2)\rho g h^2\}dx,$$

is the total perturbation energy per unit length in the y direction in the region $|x|<X$, and $F(x,t) = \rho g Huh$ is the rate per unit length in the y direction of transfer of energy in the x direction at point x.

Now, the perturbation potential energy per unit area is $(1/2)\rho g h_0^2$ in the undisturbed part of the fluid where the kinetic energy is zero. But, after the perturbation moves away, the potential energy drops to zero but reappears as kinetic energy. It is shown below that the kinetic energy so generated is exactly equal to the potential energy lost. To show this, we first solve the wave equation (15.8.4) with $f=0$, by using the initial condition at rest, i.e., $h = G(x)$, where G is a function of x, and arrive at the solution

$$h = (1/2)\{G(x+ct)+(G(x-ct)\} \qquad (15.8.25)$$

where c is the phase velocity of the gravity wave($=(gH)^{1/2}$).

15.8 Dynamics of Shallow Water Gravity Waves

We then use (15.8.25) to integrate the momentum equation (15.8.17) with respect to time and obtain

$$u = gh_0/c \tag{15.8.26}$$

where we have used the initial condition, $h = -h_0 \, \text{sgn}(x)$.

Substitution for u from (15.8.26) in the expression for kinetic energy shows that Kinetic energy $(1/2)\rho H u^2 = (1/2)\rho H(gh_0/c)^2 = (1/2)\rho g h_0^2$ (potential energy).

Thus, there is no change in total energy, but only a conversion of potential energy into kinetic energy that occurs at the time the wave front passes a point. In this way, after a sufficiently long time, all the perturbation potential energy over a fixed region will be converted into the kinetic energy of the steady current that remains after the wave front has passed. The final state is, therefore, one of rest with a horizontal free surface and all the energy initially present is said to have been lost by 'radiation' in the form of gravity waves.

(b) Energetics of shallow water motion in a rotating frame

We can now derive the energy equations in the rotating case in the same way as in the nonrotating case by multiplying (15.61) by $\rho H u$ and (15.62) by $\rho H v$ and adding. The sum eliminates the terms that come from the Coriolis acceleration, so that the rotation terms do not appear explicitly in the energy equations which, therefore, have exactly the same form as in the nonrotating case. The energy conversions in the rotating case, however, are very different from those considered above in the nonrotating case, since the adjustment process is drastically affected by rotation.

Consider first the perturbation potential energy. This is infinite initially and remains so even when a steady equilibrium solution is established (assuming such an equilibrium does occur). However, there is a finite loss of potential energy in attaining the equilibrium position. The loss of potential energy per unit length is given by

$$\text{Loss of potential energy} = 2(1/2)\rho g h_0^2 \int_0^\infty [1 - \{1 - \exp(-x/a)\}^2] \, dx$$

$$= (3/2)\rho g h_0^2 a \tag{15.8.27}$$

In the nonrotating case, all the potential energy available in the initial perturbation is converted into kinetic energy. In the rotating case, only a finite amount of the available potential energy is transformed into kinetic energy.

The amount of kinetic energy per unit length found in the equilibrium solution is given by

$$\text{Kinetic energy per unit length} = 2(1/2)\rho H(gh_0/fa)^2 \int_0^\infty \exp(-2x/a) \, dx$$

$$= (1/2)\rho g h_0^2 a \tag{15.8.28}$$

Thus, only one-third of the potential energy lost in attaining equilibrium is converted into kinetic energy. What happens to the other two-thirds? Rossby (1938) suggests that a fluid particle must "continue its displacement beyond the equilibrium point until an excessive pressure gradient develops which forces it back." The result is an inertial oscillation around the equilibrium position.

It is now known that these so-called 'inertia' oscillations which occur around the equilibrium position during the process of geostrophic adjustment are only transient features in a rotating fluid. The energy analysis carried out above, even though only partially completed, throws light on the behaviour of rotating fluids under influence of gravitational forces. It shows that energy is hard to extract from a rotating fluid. In the problem considered, only one-third of the available potential energy could be converted into kinetic energy, the reason being that a geostrophic equilibrium was established and that such equilibrium retains potential energy- an infinite amount in the case studied. It also shows that the steady equilibrium solution is not one of rest, but only a geostrophic balance, i.e., a balance between the Coriolis force and the pressure gradient force.

15.8.4 Transient Oscillations – Poicaré Waves

A complete solution of the adjustment problem (15.8.10) requires that to the solution of the homogenous equation

$$\partial^2 h/\partial t^2 + c^2(\partial^2 h/\partial x^2 + \partial^2 h/\partial y^2) + f^2 h = 0 \qquad (15.8.29)$$

we add a particular solution, which can be taken as the steady solution $h_{steady}(x)$ given by (15.8.14). This means that the solution of (15.8.29) must satisfy the initial condition

$$h = -h_0 \, \text{sgn}(x) - h_{steady}$$

i.e., $$h = -h_0 \exp(-|x|/a) \, \text{sgn}(x) \text{ at } t = 0 \qquad (15.8.30)$$

We assume that (15.8.29) has a wave-like solution of the form

$$h \propto \exp\{i(kx + ly - \mu t)\}$$

On substitution in (15.8.29), we get the dispersion relation

$$\mu^2 = f^2 + (\kappa_H c)^2$$

Or, $$(\mu/f)^2 = 1 + (\kappa_H a)^2 \qquad (15.8.31)$$

where μ is frequency, and $\kappa_H = (k^2 + l^2)^{1/2}$ is the horizontal wave number.

15.8 Dynamics of Shallow Water Gravity Waves

Waves having this dispersion relation (both k and l real) are referred to as Poicaré waves (Gill, 1982). In meteorology, they are usually called simply as gravity waves in a rotating frame.

Thus, the properties of Poicaré waves depend on how the wavelength compares with the Rossby radius of deformation. The limiting cases are as follows:
(a) For short waves ($\kappa_H a \gg 1$), (15.8.31) approximates to

$$\mu \sim \kappa_H c \tag{15.8.32}$$

This means that short waves are just ordinary nondispersive shallow water waves. However, since the shallow water theory requires that the horizontal scale of the waves should be large compared with the depth of the fluid, it follows that the Rossby radius of deformation of the waves should be still larger compared to the depth. This condition is usually satisfied in the atmosphere and ocean.
(b) For long waves ($\kappa_H a \ll 1$), (15.8.31) approximates to

$$\mu = f \tag{15.8.33}$$

which means that the frequency is approximately constant and equal to f.

Therefore, in this limit, gravity has no effect and the fluid particles move under their own inertia. For this reason, f is often called the 'inertial' frequency.

The group velocity c_g of the Poicaré waves has a maximum value of c in the short wave limit, whereas it tends to be zero in the longwave limit.

15.8.5 Importance of the Rossby Radius of Deformation

The importance of the Rossby radius of deformation in adjustment problems can hardly be over-emphasized. In the early stages of adjustment from an initial discontinuity, the change of level is confined to a short distance, the pressure gradient is large and gravity dominates the behaviour. In other words, at scales small compared to the Rossby radius, the adjustment process is approximately the same as in a nonrotating system. Later, however, as the change of level spreads to a distance comparable to the Rossby radius, the rotational effect becomes as important as the effect of gravity. This is the stage of geostrophic adjustment between gravity and rotational effects. When the change of level spreads still farther to a distance large compared to the Rossby radius, the pressure gradient becomes unimportant and rotational effect dominates the behaviour.

In geostrophic flow, the Rossby radius of deformation provides a scale for which the two terms in (15.8.6) contributing to the perturbation potential vorticity Q', are of the same order. This can be seen from the following considerations of proportionality between the two terms for Q', one representing the vorticity ζ of the flow and the other the height h of the surface of discontinuity in the case of a sinusoidal variation of the surface of discontinuity of wavenumber κ_H:

$$(-\zeta/f)/(h/H) = (\kappa_H a)^2/1 \tag{15.8.34}$$

where the vorticity $\zeta = (g/f)(\partial^2 h/\partial x^2 + \partial^2 h/\partial y^2)$ is obtained with the aid of (15.8.1) and (15.8.2) under the assumption that the steady state solution besides being geostrophic is also nondivergent. According to (15.8.34), when $\kappa_H a \gg 1$, i.e. when the scale of the perturbation κ_H^{-1} is much smaller than the Rossby radius of deformation a, the vorticity term dominates, For perturbation of scale much larger than the Rossby radius of deformation, $\kappa_H a \ll 1$, and the surface elevation term dominates.

Thus, the ratio (15.8.34) gives not only the partition of perturbation potential vorticity, but also the partition of energy between kinetic and potential, total energy remaining invariant. This may be seen by multiplying the terms on the left-hand side of (15.8.34) by $(1/2)\rho g H h$ and integrating over a wavelength, using for vorticity the expression $\zeta = (g/f)(\partial^2 h/\partial x + \partial^2 h/\partial y^2)$.

The results are:

$$\text{The first term} = -(1/2)\rho(g/f)^2 H \int\int h(\partial^2 h/\partial x^2 + \partial^2 h/\partial y^2) dx\, dy$$

$$= (1/2)\rho(g/f)^2 H \int\int \{(\partial h/\partial x)^2 + (\partial h/\partial y)^2 - \partial(h\partial h/\partial x)/\partial x - \partial(h\partial h/\partial y)/\partial y\} dx\, dy$$

$$= (1/2)\rho H \int\int (u^2 + v^2) dx\, dy = \text{Kinetic energy(K.E)}$$

The second term $= (1/2)\rho g \int\int h^2\, dx\, dy =$ Potential energy (P.E)
Thus,

$$\text{K.E./P.E.} = (\kappa_H a)^2/1 \tag{15.8.35}$$

which means that short wavelength geostrophic flow contains mostly kinetic energy, whereas in long wavelength geostrophic flow potential energy dominates.

In view of the above considerations, a distinction can be made between the processes of adjustment between the mass and the velocity fields at different length scales. Thus, it may be stated that at small scales ($\kappa_H a \gg 1$), the mass field adjusts to be in equilibrium with the velocity field which generates the vorticity. At large scales ($\kappa_H a \ll 1$), the velocity field adjusts to be in equilibrium with the mass field which produces the surface elevation.

Chapter 16
Equatorial Waves and Oscillations

16.1 Introduction

Internal gravity waves that we discussed in the previous chapter are short-period small-scale waves which are mainly due to buoyancy and the propagation of which is influenced by gravity. There are, however, some longer-period larger-scale waves in the atmosphere the propagation of which is affected by not only gravity but also the earth's rotation. These are called inertia-gravity waves. However, in the case of such waves, it can be shown that not all can propagate vertically in the atmosphere. The criterion for such vertical propagation is that the frequency of the waves must be greater than the Coriolis frequency. In midlatitudes, where the Coriolis frequency is large, many low-frequency long-period planetary-scale waves fail to meet this criterion and the waves are, so to say, trapped by the atmosphere. But the situation is different near the equator where there is little Coriolis control and the decreasing Coriolis frequency allows these long-period planetary-scale waves to be untrapped and move vertically. According to theory, these long-period vertically propagating waves generally must have very short vertical wavelengths.

There are, however, two types of vertically propagating waves whose wavelengths are large enough to be easily detected. These are the eastward-propagating Kelvin wave, and the westward-traveling mixed Rossby-gravity wave.

The Kelvin wave has a distribution of pressure and velocity which is symmetric about the equator and has little meridional velocity, whereas the mixed Rossby-gravity wave has a distribution of pressure and velocity which is antisymmetric about the equator but has a distribution of meridional velocity which is symmetric about the equator.

The horizontal distributions of pressure and velocity characteristic of these waves are shown in Fig. 16.1.

The other interesting low-frequency equatorial waves and oscillations that we examine in this chapter are the Quasi-Biennial Oscillation (QBO), the Madden Julian Oscillation (MJO) and the El Niño Southern Oscillation (ENSO). It is felt that a familiarity with tropical circulation systems and their properties will facilitate an understanding of these waves.

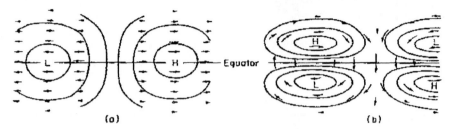

Fig. 16.1 Pressure and velocity distributions in the horizontal plane associated with: (**a**) Kelvin waves, and (**b**) mixed Rossby-gravity waves (After Matsuno, 1966, published by the Meteorological Society of Japan)

The dynamics of all these waves and oscillations can be deduced theoretically by using the log-pressure co-ordinate system in the governing equations and applying the linear perturbation technique introduced in Sect. 15.5. An introduction to the waves is furnished here, following a treatment given by Holton (1979).

16.2 The Governing Equations in Log-Pressure Co-ordinate System

In the log-pressure co-ordinate system, the vertical height z^* is defined by

$$z^* \equiv H \ln(p_0/p) \qquad (16.2.1)$$

where H is the scale height given by $H = RT_0/g$ with T_0 a global average temperature, and p_0 is a standard reference pressure usually taken to be 100 kPa. For an isothermal atmosphere at temperature T_0, z^* is exactly equal to the geometric height. For a variable temperature, they are only approximately equal. In this system, the vertical velocity w^* is given by

$$w^* \equiv dz^*/dt \qquad (16.2.2)$$

The governing equations in the log-pressure system are given by:

16.2.1 The Horizontal Momentum Equations

$$d\mathbf{V}/dt + f \mathbf{k} \times \mathbf{V} = -\nabla \Phi \qquad (16.2.3)$$

where the operator d/dt is now defined as

$$d/dt \equiv \partial/\partial t + \mathbf{V}.\nabla + w^* \partial/\partial z^*$$

16.2.2 The Hydrostatic Equation

$$\partial \Phi / \partial z^* = RT/H \qquad (16.2.4)$$

which is obtained with the aid of the ideal gas law (2.4.12).

16.2.3 The Continuity Equation

The continuity equation in the log-pressure system is obtained by transforming from the isobaric co-ordinate system as follows:

In the isobaric co-ordinate system, $\partial u/\partial x + \partial v/\partial y + \partial \omega/\partial p = 0$, where ω is the vertical p-velocity($= dp/dt$).
Since $w^* = dz^*/dt = -H\omega/p$, we can write

$$\partial \omega/\partial p = -\partial(pw^*/H)/\partial p = \partial w^*/\partial z - w^*/H$$

Thus, in the log-pressure system, the continuity equation is

$$\partial u/\partial x + \partial v/\partial y + \partial w^*/\partial z^* - w^*/H = 0 \qquad (16.2.5)$$

16.2.4 The Thermodynamic Energy Equation

The thermodynamic energy equation (11.9.1) in isobaric co-ordinates

$$dQ/dt = c_p \, dT/dt - \alpha \, dp/dt$$

when transformed into log-pressure co-ordinates, with the aid of (16.2.2), (16.2.3) and (16.2.5), takes the form

$$(\partial/\partial t + \mathbf{V} \cdot \boldsymbol{\nabla})(\partial \Phi/\partial z^*) + w^* N^2 = (\kappa/H) \, dQ/dt \qquad (16.2.6)$$

where $N^2 \equiv (R/H)(\partial T/\partial z^* + \kappa T/H)$, and $\kappa = R/c_p$.

In the stratosphere, N^2, the buoyancy frequency squared, is approximately constant with a value around $4 \times 10^{-4} \, s^{-2}$.

16.3 The Kelvin Wave

We now use Eqs. (16.2.3.), (16.2.4), (16.2.5) and (16.2.6) referring them to an equatorial β plane with the Coriolis parameter approximated by

$$f = 2 \, \Omega \, y/a \equiv \beta y$$

where y is the distance from the equator and a is the mean radius of the earth. This means that $\beta = \partial f/\partial y$.

We assume a basic state of the atmosphere which is at rest (Note that adding a constant zonal velocity simply Doppler-shifts the frequency) and has no diabatic heating.

The perturbations are assumed to be zonally propagating waves and may be written as

$$\begin{aligned} u &= u'(y,z^*) \exp(i(kx - \mu t)) \\ v &= v'(y,z^*) \exp(i(kx - \mu t)) \\ w^* &= w^{*\prime}(y,z^*) \exp(i(kx - \mu t)) \\ \Phi &= \Phi'(y,z^*) \exp(i(kx - \mu t)) \end{aligned} \quad (16.3.1)$$

where k and μ are respectively the wave number and frequency of the wave.

Substituting (16.3.1) in (16.2.3, 16.2.5 and 16.2.6) and linearizing, we get the perturbation equations

$$-i\mu u' - \beta y v' = -ik\Phi' \quad (16.3.2)$$
$$-i\mu v' + \beta y u' = -\partial \Phi'/\partial y \quad (16.3.3)$$
$$iku' + \partial v'/\partial y + (\partial/\partial z^* - 1/H)w^{*\prime} = 0 \quad (16.3.4)$$
$$-i\mu \partial \Phi'/\partial z^* + N^2 w^{*\prime} = 0 \quad (16.3.5)$$

Since, in a Kelvin wave, there is no meridional velocity, we put $v' = 0$ and eliminate $w^{*\prime}$ between (16.3.4) and (16.3.5) to get

$$-i\mu u' = -ik\Phi' \quad (16.3.6)$$
$$\beta y u' = -\partial \Phi'/\partial y \quad (16.3.7)$$
$$(\partial/\partial z^* - 1/H)\partial \Phi'/\partial z^* + (k/\mu)N^2 u' = 0 \quad (16.3.8)$$

We use (16.3.6) to eliminate Φ' from (16.3.7) and (16.3.8) and get two independent equations which the field of u' must satisfy. These are:

$$\beta y u' = -(\mu/k)\partial u'/\partial y \quad (16.3.9)$$

and

$$(\partial/\partial z^* - 1/H)\partial u'/\partial z^* + (k^2/\mu^2) N^2 u' = 0 \quad (16.3.10)$$

Equation (16.3.9) determines the meridional distribution of u' and (16.3.10) determines the vertical distribution.

It is easily verified that (16.3.9) has the solution

$$u' = u_0(z^*)\exp\{-(\beta k/2\mu)y^2\} \quad (16.3.11)$$

If we assume that $k > 0$, then $\mu > 0$ corresponds to an eastward propagating wave the amplitude of which is maximum at the equator and falls off exponentially with

16.4 The Mixed Rossby-Gravity Wave

y on either side. The field of u' in that case has a Gaussian distribution about the equator with an e-folding width given by

$$Y_L = |2\mu/\beta k| \qquad (16.3.12)$$

If we had taken $\mu < 0$, it would have resulted in a westward moving wave the amplitude of which would increase exponentially away from the equator. However, such a result would violate reasonable boundary conditions at the poles and must, therefore, be rejected. Thus, there exists only an eastward propagating atmospheric Kelvin wave.

According to Holton and Lindzen (1968), Kelvin waves may be defined as shallow water gravity waves which propagate parallel to a coastline and have no velocity component normal to the coastal boundary. The latter condition implies that the pressure gradient normal to the coastline be in geostrophic balance with the velocity field, which in turn requires that the amplitude of the wave decay exponentially away from the coast. In view of the similarity of the present solution (16.3.11) to Kelvin waves, it seems reasonable to call the wave an atmospheric "Kelvin" wave, noting that the equator plays the same role as a coastal boundary.

The solution of (16.3.10) which gives the vertical structure of the wave may be written in the form

$$u'(z^*) = u_0(z^*) \exp(z^*/2H)\{C_1 \exp(i\lambda z^*) + C_2 \exp(-i\lambda z^*)\} \qquad (16.3.13)$$

where $\lambda^2 \equiv (N^2 k^2/\mu^2) - (1/4H^2)$,
and C_1, C_2 are constants which are to be determined from appropriate boundary conditions.

For $\lambda^2 > 0$, the solution (16.3.13) is in the form of a vertically propagating wave and identical to an eastward propagating internal gravity wave that we discussed earlier in Sect. 15.7.1. Its eastward phase velocity has a downward component; hence the constant C_1 in (16.3.13) must be zero.

However, its group velocity, i.e., the direction of energy propagation, has an upward component. The Kelvin wave thus has a structure in the x, z plane which is shown in Fig. 16.2, which is identical to that shown earlier in Fig. 15.7.

16.4 The Mixed Rossby-Gravity Wave

We can make a similar analysis for the mixed Rossby-gravity wave. But, for this case, we must use the full perturbation equations (16.3.2–16.3.5). The solution corresponding to the pattern shown in Fig. 16.1(b) is

$$\begin{pmatrix} u' \\ v' \\ \Phi' \end{pmatrix} = \Psi(z^*)(1) \exp[\{-(1+k\mu/\beta)\beta^2 y^2\}/2\mu^2] \begin{pmatrix} +i\beta y(1+k\mu/\beta)/\mu \\ \\ +i\mu y \end{pmatrix} \qquad (16.4.1)$$

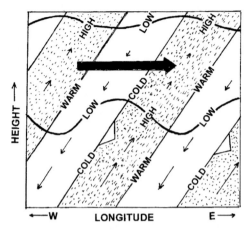

Fig. 16.2 Longitude-height section along the equator showing pressure, temperature and wind perturbations for a thermally damped Kelvin wave. Heavy wavy lines indicate material lines, short blunt arrows show phase propagation. Areas of high pressure are shaded. Length of small thin arrows is proportional to the wave amplitude which decreases with height due to damping. The large shaded arrow indicates the net mean flow acceleration due to the wave stress divergence

where the vertical structure $\Psi(z^*)$ of the three variables is given by

$$\Psi(z^*) = \exp(z^*/2H)\{C_1 \exp(i\lambda_0 z^*) + C_2 \exp(-i\lambda_0 z^*)\} \quad (16.4.2)$$

where
$$\lambda_0^2 \equiv (N^2 k^2/\mu^2)(1+\beta/k\mu)^2 - (1/4H^2),$$

and C_1, C_2 are constants which are to be determined by the boundary conditions.

Equation (16.4.1) shows that the mixed Rossby-gravity wave mode v' has a Gaussian distribution about the equator with an e-folding width given by

$$Y_L = |2\mu^2/\{\beta^2(1+\mu k/\beta)\}|^{1/2} \quad (16.4.3)$$

Equation (16.4.3) is valid for westward propagating waves ($\mu < 0$) provided that

$$(1+\mu k/\beta) > 0,$$

which condition can also be written in the form

$$|\mu| < 2\Omega/ak \quad (16.4.4)$$

In the case of frequencies which do not meet the condition (16.4.4), the wave amplitude will not decay away from the equator and hence will not satisfy boundary conditions at the pole.

For upward energy propagation, the mixed Rossby-gravity waves must have downward phase propagation just as in the case of Kelvin waves. Thus, the constant C_2 in (16.4.2) must be zero. The resulting wave structure in the x, z plane at a latitude north of the equator is shown in Fig. 16.3.

16.5 Observational Evidence

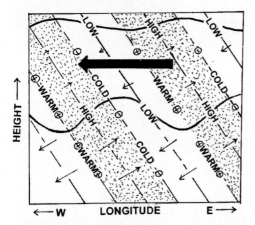

Fig. 16.3 Zonal-vertical (x, z) section along a latitude circle north of the equator showing the distribution of pressure, temperature and wind perturbations for a thermally damped mixed Rossby-gravity wave. Areas of high pressure are shaded. Small arrows indicate zonal and vertical wind perturbations with length proportional to wave amplitude. Meridional wind perturbations are shown by arrows pointed into the page (northward) and out of the page (southward). The large shaded arrow shows the direction of net mean flow acceleration due to the wave stress divergence

It is interesting to note from Figs. 16.1(b) and 16.3 that the mixed Rossby-gravity wave removes heat from the equatorial region to higher latitudes, since the poleward moving air is correlated with positive temperature perturbations so that the eddy heat flux $\langle v'T' \rangle$ is positive.

16.5 Observational Evidence

Both Kelvin and mixed Rossby-gravity wave modes have been identified in observations from the equatorial stratosphere. The observed Kelvin wave appears to have a period in the range 12–20 days, a zonal wave number 1, (i.e., one wave appears to span the whole latitude circle) and a phase speed of about $30\,\text{m s}^{-1}$ relative to the ground. If we assume a mean zonal wind of $-10\,\text{m s}^{-1}$ so that the Doppler-shifted phase velocity $c \sim 40\,\text{m s}^{-1}$, then we find from (16.3.12) that Y_L is about 2000 km. This agrees well with the observation that the significant amplitude of the Kelvin wave is largely confined within about 20° of latitude of the equator. Further, knowledge of the phase speed allows us to compute the vertical wavelength of the Kelvin wave. Assuming a value $N^2 = 4 \times 10^{-4}\,\text{s}^{-2}$ and using (16.3.13), we get vertical wavelength $\sim 2\pi/\lambda \sim 2\pi c/N \sim 12\,\text{km}$, which is in good agreement with the value of the vertical wavelength deduced from observations. An example of zonal wind oscillations caused by passage of Kelvin waves at Kwajalein, a station near the equator in western north Pacific, is shown by a time-height section in Fig. 16.4.

The existence of mixed Rossby-gravity mode has also been confirmed in the observational data over the equatorial Pacific. This mode is most easiliy identified

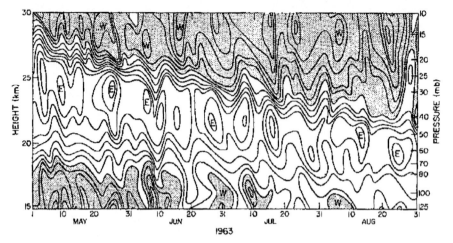

Fig. 16.4 Time-height section of zonal wind at Kwajalein (near 9N). Isotachs at intervals of $5\,\mathrm{m\,s^{-1}}$. Westerlies are shaded. (After Wallace and Kousky, 1968) (Reproduced with permission of the American Meteorological Society)

by the meridional wind component, since, according to (16.4.1), v' is a maximum at the equator ($y = 0$) for the mixed Rossby-gravity mode. The observed waves of this mode have periods in the range 4–5 days and propagate westward at about $20\,\mathrm{m\,s^{-1}}$. The horizontal wavelength appears to be about 10,000 km. The observed vertical wavelength is about 6 km which agrees closely with the theoretically derived wavelength from (16.4.2). These waves also appear to have significant wavelength within about $20°$ of the equator, which is consistent with the e-folding width Y_L derived from (16.4.3).

Theory and observations appear to suggest that both the Kelvin wave and the Rossby-gravity wave are excited by oscillations in the large-scale convective heating pattern in the equatorial troposphere. Though these waves do not contain much energy compared to other tropical disturbances such as storms and cyclones, they are the predominant disturbances of the equatorial stratosphere, and through their vertical energy and momentum transport play an important role in the maintenance of the general circulation of the stratosphere.

16.6 The Quasi-Biennial Oscillation (QBO)

Among the various types of atmospheric oscillations with which we are by now familiar from observations, the quasi-biennial oscillation of the zonal wind in the equatorial stratosphere is, perhaps, closest to one exhibiting true periodicity. It has the following observed features: Zonally-symmetric easterly and westerly wind regimes appear alternately with a period between 24 and 30 months. Successive regimes first appear above 30 km, but their phases propagate downward at a rate of

16.6 The Quasi-Biennial Oscillation (QBO)

about 1 km per month. The downward propagation occurs without change of amplitude between 30 and 23 km but there is rapid loss of amplitude below 23 km. The oscillation is symmetric about the equator with maximum amplitude of about 20 m s^{-1} and a half-width of about 12° of latitude. It is well depicted by a time-height section of mean zonal wind components (m s^{-1}) for Canton Island during the period February 1954 to October 1960, as shown in Fig. 16.5. One may note from Fig. 16.5 that the vertical shear of the wind is quite strong in the region where one regime is replacing the other. Since the oscillation is zonally symmetric and symmetric about the equator and has very small meridional as well as vertical motions associated with it, the zonal wind must be in geostrophic balance nearly all the way to the equator. Thus, there needs to be a strong meridional temperature gradient in the vertical shear zone to satisfy the thermal wind balance.

Several observed features of the quasi-biennial oscillation, such as its approximate biennial period, downward phase propagation without loss of amplitude, the occurrence of zonally symmetric westerlies at the equator, require theoretical explanations and factors responsible for them need to be identified. The occurrence of westerlies at the equator signifies that air is moving eastward faster than the earth's surface and, therefore, cannot be accounted for by advection of angular momentum from higher latitudes where the earth's angular momentum is less than that at the equator. So, there must be some eddy momentum source to create the westerly accelerations in the downward-moving westerly shear zone.

Both observational and theoretical studies have confirmed that vertically propagating equatorial waves – the Kelvin wave and the mixed Rossby-gravity waves - provide the zonal momentum sources needed to drive the quasi-biennial oscillation. Fig. 16.2 shows that eastward-propagating Kelvin waves with upward energy propagation transfer westerly momentum upward, i.e., $\underline{u'w'} > 0$. This means that u' and w' are positively correlated in the case of these waves. Fig. 16.3 shows that $\underline{u'w'} > 0$, also in the case of mixed Rossby-gravity waves. However, the mixed Rossby-gravity

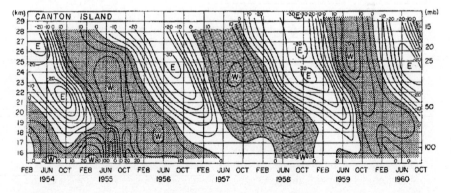

Fig. 16.5 Time-height section for Canton island (2°46'S, 171°43'W), February 1954–October 1960. Isopleths are monthly mean zonal wind components in m s^{-1}. Negative values denote easterly winds (After Reed and Rogers, 1962 Reproduced with permission of the American Meteorological Society)

mode has a strong positive horizontal heat flux, $\overline{v'T'} > 0$, which sets up a mean meridional circulation which through the Coriolis acceleration $fv' < 0$ produces a net easterly acceleration.

In this context, the two heavy wavy lines shown in Figs. 16.2 and 16.3 are highly significant. If we take the lower line as representing the vertical displacement of material particles at the tropopause level at different longitudes, the upper line shows that the amplitude of the displacement decreases with height in the stratosphere.

Theoretical calculations indicate that the equatorial stratospheric waves are thermally damped by infrared radiation to space so that the amplitude of the temperature wave decreases with height. Further, the extent of damping depends strongly on the doppler-shifted frequency of the waves. As the doppler-shifted frequency decreases, the vertical component of the group velocity also decreases, so a much longer time is available for the energy to be damped for a rise through a given vertical distance. Thus, the westerly Kelvin waves tend to be damped rapidly in westerly shear zones below their critical levels (a critical level is the altitude at which the relative horizontal phase speed of an internal gravity wave equals the mean wind speed). As a wave approaches this critical altitude, from below or above, the vertical component of the group velocity becomes zero and its energy is absorbed and transferred to the mean flow. This causes the westerly shear zone to descend with time. Similarly, the easterly mixed Rossby-gravity waves are damped in easterly shear zones, thereby causing an easterly acceleration and lowering of the easterly shear zone.

Thus, we may conclude that the quasi-biennial oscillation is, indeed, excited by the vertically propagating equatorial waves through the mechanism of radiative damping which causes the waves to decay in amplitude with height and thus transfer their energy to the mean zonal flow.

16.7 The Madden-Julian Oscillation (MJO)

This oscillation, named after its co-discoverers, Madden and Julian(1971, 1972), is a broad band intraseasonal tropical oscillation which was first detected in the co-spectrum of the 850- and 150-mb zonal wind (u) components in which a very pronounced negative coherence extreme was noted in the frequency range 0.0245–0.0100 day^{-1} (period 41–53 days). Their analysis based on nearly ten years of daily rawinsonde data for Canton Island (3S, 172W) revealed the following structure of this oscillation in the fields of wind, pressure and temperature:

(a) In the wind field, peaks in the variance spectra of the zonal wind were found to be strong in the lower troposphere, weak or non-existent in the 700–400 mb layer, and strong again in the upper troposphere. No evidence of this feature could be found above 80 mb, or in any of the spectra of the meridional component.
(b) In the pressure field, the spectrum of station pressure showed a peak in this frequency range and the oscillation was in phase with the lower-tropospheric zonal wind oscillation but out of phase with that in the upper troposphere.

16.7 The Madden-Julian Oscillation (MJO)

(c) The tropospheric temperatures exhibited similar peak and were highly coherent with the station pressure oscillation, positive station pressure anomalies being associated with negative temperature anomalies throughout the troposphere. Thus, the lower-middle troposphere appears to act as a nodal surface with u and surface pressure oscillating in phase, but 180° out of phase above or below.

Their study (Madden and Julian, 1972) also appears to indicate that the detected oscillation is of global scale but restricted to the tropics and is the result of an eastward movement of large-scale circulation cells oriented in the zonal plane. Fig. 16.6 shows (upper panel) the co-spectrum between the 850- and 150-mb u components, along with the co-spectrum between the station pressure and the 850-mb u component, as originally presented by Madden and Julian (1971). In both the co-spectra, one can see peaks in the frequency range of the MJO.

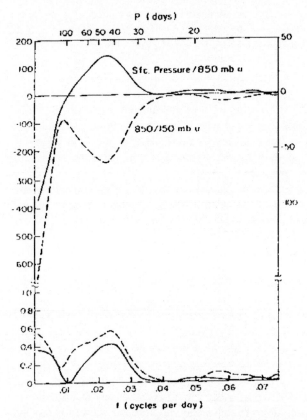

Fig. 16.6 (*Top*) The co-spectrum of the 850- and 150-mb zonal wind u(*dashed*, and *left* ordinate values), and co-spectrum of the station(sfc) pressure and the 850-mb zonal wind(*solid*, and *right* ordinate values) for Canton Island, June 1957 through March 1967. The ordinate is co-spectral density normalized to unit bandwidth in unit of $(m^2 \ s^{-1} \ day^{-1})$. (*Bottom*) The coherence-squared statistic for the 850- and 150-mb zonal wind and the station pressure and 850-mb u -series (Reproduced from Madden and Julian, 1971, with permission of American Meteorological Society)

The Meteorological glossary of the American Meteorological Society (2000) describes a Madden-Julian oscillation as follows: It is "a quasi-periodic oscillation of the near-equatorial troposphere, most noticeable in the zonal wind component of the boundary layer and in the upper troposphere, particularly over the Indian ocean and the western equatorial Pacific. It appears to represent an eastward propagating disturbance with the structure of a Kelvin wave with a vertical half-wavelength of the depth of the troposphere, but with a phase speed of only about 8m s^{-1}, much less than that of an adiabatic Kelvin wave in the stratosphere. The disturbance is accompanied by strong fluctuations of deep convection, easily detectable using satellite observations, and is a major contributor to intraseasonal weather variability in equatorial regions from eastern Africa eastward to the central Pacific." As is well-known, this region of the equator is dominated by monsoons.

Both theoretical and observational studies during the last three decades have concentrated on an understanding of the mechanism of excitation of the MJO by deep convection and how its eastward propagation as a Kelvin wave is affected by convection. Madden (1986), using a technique which he termed seasonally varying cross-spectral analysis concluded that the energy source for the 40–50 day oscillation is the large-scale convection associated with the seasonally-migrating intertropical convergence zone (ITCZ). Madden found that the variance of the zonal wind in a relatively broad band centered on a 47-day period was maximum during December, January and February (DJF) and at stations in the Indian and western Pacific oceans during all seasons. The lower- and upper-tropospheric zonal winds tend to be most coherent and out-of-phase at near-equatorial stations in the summer hemisphere. It is likely that this results from the dependence of the oscillation on the deep convection associated with the seasonal migration of the ITCZ.

An interesting aspect of this oscillation, as found by his study, is the involvement of the meridional v-component of the wind in transferring mass from the convective region of the summer hemisphere ITCZ to the winter hemisphere, as shown schematically in the upper part of Fig. 16.7.

Hendon and Salby (1994) constructed a composite life cycle of the MJO from the cross covariance between outgoing longwave radiation (OLR), wind and temperature, using episodes when a discrete signal in OLR is present. It was found that the composite convective anomaly possesses a predominantly zonal wavenumber 2 structure that is confined to the eastern hemisphere, propagates eastward at about 5m s^{-1} and evolves through a systematic cycle of amplification and decay. Unlike the convective anomaly, the circulation anomaly is not confined to the eastern hemisphere and travels faster. According to these workers, the circulation anomaly exhibits characteristics of both a forced response, coupled to the convective anomaly as it propagates across the eastern hemisphere, and a radiating response which propagates away from the convective anomaly into the western hemisphere at about $8–10 \text{m s}^{-1}$. The forced response appears as a coupled Rossby-Kelvin wave, while the radiating response displays predominantly Kelvin wave features. During amplification, the convective anomaly is positively correlated to the temperature perturbation, which means production of eddy available potential energy

16.7 The Madden-Julian Oscillation (MJO)

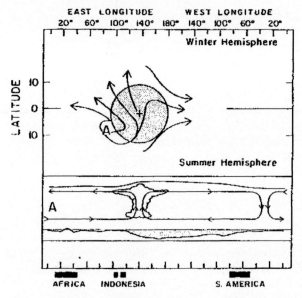

Fig. 16.7 Schematic of the structure of a large-amplitude 40–50 day oscillation in the equatorial plane The shading at the bottom of the lower panel represents the negative pressure anomalies. The line at the top of lower panel represents the tropopause. The upper panel is a plan view of the disturbance in the upper troposphere. The shaded area there corresponds to a positive anomaly in convection centered in the summer hemisphere. The divergence of the anticyclonic circulation transfers mass from summer to the winter hemisphere (Reproduced from Madden, 1986, with permission of American Meteorological Society)

(EAPE). A similar correlation between upper tropospheric divergence and temperature implies conversion of EAPE to eddy kinetic energy during this time. When it is decaying, temperature has shifted nearly into quadrature with convection, so their correlation and production of EAPE are then small. The same correspondence to the amplification and decay of the disturbance is observed in the phase relationship between surface convergence and convective anomaly. The correspondence of surface convergence to the amplification and decay of the convective anomaly suggests that frictional wave-CISK plays a key role in generating the MJO.

A numerical study by Chang and Lim (1988) shows that two types of CISK modes may arise from an interaction of two vertical modes of convection, shallow and deep. With maximum convective heating in the lower troposphere, the instability is due to the lower internal mode which gives a stationary east-west symmetrical structure. However, when heating is maximum in the midtroposphere, eastward propagating CISK modes resembling the observed and numerically-simulated oscillations occur. These modes arise from the interaction of the two internal modes which are locked in phase vertically. A time-lagged CISK analysis suggests that the shallower mode, with its stronger influence on the low-level moisture convergence, slows down the deeper mode resulting in a combined mode which has a deep vertical structure but a relatively slow propagating speed.

16.8 El Niño-Southern Oscillation (ENSO)

16.8.1 Introduction

There is yet another large-scale low frequency oscillation in the equatorial atmosphere and ocean which has an average life span of 2–7 years and may be found in varying intensity over all the global oceans. It shows itself up most prominently in the Pacific Ocean sector. In some longitudinal segments, it also influences events at higher latitudes through the tropical Hadley circulation. The variability is called the El Niño-Southern Oscillation (ENSO), with specific reference to the equatorial Pacific where the oscillation develops most strongly and has been the subject of intensive investigations. Prior to the 1960s, El Niño and Southern Oscillation were regarded as two separate phenomena, one (El Niño) involving the ocean and the other the atmosphere.

For ages, the Peruvian fishermen have known El Niño as a spectacular warm ocean current which flows southward along the coast of Peru for a few weeks every year replacing the usual northward-flowing cold upwelling ocean current. However, they noted a spectacular large-scale invasion of warm water in some years which lasted over a long period of a year or so before it was replaced by the usual cold current. The Southern Oscillation, on the other hand, was originally described by Gilbert Walker (1923, 1924, 1928) and his collaborators in the early part of the twentieth century when they detected a seesaw-type oscillation in atmospheric surface pressure between the eastern and western parts of the tropical oceans, especially in the Pacific, e.g., between Indonesia and Easter Island, which they sought to correlate with Indian monsoon rainfall.

However, it was only in the 1960s that Bjerknes (1966, 1969) and others demonstrated that the two oscillations are not separate but coupled to each other intimately via ocean-atmosphere interaction in which each constantly tries to adjust to the other across the interface in a bid to remain in thermodynamic equilibrium, inspite of their well-known differences in heat capacity and response times. They coined the term ENSO for the coupled oscillations. ENSO develops most strongly in the equatorial Pacific Ocean because of a pronounced zonal anomaly in ocean surface temperature between its eastern and western parts. In some years, the Pacific ENSO develops so strongly that its signal reaches distant parts of the globe. In what follows, we first describe El Niño and its opposite La Niña and then the SO separately and then describe their coupled effects and behaviour in the atmosphere as well as the ocean.

16.8.2 El Niño/La Niña

The climatological distribution of mean ocean surface temperature in the Pacific(see Fig. 2.2) reveals the existence of an extensive area of warm water called the warm pool (SST 28–30°C) in the western part of the equatorial Pacific (west of about

16.8 El Niño-Southern Oscillation (ENSO)

the dateline) as against its eastern part where the combined effects of the Humbolt current and intense upwelling along the coast of Peru maintain an SST which can be as low as 19°C.

However, notwithstanding this normal zonal distribution of SST, almost every year, a narrow warm ocean current known as the equatorial countercurrent flows eastward from the Western Pacific warm pool a few degrees north of the equator(5N–10N) between the westward flowing north and south equatorial currents and shifts southward closer to the equator during the months January to April and ushers in, though temporarily for a few weeks only, an influx of warm water to the coast of South America. During that short period, a branch of the warm current flowing southward along the coast of Peru overrides the usual northward-flowing cold water to a latitude of about 5S. The people of Peru especially its fishermen call this warm countercurrent 'El Niño', meaning in Spanish 'the male child', indirectly referring to 'Christ child', since it comes around Christmas time.

This usual pattern of onset of El Niño, however, is upset once every 2–7 years, by a major disturbance which appears to be related to changes in atmospheric and oceanic circulation and the flow of warm water along the coast of Peru is enhanced many times and extended further south along the coast to the latitude of Lima Callao at 12S or even beyond with disastrous effects on atmospheric and oceanic environment of the region. The surface temperatures of the coastal waters on such occasions rise at places by as much as 7°C. The gradual decrease of average water temperatures towards the south on such occasions indicates that there is considerable mixing of the warm waters with the ordinary cold coastal waters, and during this mixing the organisms in the coastal current, from plankton to anchovy fish which are used to cold temperatures, are destroyed on a very large scale. Dead fish floating in the water and thrown on the beaches decompose and befoul both the water and the air. The decomposition releases so much of hydrogen sulphide that it blackens the paint of the ships, a phenomenon known as the 'Callao painter'. The wholesale loss of fish deprives the guano birds their favorite food and they die of disease or starvation or leave their nests, so that the young ones die, triggering enormous losses to the guano industry.

Table 16.1, compiled by Schott (1931) during the disturbed El Niño period of 1925, shows how and to what extent El Nino affects the water surface temperatures off the coast of Peru.

The trend shown in Table 16.1 has been substantiated by several subsequent studies of major El Niño events in the equatorial Pacific.

Table 16.1 Surface temperatures (°C) of the water, off some coastal stations of Peru in March 1925, as compared to average March temperature (Schott, 1931)

Station	Latitude	Average temp	Temp (°C) in 1925
Lobitos	4°20'S	22.2	27.3
Puerto Chicana	7°40'S	20.3	26.9
Callao	12°20'S	19.5	24.8
Pisco	13°40'S	19.0	22.1

The meteorological phenomena which accompany a severe El Niño event are no less disastrous. Concurrent with the southward shift of the warm ocean current, the tropical rain-belt of the eastern Pacific moves south and pours torrential rain over a wide coastal belt of Peru. For example, in March 1925, rainfall at Trujillo at 8°S amounted to 395 mm, as against an average precipitation in March of the previous eight years of only 4.4 mm. These bursts of heavy rainfall cause damaging floods and erosion of land.

A recent study(Couper-Johnston, 2000) reveals that in the 473 years since 1525, catastrophic El Niños struck the equatorial eastern Pacific seaboard as many as 116 times, giving an average frequency of one such El Niño every four years. However, there appears to be no periodicity and the gap between occurrences of two disastrous El Niño's may vary from one to thirty years. This means that in majority of the years, the water surface off the Peruvian coast remains cold with little or no rain over the region. The periods of these cold ocean-surface temperatures with drought conditions have been called 'La **Niña**' or 'girl child' years.

16.8.3 Southern Oscillation (SO)

In 1877, there was a severe drought in many countries of the world, especially those bordering the Pacific and the Indian oceans. The failure of the monsoon rains in India that year caused untold suffering to the people of the country which led the then Government meteorologist, Henry Blanford, to take up a study of the problem with the ultimate objective of issuing, if possible, advance warning of the occurrence of such calamities. As a first step, Blanford wrote to meteorologists of several countries giving details of past droughts in India and enquiring if similar droughts had occurred in their countries and was surprised when he received a response from Charles Todd, South Australian Government astronomer and meteorologist, to the effect that severe droughts had occurred in Australia exactly in the same years as in India.

Encouraged by Todd's reply, Blanford and later his successor Gilbert Walker and their co-workers in India launched a world-wide search for atmospheric parameters which might show promising relationship with the Indian monsoon rainfall and which they could statistically use for prediction.

In the course of his extensive investigations, Walker (1923. 1924, 1928) found quite a few such parameters around the global tropics. The most promising of these was a relationship which he called the southern oscillation (SO) which he describes as follows (Walker,1924): "By the southern oscillation is implied the tendency of (surface) pressure at stations in the Pacific(San Francisco, Tokyo, Honolulu, Samoa, and South America), and of rainfall in India and Java ... to increase, while pressure in the region of the Indian ocean (Cairo, N.W. India, Port Darwin, Mauritius, S.E. Australia, and the Cape) decreases ... ". He wrote later (Walker, 1928): "We can, perhaps, best sum up the situation by saying that there is a swaying of pressure on a big scale backwards and forwards between the Pacific and the Indian oceans ...". A schematic (Fig. 16.8, due to Berlage, 1966) shows regions of the globe encompassed by the SO.

16.8 El Niño-Southern Oscillation (ENSO)

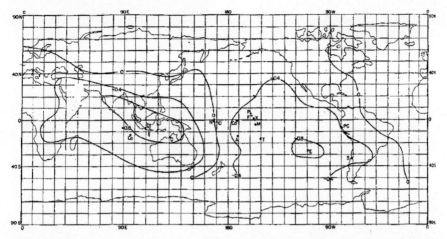

Fig. 16.8 Map showing isopleths of correlation of monthly mean station pressure with that of Djakarta (Dj), Indonesia. Other stations shown are Cocos Island(CO), Port Darwin(D), Nauru(N), Ocean island(O), Palmyra(P), Christmas Island(X), Fanning(F), Malden Island(M), Apia(A) in Samoa, Tahiti(T), Easter Island(E), Puerto Chicama(PC), Lima(L), and Santiago(S) (Reproduced from Berlage, 1966, with kind permission of Royal Netherlands Meteorological Institute)

According to Fig. 16.8, the isopleth of zero correlation with Djakarta pressure runs along about 170E with positive correlation values to the west and negative to the east. However, it may be noted that the isopleth of maximum positive correlation (+0.8) is not zonally oriented along the equator but inclined to it in an approximately NW-SE direction extending from Australia to central India across the Maritime continent. A similar deviation from the equatorial orientation is noticeable in the case of the isopleth of negative correlation (say, between 0 and –0.4) across the Americas. It would seem that this deviation from equatorial orientation occurs under the influence of the regional monsoons which causes a seasonal shift of the equatorial trough of low pressure and associated intertropical convergence zone (ITCZ) into the summer hemisphere across the respective connecting landmass. The orientation appears to represent the resultant of the zonal and meridional components of the correlation.

Julian and Chervin (1978) who computed the coherence-square statistic between station pressure at Port Darwin and Santiago emphasized two aspects of the maximum in the coherence-square. First, the phase angle of the cross spectrum is almost exactly $\pm\pi$ indicating that the pressures at the two stations are out of phase. Secondly, the bandwidth of the phenomena is rather large, giving an estimate of the period range of 87–27 months (7.2–2.2 years).

These findings emphasize the fact that the Southern Oscillation is not periodic, but certainly oscillatory, as found by the Peruvian fishermen in the oceanic case. Various combinations of stations have been used to compute an index for the SO from differences in station pressure. Most commonly used have been Djakarta (Dj), Port Darwin (D), Santiago (S), Apia (A) Samoa, Tahiti (T), and Easter Island (E), the locations of which are shown in Fig. 16.8. In some recent studies, the normalized pressure difference between Tahiti and Darwin has been used to compute a SO Index (SOI); a large negative value of the index indicates El Niño and positive value La Niña.

16.8.4 The Walker Circulation – ENSO

Examining monthly mean sea surface temperature (SST) and air temperature and precipitation data at Canton island ($2°48'S$, $171°43'W$) during a period of 5 (1963–1967) years, Bjerknes (1969) found that whenever SST was higher (lower) than air temperature at the station, there was a large increase (decrease) in rainfall, the two anomalies being positively correlated. Canton-type rainfall regime is known to prevail along the equatorial Pacific from about $165°E$ eastward to the coast of South America. He also found that temperature and pressure were negatively correlated, which means that a warm (cold) anomaly was associated with a fall (rise) of pressure. The zero isallobar was found to lie close to the dateline. One of his most important findings was that warm and cold water regimes and pressure and rainfall patterns associated with them alternated between the western and the eastern parts of the equatorial Pacific with the period of about two years(1964 and 1966 were El Niño years; 1963, 1965 and 1967 were La Niña years).

Bjerknes (1969) interpreted the findings of his study at Canton Island in terms of a vertical circulation the rising (sinking) branch of which would be located near the island whenever it had the warm (cold) SST anomaly. A warm anomaly epoch at the island would be associated with lower pressure to which air from the region of higher pressure would converge, rise in convection carrying moisture evaporated from the warm surface to higher levels where it would condense and produce cloud and precipitation. On cooling at higher levels, the rising currents would diverge and after long travel sink over region of higher surface pressure from where they would flow back again towards the island to complete a vertical circulation. On the other hand, during the period of a cold anomaly, the surface pressure at the island would be higher with cold air diverging, so it would come under the sinking branch of the vertical circulation and there would be little or no cloud or precipitation. Bjerknes called this vertical circulation the "Walker Circulation", as its east-west movement was consistent with Walker's southern oscillation in the Pacific.

Thus, Southern oscillation was shown to be directly coupled to El Niño/La Niña events through ocean-atmosphere interaction. Hence the combined oscillation is termed ENSO.

16.8.5 Evidence of Walker Circulation in Global Data

The Walker circulation is clearly evidenced by the field of upper-air divergence computed from the annual mean wind fields which have little or no monsoonal effects (e.g., Krishnamurti, 1975, 1979; Murakami, 1987). The study by Murakami who computed the annual mean wind vectors and corresponding velocity potential fields at 200 hPa during the FGGE year (December 1978–November 1979) highlights three prominent centers of divergence (D +) along the equator, one over Brazil (60°W), the second over Africa (20°E) and the third over the Maritime continent (170°E). Of these, the divergence center over the Maritime continent which reflects the dominance of the Southwest Pacific Convergence Zone (SPCZ) over the area appears to be the most prominent and strongest of the three to have within its domain a fourth divergence center near Malayasia (100°E). The annual mean winds are nearly symmetric about the equator as they become predominantly zonal on approaching the equator.

The near-symmetry of the equatorial winds between the two hemispheres is not surprising, as it is strongly suggested by several theoretical studies (e.g., Matsuno, 1966; Gill, 1980; Lim and Chang, 1983). Using a linear diagnostic model with equatorial heat sources and sinks prescribed as external forcing, the studies showed that the atmospheric response is generally confined to the equatorial latitudes, where the Rossby radius of deformation is approximately 1000 km, with its structure resembling a Walker circulation. Thus, the Walker circulation is of Kelvin wave type, but stationary. Relative to the divergence center near New Guinea (170°E), equatorial winds are westerlies to the east and easterlies to the west. These equatorial easterlies blow between the zonally-oriented subtropical ridge axes of the two hemispheres along about 10°N and 10°S. Poleward of these ridge axes, winds are westerly with two jetstreams, one near Japan in the northern hemisphere and the other over Australia in the southern hemisphere. There is little doubt that these northern and southern hemispheric W'ly jetstreams are strengthened by the local Hadley circulations between the equator and the higher latitudes.

The equatorial westerlies over the Pacific appear to be the strongest near 140°W. This may be partly due to the equatorward extension of two mid-Pacific upper-air troughs in the westerlies, one to the north and the other south of the equator. Thus, the annual mean wind fields are symmetric not only over the equatorial latitudes but also in higher latitudes, especially in regard to the locations of the subtropical ridge axes and troughs and Jetstreams in westerlies.

16.8.6 Mechanism of ENSO?

No one knows for sure what causes the ENSO and how it is caused. From time to time, scientists have pointed at a variety of possible influences from far-off places, such as excessive snowfall over Eurasia, changes in Antarctic pack ice, seismic activity following volcanic eruptions, etc., but it is believed that these could only have some marginal effects, if any, and that the main cause may be lurking closer to home.

Current thinking is that whatever the real cause a self-perpetuating loop between the ocean and the atmosphere is initiated by it. For example, a perturbation in the form of an anomalous burst of westerly winds along the equator will move the warm pool water a little to the east. This will cause further relaxation of the trade winds allowing more warm water to move towards the east. The movement of warm water flattens the thermocline and less cold water is upwelled in the east along the entire equator. This decreases the temperature difference and hence reduces the pressure gradient between the east and the west. With further weakening of the trades, more warm water flows to the east, preparing the east to have a full-fledged El Niño.

Thus, at the peak of an El Niño, the broad picture of the Pacific is very different from the norm and is described by Couper-Johnston (2000), as follows: "The thermohaline has flattened out considerably. So the deep, cold water normally close to the surface off South America is up to 30 m deeper than usual. The sea levels on both sides of the ocean are comparable. And the pressure difference between east and west has disappeared and, at times, even reversed. This is the flipside of the atmospheric seesaw that Gilbert Walker first noticed when he coined it the words "southern oscillation". Because the pressure difference drives the winds, the trade winds disappear and are replaced by westerlies that can blow nearly all the way to the Americas. The zone of cloudiness and heavy rain that characterizes the western Pacific is now located across much of the equatorial Pacific, and in its place across Australasia the abnormally high pressures bring warmer drier conditions. The clear skies associated with the usual zone of high pressures off the western coast of South America are replaced by rainclouds, as the warmer water off the coast heats up the air above it."

The above-mentioned physical processes at the ocean surface are reinforced by two other very important and related processes a few tens of metres below the ocean surface. The first is the excitation of a series of massive eastward-propagating Kelvin waves. Starting at the western boundary of the ocean, these waves as they move have the effect of lowering the thermocline in the east and thereby moving warm water eastward which may surface later. The second, which is sparked off by the first, is the generation of a series of westward-propagating Rossby waves which have the effect of raising the thermocline in the west and bringing it closer to the surface. The process goes on till the Kelvin waves, traveling at about 100 km $(\text{day})^{-1}$ and depending upon their point of origin, reach the coast of South America in about two to three months to start a full-fledged El Niño phase. There, they get partially reflected up and down the coast to advect heat to higher latitudes and partially reflected back across equatorial Pacific as Rossby waves. Now, Rossby waves travel at one-third the speed of the Kelvin waves, so they take any time between six and twelve months to reach the western boundary of the ocean. From there, they are reflected eastward as Kelvin waves, but this time as upwelling waves, i.e., raising the thermocline in the east and lowering it in the west. The process goes on till the reflected Kelvin waves reach the South American coast to end the El Niño and replace it by a La Niña phase. Thus, as Kelvin and Rossby waves keep shuttling across the equatorial Pacific Ocean from one side to the other, they would appear to provide the necessary mechanism to start or end an El Niño or a La Niña.

In 1997–1998, there was a pronounced El Niño not only in the equatorial eastern Pacific but also in the other global oceans, which lasted about a year and a half. Fig. 16.9 presents equatorial depth-longitude sections of ocean temperature

16.8 El Niño-Southern Oscillation (ENSO)

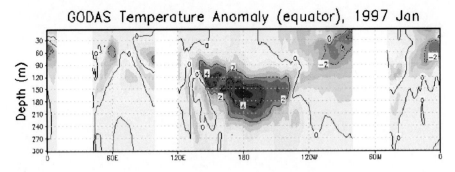

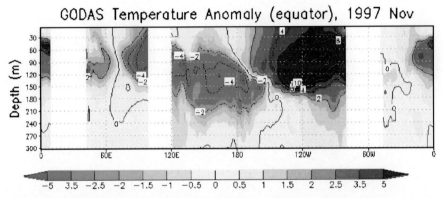

Fig. 16.9 Equatorial depth-longitude section of ocean temperature anomalies during (**a**) January 1997, and (**b**) November 1997 Negative anomalies indicate cold, positive warm. Data are derived from a Global Ocean Data Analysis System (GODAS) which assimilates oceanic observations into an oceanic GCM. Contour interval is 0.5°C. Anomalies are departures from the 1992–2003 base period means (Courtesy: NCEP/NWS, Washington, D.C.)

anomalies in (a) January 1997, and (b) November 1997, computed from a Global Ocean Data Assimilation System (GODAS) which show the underwater thermal structure of the different oceans before and during the onset of the El Niño. It clearly shows the seesaw-type swing of warm and cold water and change in the inclination of the thermocline between the western and the eastern sides of the equatorial oceans not only in the equatorial Pacific but also in the Atlantic and Indian oceans between January 1997 and November 1997. By January 1999, the La Niña condition returned to all the oceans, thus giving the oscillation a period of a little less than two years.

Chapter 17
Dynamical Models and Numerical Weather Prediction (N.W.P.)

17.1 Introduction – Historical Background

It was probably V. Bjerknes of Norway who first conceived of the idea of using the time-dependent hydrodynamical equations of motion, presented in Chap. 11, for atmospheric prediction. In 1904, in a German meteorological journal, he wrote: "If it is true, as every scientist believes, that subsequent atmospheric states develop from the preceding ones according to physical law, then it is apparent that the necessary and sufficient conditions for the rational solution of forecasting problems are the following: (1) A sufficiently accurate knowledge of the state of the atmosphere at the initial time, and (2) A sufficiently accurate knowledge of the laws according to which one state of the atmosphere develops from another." However, the idea remained in theory only till some years later when the British scientist, L.F. Richardson (1922), gave a practical shape to it and demonstrated its feasibility by actually carrying out an elaborate and arduous set of calculations by a numerical process using finite-difference techniques. However, for a variety of reasons to be discussed later in this chapter, the results of Richardson's experiment were unrealistic and discouraging. For example, his predicted value of surface pressure change in a forecast for 6 h turned out to be an order of magnitude higher than that observed.

There was no follow-up of Richardson's pioneering work till 1950, when it was revived by two important developments in meteorology and brought back to the center-stage. The first was a critical diagnosis of the reasons why Richardson's experiment failed to deliver a reasonable forecast. Richardson himself tried to analyze the cause of failure of his experiment but could not identify any particular reason, except the inadequacy of data, especially upper-air data over the observational grid. Evidently, he was unaware of the important 'Courant-Friedrichs-Lewy criterion' which requires any prediction scheme using finite-differencing technique to maintain computational stability.

It was left to Jule Charney of the Massachusetts Institute of Technology to point out in 1948 that lack of adequate data might have been only partly responsible for Richardson's failure. The main culprit was the presence of high-speed sound and gravity waves in his initial data, which contaminated the result by producing computational instability. He, therefore, stressed the need for elimination of these unwanted waves from the primitive equations in some way, so that they might not contaminate the result. He succeeded in simplifying the equations by devising methods to filter them out, thus clearing the way for further experiments. Another development which encouraged further experiments in numerical forecasting with the time-dependent hydrodynamical equations was the invention of the electronic computer which could make millions of computations at incredibly fast speed and thereby enable forecasts to be made in the shortest possible time to be useful.

The method suggested by Charney to filter out the sound and gravity waves was simplification of the governing equations by systematic use of the geostrophic and hydrostatic approximations and neglecting friction. The filtered equations came to be known as the quasi-geostrophic equations or a quasi-geostrophic model. A special version of this model known as the equivalent barotropic model was used in 1950 to make the first numerical forecast. It predicted the geopotential height field at 500 mb. It was a highly simple model and was not designed to predict any weather as such, like cloud and/or precipitation, since vertical motion which is essential for formation of such weather could not be computed with a single-level model. Later, multi-level quasi-geostrophic models and divergent models came to be used which enabled vertical motion to be forecast. However, simplified models had their obvious limitations and met with only limited success.

After the Second World War, the data situation improved a great deal, especially with the introduction of remote-sensing satellites in the 1960s. There had also been rapid advance in computer technology, bringing within reach of the modelers large-memory high-speed computers to handle much more complicated numerical models. The net result has been a rapid advance in the field of numerical weather prediction during the last few decades. Full-scale primitive equation models with a lot of complicated physics and computational skill are now used at many national and international centers for short-, medium- and long-range weather forecasts.

17.2 The Filtering of Sound and Gravity Waves

It is well-known that meteorological observations recorded at a station at any fixed time are not instantaneous values but the mean of a variable over an interval of time, say a minute or so, and, as such, have sometimes large built-in errors in them. So if raw data of different elements observed independently of each other at a time are used for numerical computations, the errors are likely to grow during computation and generate high-speed sound and gravity waves which may interfere with the meteorological waves which we want to forecast. Normally, the vertical acceleration in the atmosphere is small as may be seen from the result of scale analysis in Table 11.3, since the pressure gradient force and the gravity force

17.2 The Filtering of Sound and Gravity Waves

are more or less in balance. However, when observations of pressure and density are taken independently of each other, there is no guarantee that the difference between the two forces will be small enough. Similarly, the horizontal acceleration is generally small in the atmosphere, as may be seen from the scale analysis in Table 11.2, because the pressure gradient force and the Coriolis force are more or less in geostrophic balance. But when observations of pressure and velocity are taken independently of each other, there is no guarantee that the horizontal acceleration will remain small. So Charney argued that since it is the errors in observed data that generated the sound and gravity waves, the governing equations needed to be simplified with the aid of hydrostatic and geostrophic approximations so as to filter out the disturbing waves.

With the simplifications in place and using isobaric co-ordinates with the lower boundary condition, $\omega = 0$, the horizontal momentum equation, the continuity equation and the hydrostatic thermodynamic energy equation for a frictionless atmosphere may be written in vector notation

$$(\partial/\partial t + \mathbf{V} \cdot \nabla)\mathbf{V} + \mathbf{k} \times f\mathbf{V} + \omega \partial \mathbf{V}/\partial p = -\nabla \Phi \qquad (17.2.1)$$

$$\nabla \cdot \mathbf{V} + \partial \omega/\partial p = 0 \qquad (17.2.2)$$

$$(\partial/\partial t + \mathbf{V} \cdot \nabla)(\partial \Phi/\partial p) + \sigma \omega = 0 \qquad (17.2.3)$$

where $\sigma = (\partial \Phi/\partial p)\, \partial \ln \Theta/\partial p = \{\partial^2 \Phi/\partial p^2 - (1/p)(\partial \Phi/\partial p)(R/c_p - 1)\}$.

The above Eqs. (17.2.1–17.2.3) together with the lower boundary condition, $\omega = 0$, have no mechanism for the generation of sound waves. However, they are still capable of generating gravity waves, because, as discussed below, a divergent horizontal velocity field which varies in time is what causes propagation of gravity waves.

So it follows that neglecting the local variation of divergence from the horizontal momentum equation will totally eliminate gravity waves from the time-dependent prediction system. Of course, divergence does not appear explicitly in the horizontal momentum equation (17.2.1) as written, but its presence is revealed when the same equation is replaced by the vorticity and the divergence equations by operating on it with the Del operators ($\nabla \times$) and ($\nabla \cdot$), respectively. Thus, we obtain the vorticity equation

$$\partial \zeta/\partial t = -\mathbf{V} \cdot \nabla(\zeta + f) - \omega \partial \zeta/\partial p - (\zeta + f)\nabla \cdot \mathbf{V} + \mathbf{k} \cdot \{(\partial \mathbf{V}/\partial p) \times \nabla \omega\} \qquad (17.2.4)$$

where $\zeta = \mathbf{k} \cdot \nabla \times \mathbf{V}$ is the relative vorticity,
and the divergence equation

$$\partial D/\partial t = -\nabla^2[\Phi + (\mathbf{V} \cdot \mathbf{V})/2] - \nabla[\mathbf{k} \times \mathbf{V}(\zeta + f)] - \omega \partial D/\partial p - \partial \mathbf{V}/\partial p \cdot \nabla \omega \qquad (17.2.5)$$

where we have put D for divergence, $\nabla \cdot \mathbf{V}$, and used the vector identity

$$(\mathbf{V} \cdot \nabla)\mathbf{V} = \nabla(\mathbf{V} \cdot \mathbf{V})/2 + \mathbf{k} \times \nabla \zeta$$

Thus, if we set the left hand side of the divergence equation (17.2.5) equal to zero, we eliminate all solutions corresponding to the time-dependent gravity waves. This is the maximum simplification of the governing equations required to filter out the gravity waves.

17.3 Quasi-Geostrophic Models

A model suited for prediction of synoptic-scale motions in middle and high latitudes is the quasi-geostrophic model which utilizes the isobaric vorticity equation (17.2.4) and the thermodynamic energy equation (12.2.8). In this model, the following simplifications and approximations are made in the vorticity equation (17.2.4):

(a) The vertical advection and the twisting terms (the 2nd and the 4th terms on the right-hand side of the equation) are neglected;
(b) The relative vorticity is neglected compared to the Coriolis parameter in the divergence term;
(c) The horizontal velocity is replaced by the geostrophic wind velocity in the horizontal advection term;
(d) The relative vorticity is replaced by the geostrophic vorticity; and
(e) The beta-plane approximation is applied to the Coriolis parameter.

The Coriolis parameter f about a given latitude ϕ_0 is approximated by the β-plane approximation:
$$f = f_0 + \beta y,$$
where f_0 is the Coriolis parameter at the given latitude, $\beta \equiv (df/dy)$ at ϕ_0 is regarded as constant, and y is the distance from the given latitude ϕ_0.

For synoptic-scale midlatitude disturbances, $\beta y < f_0$ Hence, in the geostrophic model, f is replaced by f_0. Applying all the above-mentioned approximations, the quasi-geostrophic vorticity equation reduces to

$$\partial \zeta_g / \partial t = -\mathbf{V}_g \cdot \nabla(\zeta_g + f) - f_0 \nabla \cdot \mathbf{V} \qquad (17.3.1)$$

where $\zeta_g = \nabla^2 \Phi / f_0$ and $\mathbf{V}_g = \mathbf{k} \times \nabla \Phi / f_0$, have both constant f_0.

It is important to note that in the divergence term of (17.3.1), we have not replaced the wind by its geostrophic value. This step is necessary to retain vertical velocity in the model, since divergence of the geostrophic wind with a constant f_0 is zero. It is the departure of the wind from its geostrophic value that produces divergence for vertical motion. For, if, in isobaric co-ordinates, we take $\mathbf{V}'$ as the ageostrophic wind,

$$\nabla \cdot \mathbf{V} = \nabla \cdot (\mathbf{V}_g + \mathbf{V}') = \nabla \cdot \mathbf{V}' = -\partial \omega / \partial p \qquad (17.3.2)$$

With this substitution for divergence, (17.3.1) reduces to

$$\partial \zeta_g / \partial t = -\mathbf{V}_g \cdot \nabla(\zeta_g + f) + f_0 \, \partial \omega / \partial p \qquad (17.3.3)$$

Since ζ_g and $\mathbf{V}_g$ are both functions of Φ, (17.3.3) offers a better method of computing the ω-field from observations of Φ than the continuity equation (12.8.2), because in midlatitudes both Φ and $\partial\Phi/\partial t$ can be determined with somewhat greater accuracy than the wind at isobaric surfaces.

In a similar manner, using the quasi-geostrophic wind in the horizontal advection term and neglecting the role of diabatic heating in midlatitude synoptic-scale motion systems, we can approximate the thermodynamic energy equation (12.2.8) to the form

$$\partial(-\partial\Phi/\partial p)/\partial t = -\mathbf{V}_g \cdot \nabla(-\partial\Phi/\partial p) + \sigma\omega \qquad (17.3.4)$$

where the static stability σ is assumed constant and, on the basis of the hydrostatic approximation, $\alpha = -\partial\Phi/\partial p$, and the equation of state, $p\alpha = RT$, we have replaced T by $p(-\partial\Phi/\partial p)/R$.

Thus, the quasi-geostrophic vorticity equation (17.3.3) and the hydrostatic thermodynamic energy equation (17.3.4) each contains only two dependent variables Φ and ω, and, therefore, together constitute a closed set of prediction equations in Φ and ω.

From these two equations, we can eliminate ω and obtain an expression for the geopotential tendency, $\partial\Phi/\partial t$. We can also manipulate the two equations suitably to compute the vertical velocity, ω, yielding the so-called omega equation.

The expressions for the geopotential tendency and the vertical motion, derived in this manner, are as follows:

The geopotential tendency equation,

$$\{\nabla^2 + (f_0^2/\sigma)\partial^2/\partial p^2\}\partial\Phi/\partial t = -f_0 \mathbf{V}_g \cdot \nabla\{(1/f_0)\nabla^2\Phi + f\}$$
$$+ (f_0^2/\sigma)\partial\{-\mathbf{V}_g \cdot \nabla(\partial\Phi/\partial p)\}/\partial p \qquad (17.3.5)$$

and

The omega equation,

$$\{\nabla^2 + (f_0^2/\sigma)\partial^2/\partial p^2\}\omega = (f_0/\sigma)\partial[\mathbf{V}_g \cdot \nabla\{(1/f_0)\nabla^2\Phi + f\}]/\partial p$$
$$+ (1/\sigma)\nabla^2\{\mathbf{V}_g \cdot \nabla(-\partial\Phi/\partial p)\} \qquad (17.3.6)$$

17.4 Nondivergent Models

The basis for this type of models is a simplified version of the divergence equation (17.2.5). Elimination of gravity waves from the solution of the equations of motion by putting $\partial D/\partial t = 0$ in (17.2.5) leaves us with a complicated diagnostic relation involving Θ, $\mathbf{V}$, and ω in a balance. On scaling considerations, further simplification of this balance equation may be made. According to a theorem of Helmholtz, any velocity field $\mathbf{V}$ can be partitioned into two independent velocity components, a rotational velocity component $\mathbf{V}_\psi$, and a divergent velocity component $\mathbf{V}_\chi$. The rotational component is defined by the expression, $\mathbf{V}_\psi = \mathbf{k} \times \nabla\psi$, where ψ denotes a nondivergent stream function and the divergent component is defined by the expression, $\mathbf{V}_\chi = -\nabla\chi$, where χ denotes an irrotational velocity potential. Thus,

$$\mathbf{V} = \mathbf{V}_\psi + \mathbf{V}_\chi = \mathbf{k} \times \nabla\psi - \nabla\chi \tag{17.4.1}$$

It follows from (17.4.1) that

$$\zeta = \mathbf{k} \cdot \nabla \times \mathbf{V} = \nabla^2\psi, \text{ and } D = \nabla \cdot \mathbf{V} = -\nabla^2\chi$$

Now, it can be shown by scale analysis that D is almost an order of magnitude smaller than ζ, especially in extratropical latitudes. So, if we assume the motion to be largely nondivergent, both D and ω may be neglected compared to ζ in (17.2.5). The result is the balance equation between Φ and ψ which may be written in the form

$$\nabla^2[\Phi + (\nabla\psi)^2/2] = \nabla \cdot [(f + \nabla^2\psi)\nabla\psi] \tag{17.4.2}$$

The Eqs. (17.2.2), (17.2.3), (17.2.4) and (17.4.2) constitute a closed system of equations in Φ, ψ and ω which can be used in a forecast model. However, (17.4.2) is still rather complicated to use. The scale analysis shows that the nonlinear terms in it are much smaller than the linear terms. So the equation can be further simplified by neglecting the nonlinear terms to what is known as the linear balance equation

$$\nabla^2\Phi = \nabla \cdot (f\nabla\psi) \tag{17.4.3}$$

Likewise, we may simplify (17.2.4) by neglecting small terms to obtain a simplified vorticity equation

$$\partial\zeta/\partial t + \mathbf{V}_\psi \cdot \nabla(\zeta + f) + \mathbf{V}_\chi \cdot \nabla f + f\nabla \cdot \mathbf{V}_\chi = 0 \tag{17.4.4}$$

The third term in (17.4.4) which gives advection of planetary vorticity by the divergent wind $\mathbf{V}_\chi$ is actually small compared to the other terms of the equation and may, therefore, be neglected if we replace f by a mean value, say f_0, in (17.4.3) as well as in the last term in (17.4.4). The resulting simplified vorticity and divergence equations for midlatitude synoptic-scale motion systems are then

$$\partial\zeta/\partial t = -\mathbf{V}_\psi \cdot \nabla(\zeta + f) - f_0 \nabla \cdot \mathbf{V}_\chi \tag{17.4.5}$$

and

$$\nabla^2\Phi = f_0 \nabla^2\psi \tag{17.4.6}$$

Since the geostrophic wind is given by the expression, $\mathbf{V}_g = (\mathbf{k} \times \nabla\Phi)/f$, (17.4.6) simply states that to a first approximation, the vorticity of the motion is the vorticity of the geostrophic wind with a constant value of the Coriolis parameter. Thus, for such motion, we may put $\psi = \Phi/f_0$. This means that the geopotential field on a

17.4 Nondivergent Models

constant pressure map is approximately proportional to the streamline field and, to the same order of approximation,

$$\mathbf{V}_\psi = \mathbf{k} \times \nabla \Phi / f_0 \qquad (17.4.7)$$

The geostrophic vorticity equation and the hydrostatic thermodynamic energy equation can now be written in terms of ψ and ω as

$$\partial(\nabla^2 \psi)/\partial t = -\mathbf{V}_\psi \cdot \nabla(\nabla^2 \psi + f) + f_0\, \partial\omega/\partial p \qquad (17.4.8)$$

$$\partial(\partial\psi/\partial p)/\partial t = -\mathbf{V}_\psi \cdot \nabla(\partial\psi/\partial p) - \sigma\omega/f_0 \qquad (17.4.9)$$

Differentiating (17.4.9) with respect to p after multiplying through by f_0^2/σ, and adding the result to (17.4.8), we obtain the quasi-geostrophic potential vorticity equation,

$$(\partial/\partial t + \mathbf{V}_\psi \cdot \nabla)q = 0 \qquad (17.4.10)$$

where

$$q = \nabla^2 \psi + f + f_0^2 \partial\{(1/\sigma)\partial\psi/\partial p\}/\partial p$$

where we have again assumed that σ is a function of pressure only.

Equation (17.4.10) states that the quasigeostrophic potential vorticity q is conserved following the nondivergent wind in pressure co-ordinates. This conservation law forms the basis of most of the numerical prediction schemes discussed in this chapter.

Further, if the time derivatives are removed from (17.4.8) and (17.4.9), we obtain the diagnostic quasi-geostrophic omega equation

$$[\nabla^2 + (f_0^2/\sigma)\partial^2/\partial p^2]\omega = (f_0/\sigma)\partial[\mathbf{V}_\psi \cdot \nabla(\nabla^2\psi + f)]/\partial p - (f_0/\sigma)\nabla^2[\mathbf{V}_\psi \cdot \nabla(\partial\psi/\partial p)] \qquad (17.4.11)$$

The Eq. (17.4.11) can be used to compute the vertical velocity field when the ψ-field is known.

The basis for using (17.4.10) for prediction lies in advecting the quasi-geostrophic nondivergent potential vorticity by the nondivergent velocity field in accordance with a scheme of successive time integration. The usual method is first to compute an initial field of $\psi(x, y, p)$ and q at time t_0 at a finite number of points in a three-dimensional grid and then use these fields to compute the tendency $\partial q/\partial t$ over a short interval of time by using (17.4.10) in the form

$$\partial q/\partial t = -\mathbf{V}_\psi \cdot \nabla q \qquad (17.4.12)$$

For the computation of the derivatives, a time-differencing scheme is used which is forward at the first time step, but centered later. The space derivatives are evaluated using a centered finite-differencing scheme. After every successive time step, a new ψ-field is obtained. In this way, a forecast may be obtained for any desired period.

17.5 Hierarchy of Simplified Models

17.5.1 One-Parameter Barotropic Model

The simplest type of prediction model is the single-level barotropic model which assumes conservation of absolute vorticity and is usually applied at about 500 mb for the simple reason that the flow at this surface in midlatitudes is largely nondivergent. The simplified barotropic vorticity equation used for the model is:

$$\partial(\nabla^2\psi)/\partial t = -\mathbf{V}_\psi \cdot \nabla[(\nabla^2\psi) + f)] \tag{17.5.1}$$

The real atmosphere is not barotropic, nor is the flow nondivergent. However, it so happens that at some midtropospheric level near 500 mb, the vertical profile of divergence changes sign and the horizontal flow becomes quasi-nondivergent. It is for this reason that the nondivergent barotropic vorticity equation (17.5.1) has been found to be useful as a prediction tool to predict flow patterns at about this midtropospheric level over many parts of the globe, including the tropics and equatorial regions. In fact, the model result has been found to be more realistic over the equatorial regions than the midlatitudes, outside regions of precipitation.

However, neglect or omission of the divergence term, $f_0\,\partial\omega/\partial p$, from (17.4.10) poses a problem, since the linearized form of (17.5.1) admits solutions corresponding to westward-propagating Rossby waves. In the vorticity balance, the divergence term is needed to balance the planetary vorticity advection. For this reason, a modified version of (17.5.1) is often used by parameterizing the baroclinic term in q in the potential vorticity equation (17.4.12) as follows:

$$f_0^2 \partial\{(1/\sigma)\partial\psi/\partial p\}/\partial p = -\lambda^2\psi \tag{17.5.2}$$

where λ^{-1} is an empirical constant with the unit of length. Substituting from (17.5.2) in (17.5.1), we obtain a modified barotropic vorticity equation

$$\partial(\nabla^2\psi - \lambda^2\psi)/\partial t = -\mathbf{V}_\psi \cdot \nabla(\nabla^2\psi - \lambda^2\psi + f) \tag{17.5.3}$$

Experiments with (17.5.3) have shown that it provides a much better short-range prediction of long-wave components of the 500-mb flow than (17.5.1), if we choose λ to be $\simeq 10^{-6}\,\mathrm{m}^{-1}$.

17.5.2 A Two-Parameter Baroclinic Model

It was not long before the one-parameter model was replaced by a two-parameter model, since the vertical motion which is essential for any type of development in the atmosphere could not be computed with a single-level model. A two-parameter model allows not only the vorticity to vary in the vertical but also the thickness

17.5 Hierarchy of Simplified Models

between two isobaric surfaces which depends upon the horizontal distribution of temperature. To derive the model, we divide the atmosphere into two discrete layers, bounded by pressure surfaces marked 0, 2 and 4, as shown in Fig. 17.1.

We now apply the vorticity equation (17.4.2) at levels 1 and 3 representing isobaric surfaces 250 mb and 750 mb which mark the middle of the upper and the lower layer respectively. To do this, we have to evaluate the divergence term $\partial \omega / \partial p$ at each of these pressure surfaces using finite-difference approximation to the vertical derivatives. Thus,

$$(\partial \omega / \partial p)_1 \simeq (\omega_2 - \omega_0)/\Delta p, \quad (\partial \omega / \partial p)_3 \simeq (\omega_4 - \omega_2)\Delta p$$

where the suffixes denote the designated levels and Δp is the pressure interval between the levels 0–2, 1–3, and 2–4. If we now assume that $\omega = 0$ at the top and the bottom of the atmosphere, we can write the vorticity equations as

$$\partial (\nabla^2 \psi_1)/\partial t = -(\mathbf{k} \times \nabla \psi_1) \cdot \nabla (\nabla^2 \psi_1 + f) + (f_0/\Delta p)\omega_2 \qquad (17.5.4)$$

$$\partial (\nabla^2 \psi_3)/\partial t = -(\mathbf{k} \times \nabla \psi_3) \cdot \nabla (\nabla^2 \psi_3 + f) - (f_0/\Delta p)\omega_2 \qquad (17.5.5)$$

Next, we write the thermodynamic energy equation (17.4.9) for level 2. After evaluating $\partial \psi / \partial p$ by the finite-difference $(\psi_3 - \psi_1)/\Delta p$, the equation is

$$\partial (\psi_1 - \psi_3)/\partial t = -(\mathbf{k} \times \nabla \psi_2) \cdot \nabla (\psi_1 - \psi_3) + (\sigma \Delta p)\omega_2/f_0 \qquad (17.5.6)$$

where $\psi_2 = (\psi_1 + \psi_3)/2$ is the stream function at 500 mb, which is not a predicted variable but determined by the predicted values of ψ_1 and ψ_3. Using this value of ψ_2 in terms of ψ_1 and ψ_3, we obtain a closed set of equations in ψ_1, ψ_3, and ω_2.

We now eliminate ω_2 between (17.5.4) and (17.5.6) to obtain two equations in ψ_1 and ψ_3 alone.

First we add (17.5.4) and (17.5.5) to eliminate ω_2 and obtain

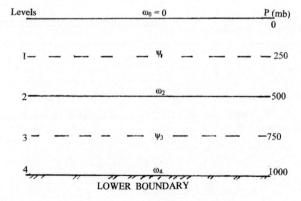

Fig. 17.1 Arrangement of variables in the vertical in a two-parameter baroclinic model

$$\partial \nabla^2(\psi_1 + \psi_3)/\partial t = -(\mathbf{k x}\nabla\psi_1)\cdot\nabla(\nabla^2\psi_1 + f) - (\mathbf{k x}\nabla\psi_3)\cdot\nabla(\nabla^2\psi_3 + f) \quad (17.5.7)$$

We now introduce a length scale λ^{-1}, where $\lambda^2 \equiv f_0^2/\{\sigma(\Delta p)^2\}$.

We next subtract (17.5.5) from (17.5.4) and add the result to $-2\lambda^2$ times (17.5.6) to obtain

$$\partial\{(\nabla^2 - 2\lambda^2)(\psi_1 - \psi_3)\}/\partial t = -(\mathbf{k x}\nabla\psi_1)\cdot\nabla(\nabla^2\psi_1 + f) + (\mathbf{k x}\nabla\psi_3)\cdot\nabla(\nabla^2\psi_3 + f)$$
$$+ 2\lambda^2(\mathbf{k x}\nabla\psi_2)\cdot\nabla(\psi_1 - \psi_3) \quad (17.5.8)$$

The physical interpretation of (17.5.7) is simple. It states that the local rate of change of the vertically-averaged vorticity between 250 mb and 750 mb is given by the vertical average of the horizontal advections of vorticity at the two pressure surfaces. Thus, it represents the barotropic part of the flow vorticity. The baroclinic part is represented by (17.5.8) which states that thickness tendency at a point is determined partly by the difference in vorticity advections between 250 mb and 750 mb and partly by the thermal advection by the mean nondivergent wind between the two pressure surfaces.

We can derive an expression for vertical motion ω_2 by combining (17.5.4) and (17.5.6) and eliminating the time tendencies. For this, we first operate on (17.5.6) with ∇^2 and add the result to the difference between (17.5.4) and (17.5.5). Rearranging the terms, we obtain

$$(\nabla^2 - 2\lambda^2)\omega_2 = (f_0/\sigma\Delta p)[\nabla^2\{(\mathbf{k x}\nabla\psi_2)\cdot\nabla(\psi_1 - \psi_3)\}$$
$$- (\mathbf{k x}\nabla\psi_1)\cdot\nabla(\nabla^2\psi_1 + f) + (\mathbf{k x}\nabla\psi_3)\cdot\nabla(\nabla^2\psi_3 + f)] \quad (17.5.9)$$

The two-level model as introduced here has not been found to be very useful in NWP, since it tended to produce stronger baroclinic development than observed. However, it remains a useful tool for the analysis of physical processes occurring in baroclinic disturbances. Multi-level baroclinic models and higher horizontal resolution used in later years led to improved forecasts.

17.6 Primitive Equation Models

The filtered and the simplified models, discussed in the foregoing sections, looked promising for a while but obviously had their limitations so far as weather forecasts were concerned, simply because the real atmosphere and its behaviour are more complicated. So, the demand for better forecast naturally called for use of more complete and original hydrodynamical equations which at one time were dubbed as primitive equations. The primitive equation model of the atmosphere consisted of the three prognostic equations (two consisting of the x and y components of the momentum equation and the third the thermodynamic energy equation) and the three diagnostic equations, viz., the hydrostatic equation, the continuity equation and the equation of state. The six equations constitute a closed set of equations in the six dependent variables u, v, ω, Φ, α and Θ and, therefore, were amenable to solution. However, the equations are nonlinear, so could not be solved by any known analytical methods. Attempts were, therefore, made to solve them by numerical methods with suitable boundary conditions. As already stated in the introduction, the first attempt to do so, though with discouraging result, was by Lewis Fry Richardson in England in 1922.

17.6.1 PE Model in Sigma Co-ordinates

Since atmospheric observations are recorded at pressure surfaces, it is advantageous to work with pressure co-ordinates. Other advantages of using a pressure co-ordinate system are that the density does not appear explicitly in the pressure gradient term and that the continuity equation has a simple form. Also, the sound waves are completely filtered.

However, there is a problem in specifying the lower boundary condition at the ground with the pressure co-ordinate. The assumption usually made is that the vertical velocity ω_0 is zero at the lower boundary z_0 where the pressure p_0 is treated as constant at 1000 mb. It is well-known that this assumption is not valid because of the uneven topography of the earth's surface. Also, pressure at the lower boundary does not remain constant but varies with time.

To overcome this difficulty, a modified version of the isobaric co-ordinate system, in which the vertical co-ordinate is pressure, p, normalized with surface pressure (p_s), is used and is known as the σ-sigma system. Thus,

$$\sigma \equiv p/p_s$$

In the sigma system, the lower boundary is defined by $\sigma = 1$, and the upper boundary by $\sigma = 0$, so the vertical σ-velocity at the lower as well as the upper boundary is zero at all times, even over a high mountain. That is,

$$d\sigma/dt = 0, \text{ at both } \sigma = 1 \text{ and } \sigma = 0.$$

We now transform the primitive equations from the p to the σ co-ordinate system as follows. The equations in the p co-ordinate system, neglecting friction, are:

Momentum equations

$$d\mathbf{V}/dt + f\mathbf{k}\mathbf{x}\mathbf{V} = -\nabla\Phi \qquad (17.6.1)$$

where $d/dt = \partial/\partial t + \mathbf{V} \cdot \nabla + \omega \partial/\partial p$,
and the horizontal Del operator refers to an isobaric surface.

Continuity equation

$$\nabla \cdot \mathbf{V} + \partial\omega/\partial p = 0 \qquad (17.6.2)$$

Hydrostatic equation

$$\partial\Phi/\partial p + \alpha = 0 \qquad (17.6.3)$$

Equation of state

$$p\alpha = RT \qquad (17.6.4)$$

Thermodynamic energy equation

$$c_p d(\ln\theta)/dt = dS/dt \qquad (17.6.5)$$

where $\theta = (p_0/p)^\kappa (p\alpha/R)$, and S is entropy

We now apply the transformation formula (11.6.8) to the momentum equation (17.6.1) and obtain

$$d\mathbf{V}/dt + f\,\mathbf{k}\mathbf{x}\mathbf{V} = -\nabla\Phi + (\sigma/p_s)\nabla p_s \partial\Phi/\partial\sigma \qquad (17.6.6)$$

where ∇ is now applied to the σ constant surface and the total differential is given by $d/dt = \partial/\partial t + \mathbf{V}\cdot\nabla + (d\sigma/dt)\partial/\partial\sigma$

Similarly, by applying (11.6.8) to the equation of continuity (17.6.2), we first transform the divergence term to the σ-constant surface

$$(\nabla\cdot\mathbf{V})_p = (\nabla\cdot\mathbf{V})_\sigma - (\sigma/p_s)(\partial\mathbf{V}/\partial\sigma)\cdot\nabla p_s \qquad (17.6.7)$$

To transform the term $\partial\omega/\partial p$ to the σ co-ordinate, we note that p_s does not depend on σ, so we may write the continuity equation in the form

$$p_s(\nabla\cdot\mathbf{V})_p + \partial\omega/\partial\sigma = 0 \qquad (17.6.8)$$

Now, ω is related to the sigma vertical velocity $d\sigma/dt$ by the relation

$$\omega = dp/dt = d(\sigma p_s)/dt = p_s\,d\sigma/dt + \sigma\,dp_s/dt$$
$$= p_s d\sigma/dt + \sigma(\partial p_s/\partial t + \mathbf{V}\cdot\nabla p_s)$$

Differentiating the above with respect to σ, we obtain

$$\partial\omega/\partial\sigma = p_s\partial(d\sigma/dt)/\partial\sigma + (\partial p_s/\partial t + \mathbf{V}\cdot\nabla p_s)_\sigma + \sigma\partial\mathbf{V}/\partial\sigma\cdot\nabla p_s \qquad (17.6.9)$$

17.6 Primitive Equation Models

The Eq. (17.6.9) combined with (17.6.7) and (17.6.8), after re-arrangement, yields the transformed continuity equation

$$\nabla_\sigma \cdot (p_s \mathbf{V}) + p_s \partial (d\sigma/dt)/\partial \sigma + \partial p_s/\partial t = 0 \tag{17.6.10}$$

With the aid of the equation of state (17.6.4) and the Poisson's relation, $T = \theta(p/p_0)^\kappa$, the hydrostatic equation (17.6.3) in sigma co-ordinate system is

$$\partial \Phi/\partial \sigma = -RT/\sigma = -(R\theta/\sigma)(p/p_0)^\kappa \tag{17.6.11}$$

where $p_0 = 1000\,\mathrm{hPa}$.

We now expand the total derivative on the left-side of the thermodynamic energy equation (17.6.5) to write it in the form

$$\partial \theta/\partial t + \mathbf{V}\cdot\nabla\theta + (d\sigma/dt)\partial\theta/\partial\sigma = (\theta/c_p)dS/dt \tag{17.6.12}$$

If we now multiply (17.6.10) by θ, multiply (17.6.12) by p_s, and add the results, we get the thermodynamic energy equation in the sigma co-ordinates in flux-form

$$\partial(p_s\theta)/\partial t + \nabla\cdot(p_s\theta\mathbf{V}) + \partial\{p_s\theta(d\sigma/dt)\}/\partial\sigma = (p_s\theta/c_p)dS/dt \tag{17.6.13}$$

A similar transformation of the x and y components of the momentum equation (17.6.6) gives them in flux forms

$$\partial(p_s u)/\partial t + \nabla\cdot(p_s u \mathbf{V}) + \partial\{p_s u(d\sigma/dt)\}/\partial t - f\,p_s v$$
$$= -p_s \partial\Phi/\partial x - R\theta(p/p_0)^\kappa \partial p_s/\partial x \tag{17.6.14}$$

$$\partial(p_s v)/\partial t + \nabla\cdot(p_s v \mathbf{V}) + \partial\{p_s v(d\sigma/dt)\}/\partial t + f\,p_s u$$
$$= -p_s \partial\Phi/\partial y - R\theta(p/p_0)^\kappa \partial p_s/\partial y \tag{17.6.15}$$

The five equations (17.6.10), (17.6.11), (17.6.13), (17.6.14) and (17.6.15) contain the six dependent scalar variables, u, v, $d\sigma/dt$, Φ, θ, p_s. This means that one more equation is required for a closed system of equations. The additional equation is the pressure tendency equation which is obtained by vertical integration of the continuity equation (17.6.10) with the boundary condition, $d\sigma/dt = 0$, at $\sigma = 1, 0$. Thus, the sixth equation is,

$$\partial p_s/\partial t = -\int_0^1 \nabla \cdot (p_s \mathbf{V}) d\sigma \tag{17.6.16}$$

The Eq. (17.6.16) states that the rate of change of surface pressure at a given point on the sigma surface is simply the total mass convergence in the overlying column of air of unit cross-section of the surface. With the inclusion of (17.6.16), we have now a complete set of prediction equations which can be put into a suitable finite-difference form and numerically integrated in time for different forecast periods.

17.6.2 A Two-Level Primitive Equation Model

In the quasi-geostrophic model that we discussed in Sect. 17.3, the static stability σ was assumed to be constant. But it is well-known that it varies in space and time depending upon the dynamical and thermodynamical processes in the atmosphere. We, therefore, need to take its variability into consideration in the primitive equation prediction models by computing $\partial \theta / \partial \sigma$ at each time step. This requires temperature to be predicted in at least two levels, instead of one as it was in the case of the two-level quasi-geostrophic model. As in the two-parameter baroclinic model (Fig. 17.1), the atmosphere is divided into two layers by boundary surfaces labelled 0, 2, and 4, corresponding to sigma surfaces 0, $1/2$ and 1 respectively The vertical differencing scheme in the case of the two-level primitive equation model, usually adopted, is illustrated in Fig. 17.2.

The momentum equations (17.6.14 and 17.4.6) and the thermodynamic energy equation (17.6.13) are applied at the levels 1 and 3, corresponding to sigma surfaces $1/2$ and $3/4$ which lie at the centers of the two layers. The vertical differencing for the momentum equations involves u_2, v_2, and $d\sigma_2/dt$, while that for the thermodynamic energy equation involves θ_2 and $d\sigma_2/dt$. The variables u_2, v_2 and θ_2 are obtained by linear interpolation. For example, $\theta_2 = (\theta_1 + \theta_3)/2$. The field of $d\sigma_2/dt$ may be obtained diagnostically by using the continuity equation (17.6.10) for levels 1 and 3 with the vertical derivatives computed by centered differencing. This means that for level 1, we take the difference between levels 0 and 2, and for level 3 we take the difference between levels 2 and 4.

The continuity equations at the two levels are then

$$\partial p_s / \partial t + \nabla \cdot (p_s \mathbf{V}_1) + 2 p_s \, d\sigma_2/dt = 0 \qquad (17.6.17)$$

$$\partial p_s / \partial t + \nabla \cdot (p_s \mathbf{V}_3) - 2 \, p_s \, d\sigma_2/dt = 0 \qquad (17.6.18)$$

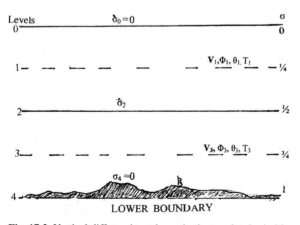

Fig. 17.2 Vertical differencing scheme in the two-level primitive equation model

17.6 Primitive Equation Models

The addition of (17.6.17) and (17.6.18) gives us the finite-difference form of the surface pressure tendency equation

$$\partial p_s/\partial t = -\nabla \cdot \{p_s(\mathbf{V}_1 + \mathbf{V}_3)/2\} \tag{17.6.19}$$

while the difference between the equations yields the diagnostic equation for $d\sigma_2/dt$ which is

$$d\sigma_2/dt = -\nabla \cdot \{p_s(\mathbf{V}_1 - \mathbf{V}_3)\}/4p_s \tag{17.6.20}$$

We now need to express Φ_1 and Φ_3 in terms of θ_1 and θ_3 using the hydrostatic relationship (17.6.11) in order to complete the specification of all the dependent variables. However, according to Arakawa (1972), the finite difference equations will have the same energy conservation properties as the original differential equations only if the variables Φ_1 and Φ_3 in (17.6.11) are expressed in a special form

$$\partial\Phi/\partial p^\kappa = -c_p\theta(p_0)^{-\kappa} \tag{17.6.21}$$

Applying (17.6.21) at level 2, we get

$$\Phi_1 - \Phi_3 = -c_p\theta_2\{(p_1/p_0)^\kappa - (p_3/p_0)^\kappa\} \tag{17.6.22}$$

Next we rewrite (17.6.11) by using the ideal gas law

$$\partial(\sigma\Phi)/\partial\sigma = \Phi - p\sigma\alpha \tag{17.6.23}$$

By integrating (17.6.23) with respect to σ from $\sigma = 1$ to $\sigma = 0$, we obtain

$$\Phi_4 = \{(\Phi_3 - p_s\sigma_3\alpha_3) + (\Phi_1 - p_s\sigma_1\alpha_1)\}/2 \tag{17.6.24}$$

where we have used the values $(\sigma\Phi) = \Phi_4$ at $\sigma = 1$, and $(\sigma\Phi) = 0$ at $\sigma = 0$.

We now solve (17.6.22) and (17.6.24) for Φ_1 and Φ_3 to obtain

$$\Phi_1 = \Phi_4 + p_s(\sigma_3\alpha_3 + \sigma_1\alpha_1)/2 + c_p\theta_2\{(p_3/p_0)^\kappa - (p_1/p_0)^\kappa\}/2 \tag{17.6.25}$$

$$\Phi_3 = \Phi_4 + p_s(\sigma_3\alpha_3 + \sigma_1\alpha_1)/2 - c_p\theta_2\{(p_3/p_0)^\kappa - (p_1/p_0)^\kappa\}/2 \tag{17.6.26}$$

This completes the formulation of the two-level primitive equation model.

17.6.3 Computational Procedure

The computational procedure in the case of the two-level primitive equation model consists of the following main steps:

1. Write suitable finite difference analogs of the momentum and thermodynamic energy equations at levels 1 and 3 and the surface pressure tendency equation at level 4.
2. Use the prediction equations to obtain the tendencies of $\mathbf{V}_1$, $\mathbf{V}_3$, θ_1, θ_3, and p_s fields.

3. Extrapolate the tendencies ahead using a suitable time differencing scheme. Usually, the first time step is forward and the subsequent ones centered.
4. Use the new values of the dependent variables to diagnostically determine $d\sigma_2/dt$, Φ_1 and Φ_3.
5. Repeat steps 2 and 4 until forecasts are obtained for the desired period.

In applying the above computational scheme, it must be borne in mind that the equations in the sigma co-ordinate system contain the mechanism to generate the fast-moving sound and gravity waves that interfere with the slow-moving meteorological waves and produce computational instability. To prevent the generation of the unwanted waves, it is necessary to keep the time increment for extrapolation small enough to satisfy the condition

$$c(\Delta t)/d = 1/\sqrt{2} \qquad (17.6.27)$$

where c is the speed of the fastest moving sound waves ($\sim 330\,\text{ms}^{-1}$), d is the horizontal grid distance and Δt is the time increment. This condition is known in literature as the Courants-Friederichs-Lewy (CFL) condition. For this reason, in order to avoid computational instability, the time increment in primitive equation prediction model must be considerably less than that in quasi-geostrophic models.

17.7 Present Status of NWP

So far, we have traced only the early developments of the subject of NWP before, say, 1970. Since then, the field has expanded so much that it is practically impossible to cover all aspects of the later developments in this brief survey. Computing machines are now millions of times faster; a development which has allowed meteorologists to take up models with much higher spatial resolution over a global domain both in the horizontal and the vertical than before. Data coverage has improved a great deal. Apart from conventional observational network, geostationary and polar-orbiting satellites now regularly report data from different parts of the global atmosphere. Four-dimensional data assimilation schemes have been devised which prepare initial data directly for integration through a 6-hr data assimilation-prediction scheme. With improved computational schemes in place (at many centers, finite-difference models have been replaced by spectral models), the global forecast centers are now able to issue extended-range forecasts of mass and flow fields over the globe for periods ranging from a day to a week or more.

Verification statistics of present-day numerical forecasts appear to suggest that by and large these improved numerical models perform better than those based on persistence or other statistical or subjective methods, especially at extended ranges. For example, according to a study by Leith (1978), quoted by Holton (1979), the root-mean-square error in a 24-hr forecast of 500 mb height was 40 m by the primitive equation model as against 65 m by persistence, the initial height error being 20 m (now less than 10 m). With extension of the forecast period to 48 h, the errors in the two methods grew to about 60 m and 95 m respectively. Experiments on how the growth of errors limits the inherent predictability of the atmosphere have been carried out by a number of workers using primitive equation forecast models. Results of these experiments appear to indicate that the theoretical limit for any useful forecasts of synoptic-scale motion is probably one to two weeks.

Actual predictive skill of present-day models falls far short of this theoretical limit due to various sources of error, some of which are computational and some physical. Among the main sources of error are the replacement of derivatives of the dependent variables by finite-differences, inadequate representation of mountain effects, boundary layer processes, frictional dissipation, cloud and precipitation, and horizontal and vertical resolution of different scales of motion. Further, it is to be understood that the above error statistics apply only to forecasts of variables of the atmosphere which are continuous functions of space and time, such as pressure, temperature, geopotential height, wind components, etc. They do not apply to forecasts of actual weather phenomena experienced in daily life, such as fog, cloud, rain, thunderstorms, tornadoes, etc, which may or may not occur during the forecast period. However, practicing forecasters give due weightage to NWP model output statistics and interpret them suitably along with other guiding materials available to them in assessing the probability of occurrence or non-occurrence of any of these weather elements.

That is how the art or science of weather forecasting stands to-day. However, there is no doubt that attempts to develop numerical models with better coverage of initial data over land and ocean, improved numerical techniques, and higher resolution in space and time have led to better understanding of the dynamics of the atmosphere during the last several decades and the trend will undoubtedly continue.

So far as long-range forecasting is concerned, there has been increasing realization in recent years of the role played by oceans in atmospheric circulation and the atmosphere in ocean circulation. This has led to development of interacting coupled ocean-atmosphere models at many of the world meteorological centers (e.g. Saha et al., 2006) and the trend holds out a great promise for future climate forecasts.

Chapter 18
Dynamical Instability of Atmospheric Flows – Energetics and Energy Conversions

18.1 Introduction

In Chaps. 3 and 4, we discussed the vertical stability of a static atmosphere by the parcel displacement method and derived the stability criteria in the cases of both dry and moist air (Eqs. 3.4.14 and 4.10.7). It was shown that the stability depends on the rate at which the potential or the equivalent potential temperature, as the case may be, varies with height. We also discussed the case of conditional instability in which an unstable surface layer is capped by a stable layer or a stable layer is capped by an unstable layer.

We know from experience, however, that the actual atmosphere is never static but always in motion such that the horizontal velocity is normally almost two orders of magnitude greater than the vertical velocity. It has also considerable horizontal and vertical shear. So, a full treatment of atmospheric stability should consider the stability of an atmosphere in three-dimensional motion with lateral and vertical shear.

However, a comprehensive treatment of that kind would obviously be very involved. So, for simplicity, the static and dynamic stability of the atmosphere are usually studied separately. The present chapter first discusses the dynamic stability of a zonal flow which has only lateral shear and is not affected by any buoyancy force which may arise from static instability. This means that we deal with a horizontal flow which has only lateral shear and call the instability 'inertial instability' because in this type of flow the pressure gradient force is balanced by the Coriolis force only. The instability that may arise due to vertical shear of the zonal wind is called 'baroclinic instability' and is treated separately. Later in the chapter, we introduce an instability mechanism called the 'Conditional Instability of the Second Kind (CISK)' which was first suggested by Charney and Eliassen (1964) to explain the development of a tropical wave disturbance into a hurricane. It involves the mutual cooperation between static and dynamic instability.

18.2 Inertial Instability

As in the case of static stability, we apply the parcel method to discuss qualitatively the stability of a zonal current with lateral shear by giving a parcel of air embedded in the current a lateral or cross-current displacement and finding out how it responds to the displacement, i.e., whether it tends to return to the original location in the current or move away farther from it. If it returns to the original location, the flow is regarded as stable. If it moves away farther, the flow is unstable. We may derive the criterion for inertial instability of a steady-state zonal flow with horizontal shear as follows:

Let the basic state zonal wind on an isobaric surface, assumed to be geostrophic, be denoted by u_g and given by

$$u_g = -(1/f)\, \partial \Phi / \partial y$$

The approximate horizontal equations of motion are then

$$du/dt = fv = f\, dy/dt \qquad (18.2.1)$$

$$dv/dt = f(u_g - u) \qquad (18.2.2)$$

Let us now give a parcel of air moving with the current at a position y_0 a displacement δy across the current. The zonal velocity of the parcel at the new position $y_0 + \delta y$ is obtained by integrating (18.2.1):

$$u(y_0 + \delta y) = u_g(y_0) + f\, \delta y \qquad (18.2.3)$$

The geostrophic wind at $y_0 + \delta y$ may be approximated as

$$u_g(y_0 + \delta y) = u_g(y_0) + (\partial u_g / \partial y)\, \delta y \qquad (18.2.4)$$

Substituting (18.2.3) and (18.2.4) in (18.2.2), we obtain

$$dv/dt = d^2(\delta y)/dt^2 = -f(f - \partial u_g / \partial y)\delta y \qquad (18.2.5)$$

The differential equation (18.2.5) has the same form as (3.4.12), the equation for vertical motion of a displaced parcel in a stable atmosphere.

Depending on the sign of the co-efficient of δy on the right-hand side of (18.2.5), the parcel will return to the original position, or stay at the displaced location, or move away farther from that position in the northern hemisphere where f is positive according as

$$(f - \partial u_g / \partial y) \begin{cases} > 0 \text{ Stable} \\ = 0 \text{ Neutral} \\ < 0 \text{ Unstable} \end{cases} \qquad (18.2.6)$$

Since $(f - \partial u_g / \partial y)$ is the absolute vorticity of the basic flow, the condition for inertial instability requires that the absolute vorticity of horizontally sheared zonal flow be negative. In the northern hemisphere, the absolute vorticity is almost always positive and the flow inertially stable. The occurrence of negative absolute vorticity over any large area would immediately set in motion inertially unstable perturbations which would mix the fluid laterally and reduce or obliterate the shear so as to restore stability again. When viewed in an absolute frame of reference, inertial instability actually results from an imbalance between the pressure gradient and the Coriolis (inertial) forces acting on a parcel displaced radially in an axisymmetric vortex.

Static instability and inertial instability are only two of the several possible ways in which the atmosphere can become unstable. A basic flow when perturbed is subject to a variety of modes of instability depending on its horizontal and vertical shear, static stability, variation of the Coriolis parameter, frictional effect, etc. In only a few cases can the stability criteria be worked out satisfactorily In the following section, we adopt a more rigorous approach to work out a criterion for baroclinic instability which arises due to vertical shear in the case of a two-level quasi-geostrophic model, following a treatment by Holton (1979).

18.3 Baroclinic Instability

18.3.1 The Model

The basic equations (17.5.4–17.5.6) for the two-level quasi-geostrophic baroclinic model may be written:

$$\partial(\nabla^2 \psi_1)/\partial t + \mathbf{V}_1 \cdot \nabla(\nabla^2 \psi_1 + f) = (f_0/\Delta p)\, \omega_2 \qquad (18.3.1)$$

$$\partial(\nabla^2 \psi_3)/\partial t + \mathbf{V}_3 \cdot \nabla(\nabla^2 \psi_3 + f) = -(f_0/\Delta p)\, \omega_2 \qquad (18.3.2)$$

$$\partial(\psi_1 - \psi_3)/\partial t + \mathbf{V}_2 \cdot \nabla(\psi_1 - \psi_3) = (\sigma \Delta p/f_0)\omega_2 \qquad (18.3.3)$$

where $\mathbf{V}_j = \mathbf{k} \times \nabla \psi_j$, for $j = 1, 2, 3$ and the arrangement of the levels are as shown in Fig. 17.1.

In order to keep the treatment simple, we adopt the perturbation technique and assume that the stream functions ψ_1 and ψ_3 consist of a basic state part which varies with y only and a perturbation part which varies with x and t. Thus, we assume

$$\begin{aligned} \psi_1 &= -U_1\, y + \psi_1'(x, t) \\ \psi_3 &= -U_3\, y + \psi_3'(x, t) \\ \omega_2 &= \omega_2'(x, t) \end{aligned} \qquad (18.3.4)$$

where the zonal velocities at levels 1 and 3 are given by U_1 and U_3 respectively and the perturbations have velocity components in the meridional and vertical directions only.

Substituting from (18.3.4) into (18.3.1–18.3.3) and linearizing, we get the perturbation equations

18 Dynamical Instability of Atmospheric Flows – Energetics and Energy Conversions

$$(\partial/\partial t + U_1 \partial/\partial x)\partial^2 \psi_1'/\partial x^2 + \beta \partial \psi_1'/\partial x = (f_0/\Delta p)\,\omega_2' \quad (18.3.5)$$

$$(\partial/\partial t + U_3 \partial/\partial x)\partial^2 \psi_3'/\partial x^2 + \beta \partial \psi_3'/\partial x = -(f_0/\Delta p)\,\omega_2' \quad (18.3.6)$$

$$[(\partial/\partial t + \{(U_1 + U_3)/2\}\partial/\partial x](\psi_1' - \psi_3')$$
$$- \{(U_1 - U_3)/2\}\partial(\psi_1' + \psi_3')/\partial x = (\sigma \Delta p/f_0)\omega_2' \quad (18.3.7)$$

where we have used the beta-plane approximation, $\beta \equiv df/dy$, and have linearly interpolated to express $\mathbf{V}_2$ in terms of ψ_1 and ψ_3. Eqs. (18.3.5–18.3.7) constitute a linear set in ψ_1', ψ_3', and ω_2', for which we assume wave-type solutions

$$\begin{aligned}\psi_1' &= A\,\exp\{ik(x-ct)\},\\ \psi_3' &= B\,\exp\{ik(x-ct)\},\\ \omega_2' &= C\,\exp\{ik(x-ct)\}\end{aligned} \quad (18.3.8)$$

Substituting the wave-type solutions (18.3.8) into (18.3.5–18.3.7), and re-arranging, we note that the amplitude factors A, B, and C must satisfy a set of simultaneous, homogeneous, linear algebraic equations

$$ik\{(c - U_1)k^2 + \beta\}\,A - (f_0/\Delta p)\,C = 0 \quad (18.3.9)$$

$$ik\{c - U_3)k^2 + \beta\}\,B + (f_0/\Delta p)\,C = 0 \quad (18.3.10)$$

$$-ik(c - U_3)A + ik(c - U_1)B - \{(\sigma \Delta p)/f_0\}C = 0 \quad (18.3.11)$$

Since the set of equations (18.3.9–18.3.11) is homogeneous it can be solved only if the determinant of the co-efficients of A, B, and C is zero. Thus, the required condition which the phase speed c must satisfy is

$$\begin{vmatrix} ik\{(-U_1)k^2 + \beta\} & 0 & -f_0/\Delta p \\ 0 & ik\{(c - U_3)k^2 + \beta\} & f_0/\Delta p \\ -ik(c - U_3) & ik(c - U_1) & (-\sigma\Delta p)/f_0 \end{vmatrix} = 0$$

Multiplying out the terms in the determinant we obtain a quadratic equation in c in (18.3.12).

$$(k^4 + 2\lambda^2 k^2)c^2 + \{2\beta(k^2 + \lambda^2) - (U_1 + U_3)(k^4 + 2\lambda^2 k^2)\}c$$
$$+ \{k^4 U_1 U_3 + \beta^2 - (U_1 + U_3)(k^2 + \lambda^2)\beta + \lambda^2 k^2(U_1^2 + U_3^2)\} = 0 \quad (18.3.12)$$

where we have put $\lambda^2 \equiv f_0^2/\{\sigma(\Delta p)^2\}$.

Solving (18.3.12) for the phase speed c, we obtain

$$c = U_m - \{\beta(k^2 + \lambda^2)/k^2(k^2 + 2\lambda^2)\} \pm \delta^{1/2} \quad (18.3.13)$$

where

18.3 Baroclinic Instability

$$\delta \equiv [\beta^2\lambda^4/\{k^4(k^2+2\lambda^2)^2\}] - U_T^2(2\lambda^2 - k^2)/(k^2+2\lambda^2),$$
$$U_m \equiv (U_1+U_3)/2, \text{ and } U_T \equiv (U_1-U_3)/2$$

Thus, U_m and U_T are respectively the vertically-averaged zonal wind and the basic-state thermal wind in the pressure interval $(\Delta p)/2$.

In the above we have shown that (18.3.8) is a solution of the equations (18.3.5–18.3.7), provided the phase velocity is given by (18.3.13). However, (18.3.13) is quite complicated, though it permits us to see immediately that for negative values of δ, the phase speed has an imaginary component and the perturbations will amplify exponentially.

18.3.2 Special Cases of Baroclinic Instability

Before discussing the general properties of (18.3.13), we consider two special cases:

Special case 1

In this case, we put $U_T = 0$ in (18.3.13), and obtain two values of c given by:

$$c_1 = U_m - \beta/k^2, \qquad (18.3.14)$$
$$c_2 = U_m - \beta/(k^2 + 2\lambda^2) \qquad (18.3.15)$$

The phase speeds c_1 and c_2 are real quantities which correspond to free stable (normal mode) oscillations of the two-level model with a vertically-averaged barotropic basic state zonal current, U_m.

It will be seen from (18.3.14) that c_1 is simply the phase speed of the barotropic Rossby wave moving westward relative to the barotropic basic state zonal wind, which was derived in (15.6.26) (l, the meridional wave number, being 0 in the present case). c_2 given by (18.3.15), however, is in a different category. It may be interpreted as the phase speed of an internal baroclinic Rossby wave mode.

However, Lindzen et al. (1968) have shown that this baroclinic mode is a spurious one, since it does not arise in the real atmosphere but appears in the two-level quasi-geostrophic model because of the top boundary condition, $\omega = 0$ at $p = 0$. This boundary condition is equivalent to putting a rigid lid at the top, whereas in the real atmosphere there is no such barrier.

Holton (1979) has shown how an analogous baroclinic mode is excited in a homogeneous incompressible ocean with a free surface and a motionless basic state when a rigid-lid boundary condition with no vertical motion at the top is specified. It is shown that a small perturbation defined by $u = u'(x, t)$, $v = v'(x,t)$ at the surface of an ocean of depth $h = H+h'$, where H is mean depth of the ocean, will move westward with a phase speed given by

$$c = -\beta/(k^2 + f_0^2/gH) \qquad (18.3.16)$$

It is obvious that the phase speed given by (18.3.16) is that of a mixed Rossby-gravity mode, i.e., a westward-moving Rossby wave under the stabilizing influence of gravity. The two expressions (18.3.15) and (18.3.16) are analogous, since in the oceanic case, $U_m = 0$ and in the denominator f_0^2/gH replaces $2\lambda^2 \{\equiv 2f_0^2/\sigma(\Delta p)^2\}$. For details of the derivation of (18.3.16), the original reference may be consulted.

Special case 2

In this case we assume $\beta = 0$.

The expression (18.3.13) then reduces to

$$c = U_m \pm U_T \{(k^2 - 2\lambda^2)/(k^2 + 2\lambda^2)\}^{1/2} \qquad (18.3.17)$$

It is easy to see that when $k^2 < 2\lambda^2$, c in (18.3.17) has an imaginary part. This means that all waves longer than a certain critical length given by,

$$L_c = \sqrt{2}(\pi/\lambda),$$

will amplify. Since $\lambda = f_0/(\sigma^{1/2}\Delta p)$, $L_c = (\Delta p)\pi(2\sigma)^{1/2}/f_0$.

For a value of σ normally obtaining in midlatitudes, say at latitude of 45°N, L_c turns out to be about 3000 km. However, it is clear from the formula above that the critical wavelength L_c for baroclinic instability increases with the static stability σ. Further, with $\beta = 0$, although L_c does not depend upon the thermal wind, the exponential growth rate α of the baroclinic wave, given by $\alpha = kc_i$, where c_i denotes the imaginary part of the phase speed c, depends upon it as given by the expression

$$\alpha = kc_i = k\, U_T \{(2\lambda^2 - k^2)/(2\lambda^2 + k^2)\}^{1/2} \qquad (18.3.18)$$

18.3.3 The Stability Criterion – Neutral Curve

If all terms are retained in (18.3.13), a stability criterion called a neutral curve which connects all values of U_T and k for the case in which δ is just about zero and for which the flow may be treated as marginally stable may be worked out. Thus, from (18.3.13), for $\delta = 0$, we have

$$\beta^2 \lambda^4 / \{k^4(k^2 + 2\lambda^2)\} = U_T^2(2\lambda^2 - k^2) \qquad (18.3.19)$$

or, solving for $k^4/2\lambda^4$, we obtain the relation

$$k^4/2\lambda^4 = 1 \pm \{1 - \beta^2/(4\lambda^4 U_T^2)\}^{1/2} \qquad (18.3.20)$$

between the thermal wind and the wavelength in the case of the marginally stable flow. In Fig. 18.1, the nondimensional quantity $k^2/2\lambda^2$, which gives a measure of the marginally stable zonal wavelength, is plotted against the nondimensional parameter $2\lambda^2 U_T/\beta$, which is proportional to the thermal wind.

18.3 Baroclinic Instability

Fig. 18.1 Neutral stability curve for the two-level baroclinic model (After Holton, 1979, published by Academic Press, Inc)

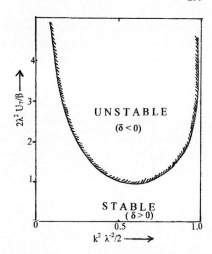

As shown in the diagram, the neutral curve separates the unstable region of the (U_T, k) plane from the stable region. It is evident that the inclusion of the β-effect stabilizes the flow, for now unstable waves can occur only when $|U_T| > \beta/2\lambda^2$. Further, the minimum value of U_T required for unstable growth depends strongly on k. This means that the β-effect stabilizes the long wave end of the spectrum. The flow is also stable for waves shorter than the critical wavelength L_c. The long wave stabilization associated with the β-effect is caused by the westward propagation of long waves which occurs only when the β-effect is included in the model. It can be shown that the baroclinically unstable waves propagate westward at a speed which lies between the maximum and the minimum mean zonal wind speeds.

To find the minimum value of U_T for which unstable baroclinic waves may exist, we differentiate (18.3.19) with respect to k and find that the required condition

$$dU_T/dk = 0$$

is met when $k^2 = \sqrt{2}\lambda^2$.

This wave number corresponds to the wave of maximum baroclinic instability. When U_T increases from zero and reaches this minimum value, the perturbation which first becomes unstable and amplifies has a wave number, $k = 2^{1/4}\lambda$, which corresponds to a wavelength of about 4000 km under normal conditions of static stability.

Observations show that the average wavelength of midlatitude synoptic-scale wave disturbances which amplify and decay is also close to this value. This may, perhaps, explain why baroclinically unstable waves are so common in midlatitude westerly currents. During amplification, the wave extracts energy from the mean thermal wind thereby decreasing the thermal wind and stabilizing the flow till the thermal wind builds up again to start another cycle of energy transformations. It is also observed that the average thermal wind in midlatitude westerlies exceeds that required for maximum baroclinic instability at wavelength near 4000 km which works out to be about $4\,\mathrm{m\,s^{-1}}$ which arises from a wind shear of $8\,\mathrm{m\,s^{-1}}$ between 250 and 750 hPa in the two-level model. We, therefore, find that the observed behaviour of the midlatitude synoptic systems is consistent with the hypothesis that the disturbances develop from infinitesimally small perturbations of a baroclinically unstable basic westerly current by deriving energy for growth from the thermal wind.

Of course, in the real atmosphere, many other sources of energy, such as those due to lateral shear of an unstable zonal current, nonlinear interaction of finite amplitude perturbations, release of latent heat due to condensation, etc., may also influence the development of synoptic-scale disturbances. However, the findings of observational studies, laboratory experiments, and theoretical models all suggest that baroclinic instability is the primary mechanism for development of synoptic-scale wave disturbances in the midlatitudes.

18.4 Vertical Motion in Baroclinically Unstable Waves

The quasi-geostrophic baroclinic model that we discussed in the foregoing sections has to satisfy two important constraints at the same time. These are: that (1) the vorticity changes are geostrophic, and (2) the temperature changes are hydrostatic. In order that both the constraints be satisfied, the vertical motion field at every instant must be adjusted so that the divergent motions keep the vorticity changes geostrophic and the vertical motions keep the temperature changes hydrostatic. These properties of the quasi-geostrophic model are clearly revealed when we use the linearized equations (18.3.5–18.3.7) to compute the vertical motion field. An omega equation for the linearized model can be obtained by taking the second derivative of (18.3.7) and eliminating the time derivatives with the aid of (18.3.5) and (18.3.6). If we neglect the β-effect for simplicity, we obtain the following omega equation

$$(\partial^2/\partial x^2 - 2\lambda^2)\omega_2' = (f_0/\sigma\Delta p)[\partial^2\{U_m(\partial\psi_1'/\partial x - \partial\psi_3'/\partial x)\}/\partial x^2 \\ - \partial^2\{U_T(\partial\psi_1'/\partial x + \partial\psi_3'/\partial x)\}/\partial x^2 \\ - \{U_1\partial(\partial^2\psi_1'/\partial x^2)/\partial x - U_3\partial(\partial^2\psi_3'/\partial x^2)/\partial x\}] \quad (18.4.1)$$

We may interpret the terms on the right-hand side of (18.4.1) as follows. The first term represents the Laplacian of the advection of the perturbation thickness by the basic state vertically-averaged mean wind. The second term is proportional to the Laplacian of the advection of the basic state thickness by the vertically-averaged perturbation meridional wind. The third term represents the differential advection of the perturbation vorticity by the basic state wind. Thus, it appears that three distinct physical processes force vertical motion in this model. However, it can be shown that the first and third terms can be combined to give an expression identical to the second term so that (18.4.1) may be written

18.4 Vertical Motion in Baroclinically Unstable Waves

$$(\partial^2/\partial x^2 - 2\lambda^2)\omega_2' = -(2f_0/\sigma\Delta p)\partial^2\{U_T(\partial\psi_1'/\partial x + \partial\psi_3'/\partial x)\}/\partial x^2 \quad (18.4.2)$$

Since $(\partial^2/\partial x^2 - 2\lambda^2)\omega_2' \propto -\omega_2'$, and the thermal wind $U_T \propto -\partial\underline{T}/\partial y$, where $\underline{T}$ is mean temperature, we have from (18.4.2)

$$\omega_2' \propto -\partial^2(v_2'\partial\underline{T}/\partial y)/\partial x^2 \propto v_2'\partial\underline{T}/\partial y \quad (18.4.3)$$

Alternatively, since the right-hand side of (18.4.2) can be expressed in terms of perturbation vorticity, we may write (18.4.2) in the form

$$\omega_2' \propto U_T \partial \zeta_2'/\partial x \quad (18.4.4)$$

where $\zeta_2' = (\zeta_1' + \zeta_3')/2$ is the vertically-averaged perturbation vorticity.

Thus, in the linearized two-level model, the net forcing of vertical motion is proportional to: (1) the advection of the basic state temperature field by the vertically-averaged perturbation meridional wind, or (2) the advection of the vertically-averaged perturbation vorticity by the basic state thermal wind.

A schematic in Fig. 18.2 depicts the phase relationships between the geopotential field and the divergent secondary motion field for a developing unstable baroclinic wave in the two-level model over the midlatitudes where U_T is usually positive.

Summarizing the above results, one may state that

(a) Cold (warm) advection forces sinking (rising) motion, or
(b) Negative (positive) vorticity advection by the thermal wind forces sinking (rising) motion.

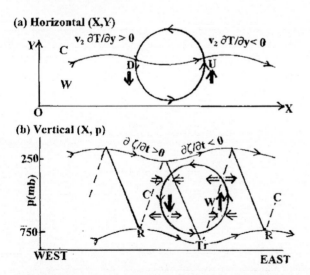

Fig. 18.2 Schematic showing (a) vertical motion forced by basic state horizontal thermal advection, and (b) vertical motion forced by divergence and vorticity changes in the field of a developing unstable baroclinic wave in the two-level model. Symbols used: C-cold, W-warm, U-upward, D-downward, Tr-trough, R-ridge, DV-divergence, CV-convergence. The circles with arrows show the directions of the secondary circulations

18.5 Energetics and Energy Conversions in Baroclinic Instability

18.5.1 Definitions

18.5.1.1 Internal Energy

Let the internal energy of a vertical section dz of a column of air of unit cross-section be denoted by dI. Then, by definition, $dI = \rho\, c_v\, T\, dz$, where ρ is the density, c_v is the specific heat at constant volume, and T is the absolute temperature of the air in the vertical section. Integrating from the earth's surface to the top of the atmosphere, the total internal energy of the column of air is given by

$$I = c_v \int_0^\infty \rho\, T\, dz \tag{18.5.1}$$

18.5.1.2 Potential Energy

On the other hand, the gravitational potential energy, dP, of the same vertical section at a height z above the earth's surface is given by

$$dP = \rho\, g\, z\, dz.$$

Integrating through the atmosphere, the total potential energy of the column is given by

$$P = g \int_0^\infty \rho\, z\, dz = -\int_{p_0}^0 z\, dp \tag{18.5.2}$$

Integrating (18.5.2) by parts and using the ideal gas law, we obtain,

$$P = -zp \Big|_{p_0}^0 + \int_0^\infty p\, dz = R \int_0^\infty \rho\, T\, dz \tag{18.5.3}$$

Thus we find that I and P are related to each other as

$$I/P = c_v/R$$

and the total potential energy of the atmosphere may be expressed as

$$I + P = (c_p/c_v)I = (c_p/R)P \tag{18.5.4}$$

18.5.1.3 Available Potential Energy and Kinetic Energy

The total potential energy as defined above (18.5.4), however, is not a suitable measure of energy in the atmosphere, because the bulk of it is unavailable for useful work for a variety of reasons. Only a tiny part is available for atmospheric circulation, as discussed below, for a model atmosphere. Let us consider two equal masses of dry air of uniform potential temperatures, θ_1 and θ_2, with $\theta_1 < \theta_2$, separated by a vertical partition (Fig. 18.3).

Horizontal dashed lines indicate approximate isobaric surfaces. Arrows show the direction of motion when the vertical partition is withdrawn. The approximate rest position of the surface of discontinuity after re-arrangement of the airmasses is indicated by an inclined dashed line.

Let the ground level pressure on each side of the partition be 1000 mb. It is obvious that when the vertical partition is removed, there will be an adiabatic re-arrangement of the air masses, the warmer air moving towards the colder air aloft, and the colder air undercutting the warmer air near the ground in opposite directions. The horizontal movement will generate a certain degree of vertical motion in the two airmasses with downward motion in the colder airmass and upward motion in the warmer airmass till a mass balance is reached along a surface of airmass discontinuity, indicated by a dashed line inclined to the vertical, as shown schematically in Fig. 18.3.

We may get an idea of the kinetic energy generated by the re-arrangement of the masses within the same volume by considering the total energy of the system before and after removal of the vertical partition. Since the total energy is conserved during an adiabatic process, we have

$$I + P + K = \text{Const}$$

where K denotes kinetic energy.

If the air masses are initially at rest, $K = 0$. Thus, if we use primed quantities to denote the final state, we have

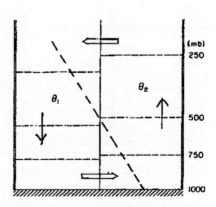

Fig. 18.3 Two airmasses of different potential temperatures separated by a vertical partition

$$I + P = K' + P' + I'$$

With the aid of (18.5.4), it can be shown that the kinetic energy that is generated by the re-arrangement of the masses is given by the relation

$$K' = (c_p/c_v)(I - I')$$

According to the above relation, K' increases with decrease in the value of I' and becomes maximum when I' attains its absolute minimum value, I'', i.e., when the warmer air lies entirely over the colder air and the surface of separation between the two air masses becomes horizontal. In this extreme case, the total potential energy that becomes unavailable amounts to $(c_p/c_v) I''$. No further reduction of potential energy is possible beyond this stage. The maximum possible kinetic energy corresponding to maximum available potential energy in the atmosphere that can be realised by adiabatic re-arrangement of two air masses is, therefore, given by the expression

$$K'_{max} = (c_p/c_v)(I - I'') \tag{18.5.5}$$

Lorenz (1960) has shown that the available potential energy (abbreviated to A.P.E.) in the earth's atmosphere is given by the volume integral of the variance of the potential temperature on isobaric surfaces over the entire atmosphere. Thus, if $\underline{\theta}$ is the average potential temperature at a given isobaric surface and θ' the local deviation from the average, the average A.P.E. per unit volume is to satisfy the proportionality

$$\text{A.P.E.} \propto (1/V) \int \{(\underline{\theta'^2})/(\underline{\theta}^2)\} dV$$

where V denotes the volume and the average value is denoted by underlining.

Observations indicate that for the atmosphere as a whole, the mean A.P.E. is only about 1/200 of the mean total potential energy and that of what is available, only about 1/10 can be converted into mean kinetic energy. This is equivalent to saying that almost 99.95% of the mean total potential energy of the atmosphere is unavailable for any useful work. Thus, from the point of view of energy conversions, the atmosphere is highly inefficient as a heat engine.

18.5.2 Energy Equations for the Two-Level Quasi-Geostrophic Model

In the two-level quasi-geostrophic model (18.3.1–18.3.3), the perturbation temperature field is proportional to $(\psi'_1 - \psi'_3)$, the 250–750 mb thickness. Thus, in conformity with the discussion in the previous sub-section, the available potential energy in this model is proportional to $(\psi'_1 - \psi'_3)^2$. To show that this is, indeed, the case, we derive the energy equations for the two-level model system as follows. We first multiply (18.3.5) by $-\psi'_1$, (18.3.6) by $-\psi'_3$, and (18.3.7) by $(\psi'_1 - \psi'_3)$. Next, we integrate the resulting equations over one wavelength of the perturbation in the

18.5 Energetics and Energy Conversions in Baroclinic Instability

zonal direction, noting that the average of any term over one wavelength will be indicated by an angle bracket as shown below.

$$<(\ )> = (1/L) \int_0^L (\ )\, dx$$

where L is the wavelength of the perturbation.

Thus, for the first term involving differentiation with respect to time in (18.3.5), the average after multiplication by $-\psi'_1$ yields

$$-<\psi'_1 \partial(\partial^2 \psi'_1/\partial x^2)/\partial t> = -<\psi'_1 \partial^2(\partial \psi'_1/\partial t)/\partial x^2>$$
$$= -<\underbrace{\partial/\{\psi'_1 \partial(\partial \psi'_1/\partial t)/\partial x\}/\partial x}_{A}> + <\underbrace{\{(\partial \psi'_1/\partial x)\partial(\partial \psi'_1/\partial x)/\partial t\}}_{B}>$$

Term A is equal to zero because it is the integral of a perfect differential in x over a complete cycle. Term B can be expressed as $<\{\partial(\partial \psi'_1/\partial x)^2/\partial t\}/2>$

which is just the rate of change of perturbation kinetic energy per unit mass and per unit time, averaged over a cycle.

Similarly, $-\psi'_1$ times the advection term on the left-hand side of (18.3.5) can be written after integration in x over a cycle

$$-U_1<\psi'_1\{\partial^2(\partial \psi'_1/\partial x)/\partial x^2\}> = -U_1<\{\partial\{\psi'_1 \partial(\partial \psi'_1/\partial x)/\partial x\}/\partial x>$$
$$+U_1<(\partial \psi'_1/\partial x)(\partial^2 \psi'_1/\partial x^2)>$$
$$= (U_1/2)<\partial(\partial \psi'_1/\partial x)^2)/\partial x>$$
$$= 0$$

The net result is that the advection of perturbation kinetic energy when integrated over a wavelength vanishes. In the same way, we can evaluate the various terms in (18.3.6) and (18.3.7) and after multiplying by $-\psi'_3$ and $(\psi'_1 - \psi'_3)$ respectively, we obtain the perturbation kinetic energy equations

$$(1/2)<\partial(\partial \psi'_1/\partial x)^2/\partial t> = -(f_0/\Delta p)<\omega'_2 \psi'_1> \qquad (18.5.6)$$

$$(1/2)<\partial(\partial \psi'_3/\partial x)^2/\partial t> = (f_0/\Delta p)<\omega'_2 \psi'_3> \qquad (18.5.7)$$

$$(1/2)<\partial(\psi'_1 - \psi'_3)^2/\partial t> = U_T<\{(\psi'_1 - \psi'_3)\partial(\psi'_1 + \psi'_3)/\partial x\}>$$
$$+ (\sigma \Delta p/f_0)<\omega'_2(\psi'_1 - \psi'_3)> \qquad (18.5.8)$$

where $U_T \equiv (U_1 - U_3)/2$.

Since the total perturbation kinetic energy is the sum of the perturbation kinetic energies at 250 and 750 mb levels, we have from (18.5.6) and (18.5.7)

$$dK'/dt = -(f_0/\Delta p)<\omega'_2(\psi'_1 - \psi'_3)> \qquad (18.5.9)$$

where we have put $K' \equiv (1/2)<\{(\partial \psi'_1/\partial x)^2 + (\partial \psi'_3/\partial x)^2\}>$

Thus, the rate of change of perturbation kinetic energy given by (18.5.9) is proportional to the correlation between the perturbation vertical motion and thickness. If we now define the perturbation available potential energy as

$$P' = (\lambda^2/2) < (\psi'_1 - \psi'_3)^2 >$$

where, as stated before, $\lambda^2 \equiv f_0^2/\{\sigma(\Delta p)^2\}$, we obtain from (18.5.8) an expression for the rate of change of perturbation available potential energy

$$dP'/dt = \lambda^2 U_T < (\psi'_1 - \psi'_3)\partial(\psi'_1 + \psi'_3)/\partial x > + (f_0/\Delta p) < \omega'_2(\psi'_1 - \psi'_3) > \quad (18.5.10)$$

So, the rate of change of available potential energy consists of two parts: the first part represents horizontal temperature advection by the perturbation meridional wind in a region where the thermal wind is positive ($U_T > 0$) and the second part a correlation between thickness and vertical motion. It should be noted that the second term on the right-hand side of (18.5.10) is exactly equal and opposite of the rate of change of perturbation kinetic energy given in (18.5.9) and thus represents a conversion between potential and kinetic energy. Physically, it states that potential energy is converted to kinetic energy when warm (cold) air is rising (sinking). Conversely, kinetic energy is converted to potential energy when cold (warm) air is rising (sinking).

Combining (18.5.9) and (18.5.10), we get

$$d(P' + K')/dt = \lambda^2 U_T < (\psi'_1 - \psi'_3)\partial(\psi'_1 + \psi'_3)/\partial x > \quad (18.5.11)$$

Thus, in areas where $U_T > 0$, the total energy of the perturbation will increase or decrease, according as the correlation between the meridional wind and temperature is positive or negative, regardless of the vertical circulation which simply converts a part of the available potential energy to perturbation kinetic energy.

18.6 Barotropic Instability

Barotropic instability arises when the horizontal shear of the wind increases beyond a certain critical limit. It can be shown by scale analysis that in the absence of condensation, vertical motion in the tropics must be small. So, the flow is governed by the barotropic vorticity equation

$$(\partial/\partial t + \mathbf{V} \cdot \nabla)(\zeta + f) = 0 \quad (18.6.1)$$

We now take a basic state zonal wind $\underline{u}$ (y) which varies with y only and superimpose on it a small barotropic perturbation, u′, so that

$$u = \underline{u}(y) + u' \quad v = v'$$

18.6 Barotropic Instability

Since the flow is quasi-nondivergent, the perturbation can be represented by a streamfunction ψ' by putting

$$u' = -\partial\psi'/\partial y, \qquad v' = \partial\psi'/\partial x$$

Substituting in (18.6.1), we can write the linearized perturbation vorticity equation in the form

$$(\partial/\partial t + \underline{u}\partial/\partial x)\nabla^2\psi' + (\beta - d^2\underline{u}/dy^2)\partial\psi'/\partial x = 0 \qquad (18.6.2)$$

where $(\beta - d^2\underline{u}/dy^2)$ is simply the latitudinal gradient of the absolute vorticity.

We now assume a wave solution of (18.6.2) and put

$$\psi'(x,y,t) = \psi(y)\exp\{ik(x-ct)\} \qquad (18.6.3)$$

where $\psi = \psi_r + i\psi_i$ is a complex function of y alone with the suffixes r and i denoting real and imaginary parts respectively. Substituting from (18.6.3) into (18.6.2) we get

$$(\underline{u}-c)(d^2\psi/dy^2 - k^2\psi) + (\beta - d^2\underline{u}/dy^2)\psi = 0 \qquad (18.6.4)$$

which is an ordinary second-order partial differential equation in $\psi(y)$.

For solution of (18.6.4), it is generally assumed that the flow is confined into a zonal channel with boundaries at $y = \pm L$, so that

$$\psi(y) = 0, \text{ at } y = \pm L \qquad (18.6.5)$$

Since the amplitudes of the perturbations are assumed to be small, the presence of these artificial boundaries at $\pm L$ should not materially affect the solutions. For a given distribution of $\underline{u}(y)$, it turns out that (18.6.4) will have solutions which satisfy (18.6.5) only for certain values of the phase speed c. In those cases where c is complex (i.e., $c = c_r + ic_i$) with a positive imaginary part, we see from (18.6.3) that the perturbation amplitude, $\exp(kc_i t)$, will grow exponentially with time.

In a given situation it is not easy to solve (18.6.4) because the co-efficients are not constant. However, it is possible to obtain the necessary condition for existence of instability by applying simple integral considerations. For this, we re-write (18.6.4) in the form

$$d^2\psi/dy^2 - [k^2 - \{(\beta - d^2\underline{u}/dy^2)/(\underline{u}-c)\}]\psi = 0 \qquad (18.6.6)$$

If the phase speed c is complex, $(\underline{u}-c)^{-1}$ is also complex and has real and imaginary parts

$$\delta_r = (u-c_r)/\{(u-c_r)^2 + c_i^2\}, \qquad \delta_i = c_i/\{(u-c_r)^2 + c_i^2\}$$

Equation (18.6.6) can then be separated into real and imaginary parts as follows:

$$d^2\psi_r/dy^2 - [k^2 - (\beta - d^2\underline{u}/dy^2)\delta_r]\psi_r - (\beta - d^2\underline{u}/dy^2)\delta_i\psi_i = 0 \qquad (18.6.7)$$

$$d^2\psi_i/dy^2 - [k^2 - (\beta - d^2\underline{u}/dy^2)\delta_r]\psi_i + (\beta - d^2\underline{u}/dy^2)\delta_i\psi_r = 0 \qquad (18.6.8)$$

Multiplying (18.6.7) by ψ_i and (18.6.8) by ψ_r, and then subtracting the latter from the former, we get

$$\psi_i\, d^2\psi_r/dy^2 - \psi_r\, d^2\psi_i/dy^2 - (\beta - d^2\underline{u}/dy^2)\delta_i(\psi_i^2 + \psi_r^2) = 0$$

which can be written as

$$d(\psi_i\, d\psi_r/dy - \psi_r\, d\psi_i/dy)/dy = (\beta - d^2\underline{u}/dy^2)\delta_i(\psi_r^2 + \psi_i^2) \qquad (18.6.9)$$

Integrating (18.6.9) with respect to y and applying the boundary conditions

$$\psi_r = \psi_i = 0, \text{ at } y = \pm L$$

we find that the terms on the left-hand side integrate to zero so that we are left with the integral condition

$$\int_{-L}^{+L} (\beta - d^2\underline{u}/dy^2)\delta_i |\psi|^2 dy = 0 \qquad (18.6.10)$$

where $|\psi|^2 = \psi_r^2 + \psi_i^2$

For an unstable perturbation to exist, δ_i should be positive, which means that $c_i > 0$.

Since $|\psi|^2 \geq 0$ everywhere in the domain, the integral condition in (18.6.10) can be satisfied for an unstable wave only when $(\beta - d^2\underline{u}/dy^2)$ changes sign somewhere in the region between $+L$ and $-L$. Thus, a necessary condition for barotropic instability is that the latitudinal gradient of absolute vorticity of the mean wind must vanish somewhere in the region, i.e.

$$\beta - d^2\underline{u}/dy^2 = 0 \text{ somewhere} \qquad (18.6.11)$$

18.7 Conditional Instability of the Second Kind (CISK)

It is well-known that the tropical atmosphere as a whole is in general conditionally unstable. The lower troposphere in which most of the moisture is confined is unstable, while the upper troposphere is relatively dry and gravitationally stable. Cloud development in such an atmosphere seldom goes beyond the stage of small-scale cumulus clouds. But when a depression forms, which usually has a cold core below and a warm core above, the situation is somewhat different, since it allows large-scale clouds and weather to develop in some favorable sectors where low-level moisture convergence is maximum. In some rare cases (perhaps, one in a thousand), a depression changes its structure and amplifies into a warm-core hurricane (or cyclone or typhoon).

The question is: What causes this transformation from a zonally-asymmetric depression into an axisymmetric cyclone and what is the triggering mechanism? In an attempt to address this question and make the perturbation analysis applicable

18.7 Conditional Instability of the Second Kind (CISK)

to the case of the tropical depression, Charney and Eliassen (1964) made two significant assumptions: that a depression is an infinitesimally small-amplitude disturbance and that surface frictional convergence causes a small-amplitude, symmetric perturbation of depression scale in a conditionally unstable atmosphere to amplify spontaneously. They were aware that these assumptions are not quite valid in the real atmosphere and will prevent a direct application of their theoretical analysis to the actual atmosphere but felt content with indicating the importance of the physical processes in a small amplitude system that they believed would be important in finite amplitude systems as well. According to them, the cyclone develops by a kind of secondary instability in which existing cumulus convection is augmented in regions of low-level horizontal convergence and suppressed in regions of low-level divergence. The cumulus – and cyclone-scale motions are thus to be regarded as co-operating rather than competing – the clouds supplying the latent heat energy to the cyclone, and the cyclone supplying the fuel, in the form of the low-level moisture, to the clouds. The amplification of the disturbance is, therefore, due to the surface frictionally induced convergence of moisture and liberation of latent heat in the center of the cyclone.

There is converging evidence that the basic concepts of CISK as a triggering mechanism for development of a tropical depression to a cyclone, hold in the real atmosphere. Following the pioneering studies of Charney and Eliassen (1964) as well as Arakawa and Schubert (1974), several studies (e.g., Krishnamurti et al., 1976; Shukla, 1978) have been undertaken to test the validity of CISK in the case of a monsoon depression using both observed and model-predicted data. According to Krishnamurti et al. (1976), who studied the structure and energetics of a monsoon depression as derived from a multi-level primitive equation model and concluded that though barotropic and baroclinic energy sources make significant contributions to development, the primary source of energy for a monsoon depression is condensation heating organized in some active sectors of the depression. The cumulus convective heating generates eddy available potential energy by releasing heat at appropriate levels above the cold core of the depression where a slight warm core is present. Here, the rising of relatively warm air contributes significantly to the generation of eddy kinetic energy of the depression.

Shukla (1978) who used a three-layer quasi-geostrophic model and the Arakawa and Schubert (1974) cumulus parameterization scheme to deduce heating due to condensation also computed the structure and energetics of a perturbation for a wave-length corresponding to that of the observed monsoon depression and found that the dominant energy transformation for the computed as well as the observed perturbations was from eddy available potential energy to eddy kinetic energy. According to his study, the primary source of heating is condensational heating. He finds reasonable agreement between the structure and energetics of the computed perturbation and the observed depression and suggests that CISK may act as the primary driving mechanism for the growth of the perturbation into a depression.

Chapter 19
The General Circulation of the Atmosphere

19.1 Introduction – Historical Background

Historically, there must been a time when people had little idea of a circulation in the earth's atmosphere. Few were aware that the wind at their locality was related to the wind at another location on the face of the earth. It appears that the first to visualize a circulation in the atmosphere was the British scientist, Halley (1686), who made a detailed study of the wind systems over the tropical belt with the data then available and hypothesized that the observed trade winds at the surface were part of a direct thermally-driven vertical circulation between a heat source and a heat sink, which reversed direction between the lower and the upper levels and between winter and summer. About half a century later, George Hadley (1735), also a distinguished and well-known British scientist, investigated the same problem and offered an explanation for the cause of the trade winds as well as their observed direction on the basis of differential heating between the equatorial and the higher latitudes and rotation of the earth. He argued that a general equatorward drift of the tradewinds at low levels required a compensating poleward drift at high levels in order to prevent an undue accumulation of mass near the equator. Further, a general westward drag by the tradewinds due to the rotation of the earth on the earth's surface at low latitudes required a compensating eastward drag by the westerlies at high latitudes so as to prevent a general slowing down of the earth. It was found later that the general westward or eastward component of the wind could be easily explained on the basis of the principle of conservation of absolute angular momentum of the earth. A parcel of air moving equatorward from high latitudes in order to conserve the angular momentum of its original latitude would acquire an increasingly westward drift, while a poleward-moving parcel would acquire an increasingly eastward drift. This was due to the fact that the earth's surface at the equator moved faster than at higher latitudes. A change of wind direction from easterly at low levels to westerly at high levels also follows from thermal-wind considerations due to equator-to-pole horizontal temperature gradient. Hadley's idealized single-cell circulation model, shown schematically in Fig. 19.1 held ground and went unchallenged for nearly a

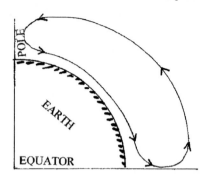

Fig. 19.1 Schematic of a single-cell Hadley circulation model

century and it was once thought that Hadley's model was representative of mean meridional circulation over all parts of the globe at all times of the year. However, later observations called for a modification of Hadley's idealized single-cell model. The new observations revealed the presence of a well-marked high pressure belt over the subtropics and a low pressure belt further poleward near 60° latitude, which suggested a meridional pressure gradient and a poleward drift of air, instead of an equatorward drift near surface, and a compensating equatorward drift at some height, over the midlatitudes. Further, the westerly wind over the midlatitudes were found to be baroclinically unstable and characterized by large-scale eddy motion. Amongst the early attempts made to modify Hadley's original scheme were those of Thomson (1857) and Ferrel (1859) who introduced a shallow indirect cell, characterized by a poleward flow near the surface and equatorward flow at some height, over the midlatitudes, within the framework of the idealized single-cell Hadley circulation model, as shown schematically in Fig. 19.2.

Further modifications to the meridional circulation model were made in the light of later observations. Of these, one proposed by Rossby (1947), is shown in Fig. 19.3.

The three-cell meridional circulation model shows a direct circulation cell over the tropical belt, an indirect circulation cell over the midlatitudes and a direct circulation cell over the polar latitudes, with a polar front located at a latitude of about 60°. The Rossby model has, by and large, stood the test of time and found general

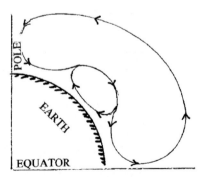

Fig. 19.2 The Hadley circulation model as modified by Thomson (1857) and Ferrel (1859)

19.2 Zonally-Averaged Mean Temperature and Wind Fields Over the Globe

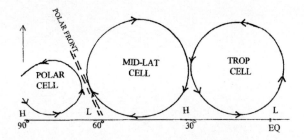

Fig. 19.3 Schematic of a three-cell meridional circulation model proposed by Rossby (1947)

acceptance by the scientific community to this day. In this model, it is the tropical cell which is identified as the classical Hadley circulation.

Palmen (1951), however, found that the use of the zonally-averaged data often tended to obscure or wipe out some of the more important features of the mean meridional circulation. He showed that the mean meridional circulation during winter differs significantly from the Rossby three-cell model. He found that over the high latitudes the meridional drifts of air were very slow compared to the zonal flow and concluded that the major part of the heat transfer over these high latitudes was effected by horizontal waves, whereas that over the tropics was accomplished by the tropical Hadley circulation.

Mintz (1951) presented a composite picture of the zonally-averaged observed mean zonal winds over the globe in a vertical cross-section extending from pole to pole and from m.s.l. to 50 mb during winter and summer, which is shown in Fig. 19.4.

The profiles of the zonal wind shown by Fig. 19.4 reveal the presence of deep easterlies over the tropics and westerlies over the middle and high latitudes in both the seasons though with some seasonal shifts in their latitudinal boundaries but considerable uncertainty appears to exist in the latitudinal extent and depth of the easterlies over the polar belt. However, the mean zonal wind over the Antarctic during southern summer appears to be easterly throughout the whole troposphere.

19.2 Zonally-Averaged Mean Temperature and Wind Fields Over the Globe

19.2.1 Longitudinally-Averaged Mean Temperature and Wind Fields in Vertical Sections

Results of a computation of the zonally-averaged fields of temperature and zonal wind in the northern hemisphere troposphere and lower stratosphere for January and July are presented in Figs. 19.5(a) and (b) respectively.

The salient features of Figs. 19.5(a, b) may be summarized as follows:

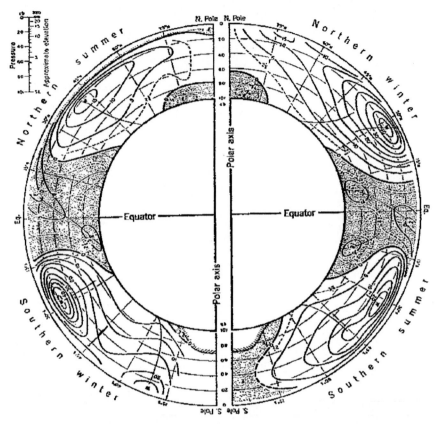

Fig. 19.4 The profile of the observed mean zonal wind in a vertical section of the global atmosphere during winter and summer: Sections with easterlies are *shaded* (Mintz, 1951)

In January (Fig. 19.5a, lower panel), the mean temperature decreases continuously from equator to pole with strongest horizontal temperature gradient over the midlatitudes and from surface upward. A minimum temperature of about $-80\,°C$ is found at the tropopause level over the equatorial belt at about 100 mb at altitude about 16 km a.s.l. The tropopause level breaks over the subtropical belt and suddenly lowers poleward of about 30N and gradually descends to about 300 mb at an altitude of about 10 km over the polar belt. The stratospheric temperature increases with height over the tropical and subtropical belts but decreases with height over the polar region.

The wind field shows two belts of light easterly winds in the troposphere, one over the tropics equatorward of 30N with depth increasing towards the equator and the other, a shallow one, over the polar belt, north of about 70N. Light easterly winds also appear over the tropical stratosphere. A strong westerly jet of velocity about $40\,\mathrm{m\,s^{-1}}$ prevails at an altitude of about 12 km over the subtropical belt and a westerly wind maximum about $20\,\mathrm{m\,s^{-1}}$ appears in the stratosphere over the latitudes between about 60N and 70N.

19.2 Zonally-Averaged Mean Temperature and Wind Fields Over the Globe

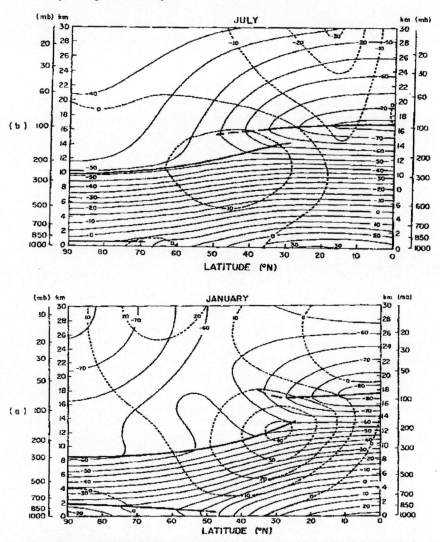

Fig. 19.5 Meridional cross-sections of longitudinally-averaged temperature in degrees Celsius (___) and zonal wind in meters per second (......) for the northern hemisphere in (**a**) January, and (**b**) July. Zonal winds positive westerly, negative easterly. *Heavy lines* denote the tropopause and the Arctic inversion. (From U.S. Navy Weather Research Facility Arctic Forecast Guide, 1962)

In July (Fig. 19.5b, upper panel), though the temperature field looks somewhat similar to that in January, there are significant differences. The warmest temperatures are now near surface over the tropical belt with temperatures decreasing both poleward and upward as in January but the horizontal temperature gradient is now much smaller.

A major change appears to have taken place in July in the thermal structure of the stratosphere over latitudes north of about 50N, with significant warming of the polar atmosphere by more than 30 °C.

Significant changes may also be noticed in the wind field in July. The westerly jet over the subtropical belt has weakened and moved poleward, while the easterly wind dominates the whole tropical belt and extends to great heights. The westerly wind maximum in the stratosphere over the high latitudes has disappeared and is replaced by light easterlies.

19.2.2 Idealized Pressure and Wind Fields at Surface Over the Globe in the Three-Cell Model

A general view of the mean sea-level pressure and wind systems consistent with the three-cell model is shown in Fig. 19.6.

Fig. 19.6 shows the tradewinds of the two hemispheres over the tropical belts blowing equatorward and converging into a well-marked trough of low pressure known as the equatorial trough and forming an intertropical convergence zone (ITCZ) at the equator, the westerly winds over the midlatitudes poleward of a well-marked ridge of high pressure over the subtropical belt; and a belt of easterlies poleward of the polar front.

The current view is that neither the original single-cell Hadley circulation model nor any of the above-mentioned modified models truly represents the actual circulation in the atmosphere which varies with time and longitude.

However, a three-cell model consisting of two direct Hadley-type cells, one over the tropics between the equator and about 30° parallels and the second over the polar belt poleward of about 60° parallels, and an indirect Thomson-Ferrel cell over the midlatitudes is, perhaps, the closest to what is observed in the real atmosphere over

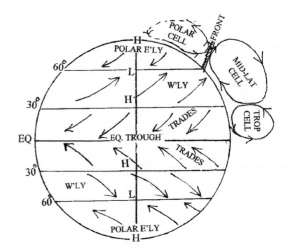

Fig. 19.6 Idealized pressure and wind belts at the earth's surface in a three-cell circulation model. Note that in this model, the tradewinds converge at the equatorial trough of low pressure along the equator

any part of the globe at any time of the year. This appears to be well borne out by a study of Starr (1968) who computed the mean northward wind component from a dataset of 700 stations spread over 5 years from all parts of the globe and later revised his earlier estimates by adding data of another 100 stations.

In this chapter, we use the term 'general circulation' to mean generally a zonally-symmetric annual-mean basic background circulation which is assumed to be present at all longitudes at all times of the year, with a clear understanding that it may be influenced by several regional or local meteorological factors, such as large thermal contrasts between continents and oceans, or between one part of the ocean and another part, or by large-scale orography.The deviations from the general circulation as defined above may be particularly large over the tropics where along with surface thermal contrasts, an extra source of diabatic heating may be introduced by release of latent heat of condensation and its effects on convection and subsidence.

19.3 Observed Distributions of Mean Winds (Streamlines) and Circulations Over the Globe During Winter and Summer

Fig. 19.7 shows maps of mean low-level winds and circulations over the globe during (a) January, and (b) July.

It is evident from Fig. 19.7 that the observed low level circulations over the globe are to a large measure consistent with the fields of surface pressures shown in Figs. 2.1. Oceans and continents appear to have their own characteristic circulation systems and that they respond differentially to the seasonal movement of heat sources and sinks across the equator, as evident from the seasonal movement of the ITCZ (indicated by a double-dashed line). For example, the ITCZ sweeps out nearly $45°-50°$ of latitude between the hemispheres over the Asia-Australia region as well as between North and South America, whereas it moves through only about $10°$ of latitude over most of the Pacific and the Atlantic oceans. Further, circulations over continents are influenced by topographic features such as high mountain ranges, while those over oceans are affected by the distribution of warm and cold ocean currents. Over broad expanses of the eastern Pacific and eastern Atlantic, the ITCZ remains confined to the northern hemisphere only, since its movement to the southern hemisphere is prevented by the effect of the powerful cold ocean currents, viz., the Humbolt or Peruvian current in the South Pacific Ocean and the Benguela current in the South Atlantic Ocean, which keep the equatorial ocean surface cold.

The dominating features of low level wind systems over both continents and oceans in both seasons are the tradewinds which diverge anticyclonically from the subtropical high pressure cells and blow towards the equator to converge at the cyclonic circulation around the equatorial trough of low pressure and form the ITCZ. In this regard, a special situation appears to exist over the Asian continent. During winter, the lofty Himalaya Mountain ranges and the Tibetan plateau divide the subtropical high pressure system over Asia into two parts, one to the north of the

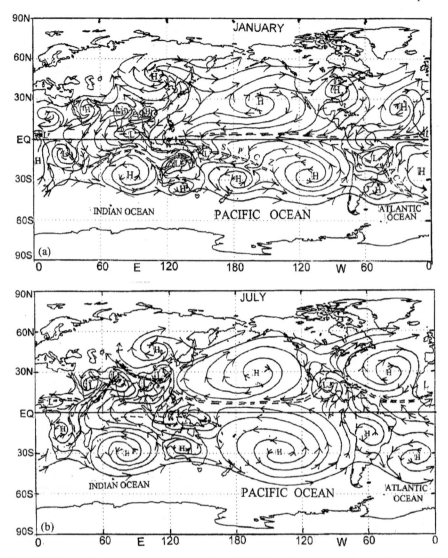

Fig. 19.7 Streamlines showing mean low level (900 hPa) circulations over the globe during January, and July. H-High, L-Low *Arrow* shows direction of airflow

mountain ranges to be centered over the Mongolian region, and the other to the south broken up into three sub-cells over southern Asia, one over the Arabian peninsula, the second over India and the third over upper Myanmar. During northern summer, the high pressure sub-cells over southern Asia are replaced by low-pressure cells to which winds diverging from the southern hemisphere subtropical high pressure cells converge, thereby ushering in the southwest monsoon over the Indian subcontinent. Poleward of the subtropical high pressure belt, winds are generally westerly to latitude about 60°.

19.4 Maintenance of the Kinetic Energy and Angular Momentum of the General Circulation

We now enquire how the observed circulation and some of its important properties, such as kinetic energy and absolute angular momentum, etc., are maintained in the atmosphere. The list is not exhaustive, for one could also investigate into the global balances of potential energy, internal energy, water vapour, etc. In this section, we review the physical processes by which the atmosphere gains or loses some of these properties and the transfer mechanisms by which a balance is achieved between the source and the sink. Let us first look into the kinetic energy balance of the atmosphere.

19.4.1 The Kinetic Energy Balance of the Atmosphere

An expression for the kinetic energy of the atmosphere can be obtained from the horizontal momentum equation (11.4.7) by multiplying it scalarly by the velocity vector $\mathbf{V}$. Thus,

$$\mathbf{V} \cdot d\mathbf{V}/dt = \mathbf{V} \cdot (-\alpha \, \nabla p - 2 \, \mathbf{\Omega} \times \mathbf{V} - \nabla \Phi + \mathbf{F})$$

Or, $\qquad \rho \, d(V^2/2)/dt = -\mathbf{V} \cdot \nabla p + \rho \, \mathbf{V} \cdot \mathbf{g} + \rho \, \mathbf{V} \cdot \mathbf{F}$ \hfill (19.4.1)

Using the relationship, $\rho d (V^2/2)/dt = d(\rho V^2/2)/dt - (\rho V^2/2\rho) \, d\rho/dt$ and the vector identity $\mathbf{V} \cdot \nabla p = \nabla \cdot p\mathbf{V} - p\nabla \cdot \mathbf{V}$, in (19.4.1), we obtain, after re-arranging,

$$d(\rho V^2/2)/dt = -(\rho V^2/2)\nabla \cdot \mathbf{V} - \nabla \cdot p\mathbf{V} + p\nabla \cdot \mathbf{V} + \rho \mathbf{V} \cdot \mathbf{g} + \rho \mathbf{V} \cdot \mathbf{F}$$

Or, $\qquad \partial E/\partial t = -\nabla \cdot E\mathbf{V} - \nabla \cdot p\mathbf{V} + p\nabla \cdot \mathbf{V} + \rho \mathbf{V} \cdot \mathbf{g} + \rho \, \mathbf{V} \cdot \mathbf{F}$ \hfill (19.4.2)

where we have put E for $\rho V^2/2$ which represents kinetic energy per unit volume and used the relationship, $dE/dt = \partial E/\partial t + \mathbf{V} \cdot \nabla E$.

If we integrate (19.4.2) over a finite volume δv, we obtain, after applying Gauss's theorem to the first two terms on the right-hand side, the following relationship:

$$\partial \left(\int E \, \delta v \right)/\partial t = -\int E V_n \, \delta a - \int p V_n \, \delta a + \int p \nabla \cdot \mathbf{V} \, \delta v + \int \rho \, \mathbf{V} \cdot \mathbf{g} \, \delta v + \int \rho \mathbf{V} \cdot \mathbf{F} \, \delta v$$
\hfill (19.4.3)

where δa is the surface area of the volume δv, and V_n is the velocity component normal to the surface.

If we now replace the volume element by the whole atmosphere and apply (19.4.3) to it, the first two terms on the right-hand side of (19.4.3) disappear and we are left with the relation

$$\partial K/\partial t = \int p\boldsymbol{\nabla}\cdot\mathbf{V}\,\delta v + \int \rho\mathbf{V}\cdot\mathbf{g}\,\delta v + \int \rho\mathbf{V}\cdot\mathbf{F}\,\delta v \qquad (19.4.4)$$

where we have replaced $\int E\,\delta v$ by K, which we call the global kinetic energy.

Further, if we consider the kinetic energy of horizontal motion only, the second term on the right-hand side of (19.4.4) which represents the work done by gravity against vertical displacements disappears and we are finally left with the relation

$$\partial \mathbf{K}/\partial t = \int p\boldsymbol{\nabla}\cdot\mathbf{V}\,\delta v + \int \rho\mathbf{V}\cdot\mathbf{F}\,\delta v \qquad (19.4.5)$$

The Eq. (19.4.5) states that if the Kinetic energy of horizontal motion is to remain constant in the atmosphere, the dissipation of kinetic energy by frictional forces (the second term on the right-hand side of the equation) must be continually replaced or balanced by the generation of kinetic energy by pressure forces. In the atmosphere, areas of high pressure with anticyclonic divergent circulation, such as those found at surface over most parts of the subtropical belt act as sources of kinetic energy, while those of low pressure with cyclonic convergent circulation, such as the equatorial trough or the polar low, act as sinks.

The frictional dissipation of kinetic energy occurs only in molecular motion and small-scale turbulent flow. In flow involving large-scale eddies and waves, however, the frictional effect may be reversed in that instead of dissipating the energy, it may actually add to the kinetic energy of the time-mean flow. This reversal is often referred to as a phenomenon of 'negative viscosity'.

19.4.2 The Angular Momentum Balance – Maintenance of the Zonal Circulation

The principle of conservation of absolute angular momentum requires that a parcel of air at rest relative to the rotating surface of the earth at a latitude acquires from, or loses to, the underlying earth's surface angular momentum according as it moves to lower or higher latitudes. Thus, the equatorward-moving easterly tradewinds over the tropics and easterly winds over the polar belt pick up angular momentum from the earth's surface, while the poleward-moving westerly winds lose angular momentum to the surface. Since all these zonal winds are maintained over long periods of time, it follows that the excess angular momentum picked up over the tropics and the polar belt must be transported so as to meet the deficit over the midlatitudes.

The following analysis shows how the zonal winds are maintained in the atmosphere.

Let u be the zonal wind velocity of a parcel of air of unit mass at a latitude ϕ relative to the earth's surface. Its absolute angular momentum denoted by M is then given by

19.4 Maintenance of the Kinetic Energy and Angular Momentum

$$M = (u + a\Omega \cos \phi) \, a \cos \phi = u \, a \cos \phi + \Omega a^2 \cos^2 \phi \tag{19.4.6}$$

where Ω denotes the angular velocity and 'a' the mean radius of the earth.

In (19.4.6), the first term stands for relative momentum and is positive or negative according as the relative velocity is eastward or westward, while the second denotes the earth's angular momentum and is called the Ω-momentum.

The forces which can change the absolute angular momentum M of a unit mass of air at latitude ϕ are those due to torques exerted by the pressure gradient and frictional forces.

Thus, according to Newton's second law, the equation of absolute angular momentum for zonal motion may be written

$$dM/dt = (-\alpha \partial p/\partial x + F_x) r \tag{19.4.7}$$

where $r = a \cos \phi$, p is pressure, α is specific volume and F_x is the zonal component of the frictional force per unit mass.

Since $\alpha = 1/\rho$, we may write (19.4.7) in the form

$$\rho \, dM/dt = (-\partial p/\partial x + \rho F_x) r$$

and simplify it, by using the flux form of the equation of continuity (12.2.6) and noting that r does not vary with x, to obtain an expression for the rate of change of absolute angular momentum per unit volume in the form

$$\partial(\rho M)/\partial t = \underbrace{-\nabla \cdot \rho M \mathbf{V}}_{(1)} - \underbrace{\partial(p \, r)/\partial x}_{(2)} + \underbrace{\rho F_x \, r}_{(3)} \tag{19.4.8}$$

The interpretation of (19.4.8) is simple. It states that the absolute angular momentum of the atmosphere poleward of a latitude can change as a result of (1) convergence of the meridional transport of absolute angular momentum across the latitude wall, (2) the torque exerted by the pressure gradient force across mountain ranges, and (3) the torque exerted by the frictional drag of the earth on the atmosphere within the polar cap. In (19.4.8), the term (1) on the right-hand side acts as a source of absolute angular momentum, while the terms (2) and (3) act as sinks. Thus, for balance, it is required that the source (1) should equal the combined effects of the sinks at (2) and (3).

Palmen (1951) has given a more convenient expression for the source term (1). If v be the poleward component of the wind velocity, the total poleward transport of the absolute angular momentum, $\eth$, through the vertical cross-section at latitude ϕ is

$$\eth = \int \int \rho M v \, (a \cos \phi \, \delta \lambda) \, \delta z$$

where $\delta \lambda$ and δz are small increments along longitude and vertical respectively.

We now integrate $\eth$ around the latitude wall with heights extending from surface to infinity and obtain

$$\eth = a^2 \cos^2 \phi \int_{Z=0}^{\infty} \int_0^{2\pi} \rho \, (u + \Omega a \cos \phi) v \, \delta\lambda \, \delta z$$

$$= [(2\pi a^2 \cos^2 \phi)/g] \int_0^{p_0} (uv + \Omega v \, a \cos \phi) \, \delta p \qquad (19.4.9)$$

where p_0 is surface pressure and we have used the hydrostatic approximation, $\rho \delta z = -\delta p/g$.

Let us now divide the observed wind components u and v into their zonally-averaged values $\underline{u}$ and $\underline{v}$ and the respective eddy components u' and v' which are deviations from the zonal average. Since, by perturbation theory, $\underline{uv} = \underline{u}\,\underline{v} + \underline{u'v'}$, (19.4.9) may be written

$$\eth = [(2\pi a^2 \cos^2 \phi)/g] \int_0^{p_0} (\Omega \underline{v} \, a \cos \phi + \underline{u}\,\underline{v} + \underline{u'v'}) \, \delta p \qquad (19.4.10)$$

The terms within the integral on the right-hand side of (19.4.10), from left to right, are: (1) the Ω-transport term, (2) the drift term, and (3) the eddy-transport term. It may be seen that if $\underline{v} = 0$, the only mechanism left to effect northward transfer of the absolute angular momentum of the atmosphere is that through the eddies. If $\underline{v} \neq 0$, the condition implies the existence of a mean meridional circulation at the latitude under consideration

However, even in this case, $\int_0^{p_0} \underline{v} dp = 0$,

since a northward drift of mass is balanced by a southward drift over a long period of time. The same remark applies to the Ω-transport term, a northward transport of which at some level or latitude will be balanced by a southward transport at some other level or latitude over a long period of time. However, the integral of the drift term $\int \underline{u}\,\underline{v} \, dp$ is positive if $\underline{u}$ and $\underline{v}$ are positively correlated in the vertical.

Palmen and Alaka (1952) applied (19.4.10) to study the angular momentum balance in the mean January atmosphere over the tropical belt between 20N and 30N. The results of their evaluation of the three terms of the equation are presented in Fig. 19.8.

The fluxes computed were of the Ω-transport term, the drift term and the eddy-flux term in each 5-deg latitude boxes shown. Since the tropical atmosphere has the NE-trades at low levels and the SW-ly flow aloft, it was divided into two layers, the lower between 1010 mb and 700 mb and the upper extending from 700 mb to the top of the atmosphere where the pressure was assumed to be 0 mb. In the Figure, the solid arrows represent the drift term, and the broken arrows the eddy-transport term. The large amounts by the side of the solid arrows represent the amount of the Ω-transport term, while the small amounts represent the momentum transfer arising from the tendency of the meridional circulation to move north or south the whole isotach pattern in the mean cross-section. The zonally-averaged values of the fluxes for each block are shown at the latitude walls for both the layers. The direction of the

19.4 Maintenance of the Kinetic Energy and Angular Momentum

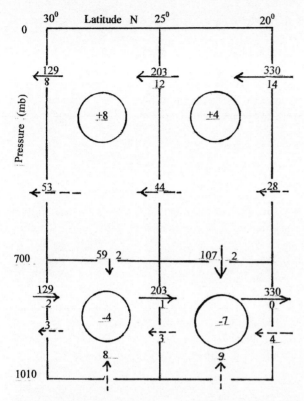

Fig. 19.8 Computed angular momentum balance for the latitude belt between 20N and 30N in January. The units of the angular momentum fluxes in the diagram are in 10^{25} g cm^2 s^{-2} (After Palmen and Alaka, 1952)

meridional transport is indicated by the arrows, while the circled numbers represent the net total flux out of the box in either the meridional or the vertical direction.

Note that the meridional circulation is responsible for the largest part of the meridional transport and this is accomplished almost entirely by the Ω-transport term. The drift term makes only a minor contribution, while the eddy-flux term becomes important only in the upper layer. An important point to note is that there is a divergence of the southward flux of angular momentum in the lower layer and convergence of northward flux in the upper layer and the balance is restored by the downward flux into the box immediately below.

The validity of the balance may be checked by comparing the circled numbers appearing on opposite sides of the 700 mb partition line. Also, it is noteworthy that a large amount of angular momentum that is transported poleward in the upper layer across the latitudinal wall at 30N is available for maintenance of angular momentum of the midlatitude westerlies. It is likely that part of this tropical contribution may be transported poleward by the breakthrough extension of the tropical cell in the vicinity of about 30N. However, it appears that over the midlatitudes the contribution made by the eddy-flux term far exceeds that due to the Ω- transport and the drift terms.

The tropical circulation cell also plays an important role in the meridional transport of water vapour in the tropics. Over most of the tradewind region, evaporation exceeds precipitation, while reverse is the case near the equator where precipitation exceeds evaporartion. The required water vapour transport to the equatorial region for net precipitation is provided by the meridional circulation.

19.5 Eddy-Transports

We now turn to the latitudes where wave activity dominates the air flow and most of the poleward transports of sensible heat, angular momentum and water vapour are effected by large-scale eddies.

Wiin-Nielsen (1967) showed that maximum kinetic energy is transported across midlatitudes by eddies of circumpolar wave numbers about six, i.e., wavelengths around 6500 km.

Starr (1968) computed the direction and magnitude of the eddy- transfer of annual mean zonally averaged heat, angular momentum and water vapour across the different latitudes. Some of the results of his computations were reviewed by Lorenz (1969). Here, we summarize some of the important findings of the studies.

19.5.1 Eddy Flux of Sensible Heat

The maximum northward eddy-flux of sensible heat occurs across midlatitudes where there is strong northward gradient of temperature and where baroclinic waves dominate. A northward flux requires that the zonal fluctuations in v and T are positively correlated, i.e., $\underline{v'T'} > 0$. [Here, the underlining denotes a zonal average].

19.5.2 Eddy-Flux of Angular Momentum

Starr (1968) computed the northward eddy-flux of angular momentum by using observed winds in the eddy-transport term of (19.4.10). About his results, Lorenz (1969) remarks as follows:

'Since the transport of angular momentum has been computed from observed winds rather than analysed maps, it presumably includes the contribution of most of the spectrum. It can be shown, however, that the transport is mainly due to the larger scales. In any band of the spectrum, the maximum possible angular momentum transport is limited by the kinetic energy. The contribution of the small scales, even if the eastward and northward components of the wind were correlated in these

19.5 Eddy-Transports

Fig. 19.9 Trough-ridge tilt by an angle, φ, from the meridian for poleward eddy-transport of angular momentum

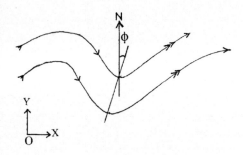

scales, could not match the observed contribution of the large scales, which results generally from correlations of about 0.2. Similar considerations apply to the transport of sensible heat, since the temperature spectrum has somewhat the character of that of kinetic energy.' The validity of Lorenz's remarks becomes evident from at least two considerations, viz., (1) asymmetry, or tilt from the meridian, of troughs and ridges in zonal flow, and (2) the Index cycle, shown in Fig. 19.9 and Fig. 19.10 respectively.

Machta showed in 1949 that a northward eddy-transport of angular momentum is proportional to the tangent of the angle of tilt φ from the meridian, i.e., when the troughs and ridges are aligned in a NE-SW direction, as shown in Fig. 19.9. A tilt in the opposite direction, i.e., in a NW-SE direction, will result in a southward eddy-transfer.

The alternation between strong and weak westerlies over the midlatitudes is usually described as an Index cycle. In the meridional distribution of the zonal

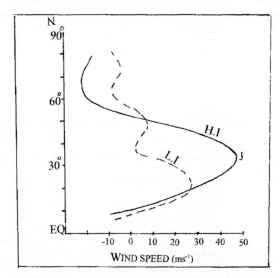

Fig. 19.10 Meridional profile of westerlies during (a) low-index (L.I), and (b) high index (H.I) phase of the Index cycle. J denotes the W'ly Jet. (After Riehl et al., 1954)

winds, a high-index phase is characterized by a single strongly peaked maximum, while in a low-index situation, the zonal winds are relatively weaker over the midlatudes and may have two maxima in winter, one at a latitude near 25N and the other at about 45N, as shown in Fig. 19.10. The two maxima, one in the form of a strong subtropical jet and the other a somewhat weaker midlatitude jet was also apparent in Palmen's model of mean meridional circulation in winter.

A high-index situation arises when a strong meridional temperature gradient develops over the midlatitudes. The thermal wind builds up a strong westerly jet in the upper troposphere. The flow becomes baroclinically unstable and breaks down into waves which transport heat rapidly northward. The result is a weakening of the meridional temperature gradient, a weakening of the jet, and the beginning of a low-index phase of the cycle. The waves draw kinetic energy from the available potential energy of the zonal flow during the transition from high-index to low-index phase and return kinetic energy to the zonal flow during the transition from the low-index to the high-index phase.

19.5.3 Eddy-Flux of Water Vapour

An estimate of the northward eddy-transport of water vapour as computed from annual mean zonally-averaged distribution of specific humidity in the northern hemisphere was made by Peixoto and Crisi (1965) from one year of data. Since the transport of water vapour implies a transport of latent heat of condensation, it plays a role similar to that of sensible heat, in the energy balance. Further, it plays an important role in the global hydrological cycle, as qualitatively suggested by Fig. 19.11 which shows the meridional distribution of zonally averaged annual evaporation and precipitation over the globe and the sense of water vapour flux required for moisture balance.

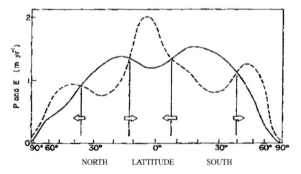

Fig. 19.11 Average annual evaporation and (—) precipitation (- - - -) per unit area expressed in meters per year. *Arrows* represent the sense of the water vapour flux in the atmosphere (Sellers, 1965)

19.5.4 Vertical Eddy-Transports

The above considerations certainly do not apply to the vertical transports of atmospheric properties by eddies. The vertical kinetic-energy spectrum is difficult to determine and there is no evidence that the energy is concentrated in the larger-scale eddies as in horizontal transfers. In fact, the observed upward and downward motions inside and outside some of the small-scale towering clouds may be many orders of magnitude larger that the averaged large-scale vertical motion computed in the atmosphere. Regarding the reliability of vertical transports evaluations, Lorenz (1969) writes:

"It is doubtful that the large-scale vertical motion fields deduced by one means or another from more readily observable quantities are sufficiently accurate for evaluating reliable vertical transports of sensible heat, angular momentum and water vapour. Thus, even if we knew the small-scale contributions, we could not readily compare them with the large-scale contributions. However, it is easy to see how convective-cloud circulations can carry significant amounts of energy upward." As an example of possible vertical transport of energy, Lorenz refers to the work of Riehl and Malkus (1958) who estimated that the entire vertical transport of energy in the equatorial zone could be effected by one giant cumulonimbus cloud per square degree. Palmen and Newton (1969) have indicated that a large part of the angular momentum also can be transported vertically by cumulonimbus clouds, or by squall lines along which they are organized. Comparing the evaluated vertical eddy-transports of energy and angular momentum by eddies of all scales; they conclude that the vertical transport of angular momentum appears to be rather small, while that of energy can be quite large.

19.6 Laboratory Simulation of the General Circulation

From about the middle of the eighteenth century, there have been several attempts by many an experimental physicist (e.g., Wilcke, 1785; Vettin, 1857, 1884; Taylor, 1921; Fultz, 1949, 1950, 1951; Long, 1951) to simulate the essential features of the general circulation of the atmosphere by using differentially heated rotating fluids in the laboratory, with encouraging results. Earlier experiments used air as the working fluid, but in most of the later experiments, it has been replaced by water, since by putting tracer elements in water one could follow fluid motions better than in air. The early experiments of the latter type were carried out with rotating dishpans and as such came to be known as 'dishpan experiments'. In these experiments, the fluid in the dishpan was supposed to represent the atmosphere over one hemisphere with the bottom curved surface of the pan representing the curved surface of the earth. The pan was heated at the rim which was supposed to correspond to the equator and cooled at the center which was assumed to correspond to the pole. One may note that in the laboratory experiments, the scales of several characteristic parameters of the fluid, such as density, velocity, length, time, pressure fluctuations, etc., were many orders of magnitude different from those in the atmosphere. For example, (a) the density of the working fluid was almost a thousand times greater than that of air; (b) gravity has the same direction throughout the fluid, whereas in the atmosphere it varies with latitude; (c) the angular velocity of the dishpan is constant horizontally, whereas in the atmosphere the vertical component of the angular velocity varies with latitude. However, despite these differences, for certain combinations of differential heating and rate of rotation, the fluid was observed to circulate forming patterns which appeared to bear close resemblance to those observed in the atmosphere. For example, Fultz found that at low speeds of rotation, there was simple overturning of the fluid with heated water rising at the rim and cooled water sinking at the center, thus forming a mean meridional circulation which closely resembled the Hadley circulation of the atmosphere. At higher rate of rotation, the rotating fluid appeared to develop jet streams and wave motions in patterns which closely resembled the baroclinic waves in the atmosphere over the midlatitudes. Another series of experiments called the 'annulus experiments' was carried out using two co-axial cylinders of different radii and the annular space between the two filled with the working fluid. Differential heating was created between the outer and the inner walls of the annulus by heating the outer surface and cooling the inner surface and maintaining a precise temperature difference across the fluid. The inner cylinder was rotated at different speeds which simulated the rotation of the earth and the motion of the fluid in the annular region was observed. It was found that for certain combinations of heating and rotation, the different flow patterns that evolved were very similar to those that were observed with the dishpan experiments.

Phillips (1963) showed that the flow patterns observed in the annular experiments could be interpreted within the framework of the quasi-geostrophic theory, since the geostrophy of a flow did not depend upon the scaling parameters, but upon the nondimensional ratio of these parameters called the Rossby number, Ro, defined in the case of the annular fluid by

$$\text{Ro} \equiv U/\{\Omega(b-a)\} \tag{19.6.1}$$

where Ω is the angular velocity of the rotating cylinder, U is a typical velocity of the fluid, and $(b-a)$ is the width of the annulus, b and a being the radii of the outer and the inner cylinders respectively.

Since water is nearly incompressible, adiabatic temperature changes are negligible following the motion. So the density ρ varies with the temperature T according to the relation

$$\rho = \rho_0\{1 - \epsilon(T - T_0)\} \tag{19.6.2}$$

where ϵ is the thermal expansion co-efficient ($= 2 \times 10^{-4}$ per degree Celsius for water) and ρ_0 is the density at the mean temperature T_0.

Using the hydrostatic approximation

$$\partial p/\partial z = -\rho g \tag{19.6.3}$$

19.6 Laboratory Simulation of the General Circulation

and the geostrophic relationship

$$\mathbf{V}_g = (\mathbf{k} \times \nabla p)/(2\Omega\rho_0) \tag{19.6.4}$$

where $\mathbf{V}_g$ is the geostrophic wind velocity,

We obtain an expression for the thermal wind

$$\partial \mathbf{V}_g/\partial z = -(g \in /2\Omega)\mathbf{k} \times \nabla T \tag{19.6.5}$$

If we now let U denote the scale of the geostrophic wind, H the depth of the fluid, and ΔT the radial temperature difference across the annulus, (19.6.5) may be written

$$U \sim (g \in H\Delta T)/\{2\Omega(b-a)\} \tag{19.6.6}$$

Substituting the value of U from (19.6.6) in (19.6.1), we obtain an expression for the thermal Rossby number

$$Ro_T = (g \in H\Delta T)/\{2\Omega^2(b-a)^2\} \tag{19.6.7}$$

For typical values of the parameters involved in (19.5.7), we may use

$$(b-a) \simeq H \simeq 10\,\text{cm}, \Delta T = 10C, \text{ and } \Omega \simeq 1s^{-1}$$

Ro_T is then found to have a value of about 10^{-1}, a value low enough to justify the application of the quasi-geostrophic theory to the problem. However, there is a difficulty in applying Ro_T directly to the annular flow problem in view of the existence of strong conduction boundary layers near the vertical walls of the annulus, in which the temperature changes rather rapidly away from the boundaries. For this reason, the thermal wind concept cannot be applied to these boundary layers but only to the interior region of the annular fluid. However, since the boundary layers are not separated from the interior by any rigid barrier, their temperatures are continually regulated by flow conditions in the interior. From this viewpoint, the temperatures measured at the walls are to be treated as externally imposed. We, therefore, talk of an imposed thermal Rossby number which is defined by

$$Ro_T^* \equiv \in gH(T_b - T_a)/\{2\Omega^2(b-a)^2\} \tag{19.6.8}$$

where T_b and T_a are the imposed temperatures at the outer and the inner walls of the annulus respectively. For reasons stated above, it follows that $Ro_T^* > Ro_T$. This means that $Ro_T^* \ll 1$. This condition fully justifies the application of quasi-geostrophic theory to the annular flow problem.

Figure 19.12 summarizes the results of Fultz's annular experiments, as presented by Phillips (1963). It shows, on a log-log plot, the Hadley and the Rossby regimes of flow for different combinations of the imposed thermal Rossby number, Ro_T^*, and the non-dimensionalized rotation rate, $(G^*)^{-1} \equiv ((b-a)\Omega^2/g$. The heavy solid line separates the axially-symmetric Hadley regime from the wavy Rossby regime.

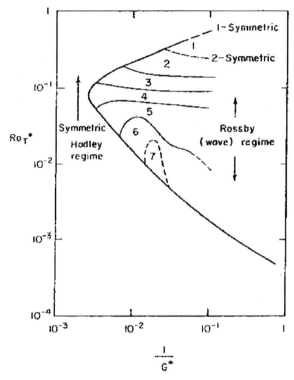

Fig. 19.12 Separation of Hadley and Rossby regimes in Fultz's annular experiments (After Phillips, 1963)

The interpretation of Fig. 19.12 is as follows: For very low values of thermal Rossby number, i.e., small differential heating across the annulus and high rate of rotation, there is only axially-symmetric Hadley-type circulation. But, as the thermal Rossby number is increased, there is an increase in the value of the thermal wind and a stage is reached when the flow becomes baroclinically unstable and baroclinic waves appear. This marks the beginning of the Rossby wave regime. However, in a baroclinic wave, heat is transported both horizontally and vertically, so that an increase in the value of Ro_T^* will tend to increase the static stability of the fluid. The strengthening of the static stability together with a lowering of the rotation rate leads to a decrease of the wave number or increase of the wavelength of the waves. Thus, with increasing Ro_T^*, the flow undergoes transitions to longer and longer waves until the static stability becomes so strong that even the longest wave cannot change the character of the flow which then returns to the axially-symmetric Hadley regime.

19.7 Numerical Experiment on the General Circulation

The invention of the electronic computer and its application to solve meteorological problems from about the middle of the last century marked the beginning of a new era in studies of the general circulation of the atmosphere. At first, because of limited memory and working speed, the computers could be used to solve only simple time-dependent mathematical models of the atmosphere, like a single-level barotropic model or two-level baroclinic models, to produce short-term forecasts of atmospheric flow patterns. But, as time passed, there was rapid advance in the design of computers and large-memory high-speed computers became available to handle the more complete primitive equation models.

These developments led to rapid advance in the field of numerical weather prediction (NWP) and attempts were made to turn out reliable forecasts for larger areas and for longer periods. However, it was soon realized that the atmosphere was much too complicated a medium involving motions at various space and time scales to be modelled adequately for long-range forecasts without a better understanding of its physics and dynamics and associated computational problems than we could do at present. This need for a better understanding of the atmosphere led several meteorologists to undertake numerical experiments to simulate the general circulation of the atmosphere.

The first successful numerical experiment in this direction was carried out by Phillips (1956) who used a two-level quasi-geostrophic model for the numerical integration, starting from a state of rest. Diabatic heating and friction were externally prescribed. However, there were several shortcomings, the most serious of which was that the static stability of the atmosphere could not be determined since the temperature was computed at one level only. It was evident from the annular experiment in the laboratory that the static stability played an important role in determining the character of the flow. Also, the actual distribution of diabatic heating in the atmosphere differed considerably from what was prescribed in the model. There were also problems of computational instability which had to be overcome. But it is remarkable that inspite of all the limitations, the model turned out thermal and flow patterns which were found to be quite realistic over extratropical latitudes. Phillips' pioneering experiment was followed by several others (for example, Smagorinsky, 1963; Mintz, 1968; Hahn and Manabe, 1975) at many global meteorological centers to simulate the general circulation of the atmosphere under more realistic conditions.

The last few decades have witnessed emergence of comprehensive numerical models to simulate the general circulation of the atmosphere under conditions which can be prescribed and controlled at will in order to study different aspects of weather and climate. To-day, several such models are in operation at meteorological research centers in many parts of the globe.

Appendices

Appendix-1(A) Vector Analysis-Some Important Vector Relations

1.1 The Concept of a Vector

In meteorology, some physical properties such as mass, time, temperature, pressure, etc., are measured and reported by their magnitudes only. They are called scalars. However, there are some others for which we need to specify a direction in addition to magnitude. These are called vectors. In the latter category, we may mention the position of a point, displacement, velocity, acceleration, force, etc. [In the appendices as well as in text, vectors are denoted by letters in bold-face.]

We illustrate the concept of a vector by taking the simple case of the position of a particle P in space. Let it be located at a distance r from a fixed point O at time t (see Fig. 1.1′, left panel). The position of P is then defined by a vector $\mathbf{R} = r\,\mathbf{r}$, where r is the magnitude of $\mathbf{R}$, indicated by $|\mathbf{R}|$, and $\mathbf{r}$ a unit vector in the direction from O to P, indicated by arrow. Here, both magnitude and direction are important to define the vector.

It is possible to resolve the position vector $\mathbf{R}$ into component vectors along the co-ordinate axes of a system of reference, by defining unit vectors along the co-ordinate axes.

Let $\mathbf{i}, \mathbf{j}, \mathbf{k}$ be the unit vectors along the axes of a rectangular co-ordinate system and x, y, z the resolved components of r along the axes OX, OY, OZ respectively.

Then we can express the vector $\mathbf{R}$ as

$$\mathbf{R} = r\,\mathbf{r} = x\,\mathbf{i} + y\,\mathbf{j} + z\,\mathbf{k} \qquad (1.1')$$

where $r = (x^2 + y^2 + z^2)^{1/2}$, and r, x, y, z are all scalars.

A unit vector has a magnitude unity in a specified direction and plays an important role in vector analysis. It helps to provide the direction in which a vector acts. As we shall see in several parts of this book, the use of a vector is advantageous for two important reasons: firstly, it greatly simplifies mathematical treatment and secondly,

Fig. 1.1′ The concept of a vector and its resolution into components

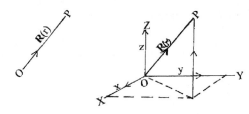

it provides a simpler physical and geometrical representation of mathematical results. Some of the standard mathematical operations with vectors are summarized in the following sections.

1.2 Addition and Subtraction of Vectors: Multiplication of a Vector by a Scalar

We have already demonstrated in (1.1′) how the vectors are added. In general, if two vectors **A** and **B** are added, their sum **C** is a vector given by the vector equation

$$\mathbf{C} = \mathbf{A} + \mathbf{B} \qquad (1.2')$$

The sum **C** is obtained by laying off the vector **B** from the end-point of **A** and then drawing a vector from the initial point of **A** to the end-point of **B** (see Fig. 1.2′).

A vector equation is equivalent to three scalar equations, since, as we saw in (1.1′), a vector is represented by its three components along the co-ordinate axes and two vectors are equal if their components are respectively equal. The geometric sum is commutative, like an ordinary arithmetic sum, i.e., it does not depend upon the order in which the vectors are added. Further, a vector sum is also associative, i.e., it does not matter how the individual vectors are grouped. For example, if **S** is the sum of three vectors **A, B** and **C**,

$$\mathbf{S} = (\mathbf{A} + \mathbf{B}) + \mathbf{C} = \mathbf{A} + (\mathbf{B} + \mathbf{C}) = \mathbf{B} + (\mathbf{A} + \mathbf{C}) \qquad (1.3')$$

The sum of two vectors having the same direction and sense is a vector the magnitude of which is the sum of the magnitudes of the two vectors, the direction remaining the same. If a vector is added n times, the vector can simply be multiplied by the number n.

A negative vector, e.g., −**B** has the same magnitude as **B** but opposite direction. Accordingly, the vector difference **A** − **B** may be written as **A** + (−**B**).

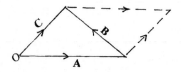

Fig. 1.2′ Sum of two vectors

1.3 Multiplication of Vectors

Two vectors can be multiplied, either scalarly or vectorially. A scalar product is indicated by a bold dot sign (**.**) placed between them, whereas a bold cross sign (**x**) indicates a vector product.

(a) The scalar product of two vectors

The scalar product, also called the dot product, of two vectors, **A** and **B**, is a scalar which is equal to the product of the magnitudes of the vectors and the cosine of the angle between them (Fig. 1.3′). Thus,

$$\mathbf{A}.\mathbf{B} = AB \cos \alpha \qquad (1.4')$$

where A and B are respectively the magnitudes $|\mathbf{A}|$ and $|\mathbf{B}|$ of the vectors and α is the angle ($\leq 180°$) between them.

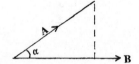

Fig. 1.3′ The scalar product of two vectors

The relation (1.4′) means that when two vectors are equal and parallel to each other, the angle between them is zero and $\cos \alpha = 1$. The scalar product is then simply the product of the magnitudes of the two vectors. But when they are at right angles to each other, $\cos 90° = 0$, and the scalar product is zero. This is an important result, for, when applied to the unit vectors, **i, j, k**, we get

$$\mathbf{i}.\mathbf{i} = \mathbf{j}.\mathbf{j} = \mathbf{k}.\mathbf{k} = 1 \text{ (for } \alpha = 0\text{); and, } \mathbf{i}.\mathbf{j} = \mathbf{j}.\mathbf{k} = \mathbf{k}.\mathbf{i} = 0 \text{ (for } \alpha = 90\text{)} \qquad (1.5')$$

From (1.5′) it follows that if we take the resolved components of two vectors **A** and **B** along the co-ordinate axes, their scalar product may be written in terms of the components as

$$\begin{aligned}\mathbf{A} \cdot \mathbf{B} &= (A_x\mathbf{i} + A_y\mathbf{j} + A_z\mathbf{k}) \cdot (B_x\mathbf{i} + B_y\mathbf{j} + B_z\mathbf{k}) \\ &= A_xB_x + A_yB_y + A_zB_z = \mathbf{B} \cdot \mathbf{A}\end{aligned} \qquad (1.6')$$

(b) The vector product of two vectors

The vector product, also called the cross product, of two vectors, **A** and **B**, is a vector **P** the direction of which is that of a right-hand screw, i.e., perpendicular to the plane determined by the vectors **A** and **B**, and the magnitude of which is equal to the area of the parallelogram formed by them, i.e., equal to $AB \sin\alpha$, where $\alpha(\leq 180)$ is the angle between them (Fig. 1.4′).

Fig. 1.4' The vector product of two vectors

The stipulation of a right-hand screw as determining the direction of the vector product denies it the property of commutativity possessed by a scalar product; for, if we reverse the order of the vectors, then the turning of **B** towards **A**, in the sense of a right-hand screw, would give the vector product a direction opposite to that of **P**. Thus,

$$\mathbf{P} = \mathbf{A} \times \mathbf{B} = -\mathbf{B} \times \mathbf{A} \tag{1.7'}$$

The most important property of the vector product is that of its distributivity with respect to addition. This means that

$$\mathbf{A} \times (\mathbf{B} + \mathbf{C} + \mathbf{D} + \cdots) = \mathbf{A} \times \mathbf{B} + \mathbf{A} \times \mathbf{C} + \mathbf{A} \times \mathbf{D} + \ldots \tag{1.8'}$$

The proof of (1.8') is a little cumbersome and will not be attempted here.

If the vectors are parallel to each other, $\sin \alpha = 0$, and the vector product is zero. If they are perpendicular to each other, $\sin \alpha = 1$, and the magnitude of the product vector is simply the arithmetic product of the individual magnitudes of the vectors, the direction being at right angles to the plane of the two vectors according to the right-hand screw rule. We can apply the vector product definition given above immediately to the unit vectors and obtain

$$\mathbf{i} \times \mathbf{i} = \mathbf{j} \times \mathbf{j} = \mathbf{k} \times \mathbf{k} = 0. \text{ (since } \sin \alpha = 0\text{); and}$$
$$\mathbf{i} \times \mathbf{j} = -\mathbf{j} \times \mathbf{i} = \mathbf{k}; \ \mathbf{j} \times \mathbf{k} = -\mathbf{k} \times \mathbf{j} = \mathbf{i}; \ \mathbf{k} \times \mathbf{i} = -\mathbf{i} \times \mathbf{k} = \mathbf{j} \tag{1.9'}$$
$$(\text{Since } \sin \alpha = 1)$$

Using these relationships, we can obtain the vector product of two vectors **A** and **B** in terms of the products of their components as follows:

$$\mathbf{A} \times \mathbf{B} = (A_x \mathbf{i} + A_y \mathbf{j} + A_z \mathbf{k}) \times (B_x \mathbf{i} + B_y \mathbf{j} + B_z \mathbf{k})$$
$$= (A_y B_z - A_z B_y) \mathbf{i} + (A_z B_x - A_x B_z) \mathbf{j} + (A_x B_y - A_y B_x) \mathbf{k} \tag{1.10'}$$

The result (1.10') can also be expressed in the determinant form

$$\mathbf{A} \times \mathbf{B} = \begin{vmatrix} \mathbf{i} & \mathbf{j} & \mathbf{k} \\ A_x & A_y & A_z \\ B_x & B_y & B_z \end{vmatrix} \tag{1.11'}$$

(c) Multiple vector/scalar products

The methods of determining scalar/vector products of vectors described above may be applied successively in multiple products of vectors. The following are some of the products which occur frequently in several branches of science.

If **A, B, C** are three vectors, then

$$\mathbf{A} \cdot \mathbf{B} \times \mathbf{C} = \mathbf{B} \cdot \mathbf{C} \times \mathbf{A} = \mathbf{C} \cdot \mathbf{A} \times \mathbf{B} = -\mathbf{A} \cdot \mathbf{C} \times \mathbf{B} = -\mathbf{B} \cdot \mathbf{A} \times \mathbf{C} = -\mathbf{C} \cdot \mathbf{B} \times \mathbf{A} \quad (1.12')$$

The products in (1.12') may also be expressed in terms of the components of the vectors. For example,

$$\mathbf{A} \cdot \mathbf{B} \times \mathbf{C} = (A_x \mathbf{i} + A_y \mathbf{j} + A_z \mathbf{k}) \cdot \{(B_y C_z - B_z C_y)\mathbf{i} + (B_z C_x - B_x C_z)\mathbf{j} + (B_x C_y - B_y C_x)\mathbf{k}\} \quad (1.13')$$

(1.13') may also be written in the form of a determinant

$$\mathbf{A} \cdot \mathbf{B} \times \mathbf{C} = \begin{vmatrix} A_x & A_y & A_z \\ B_x & B_y & B_z \\ C_x & C_y & C_z \end{vmatrix}$$

It is easy to see that all the multiple products in (1.12') represent the volume of a parallelepiped formed by the three vectors as contiguous sides.

The vector product of a vector with the vector product of two other vectors is represented as follows:

$$\mathbf{A} \times (\mathbf{B} \times \mathbf{C}) = (\mathbf{A} \cdot \mathbf{C})\mathbf{B} - (\mathbf{A} \cdot \mathbf{B})\mathbf{C} \quad (1.14')$$
$$(\mathbf{A} \times \mathbf{B}) \times \mathbf{C} = (\mathbf{C} \cdot \mathbf{A})\mathbf{B} - (\mathbf{C} \cdot \mathbf{B})\mathbf{A}$$

1.4 Differentiation of Vectors: Application to the Theory of Space Curves

Let the position vector **r** of a point P on a space curve be a function of the length of the arc s, measured from an initial point on the curve (Fig. 1.5').

Then, $|\Delta \mathbf{r}|$ is identical with Δs, and in the limit, as $\Delta s \to 0$, $\Delta \mathbf{r}/\Delta s$ is a vector of length 1 directed along the tangent to the curve. Let us denote this unit vector by **t**. Then

$$\mathbf{t} = d\mathbf{r}/ds \quad (1.15')$$

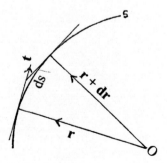

Fig. 1.5' Differentiation of a vector with respect to a scalar

The scalar products of vectors are differentiated just as ordinary scalar functions, regardless of the order of the vectors. In the case of vector products, however, the order of the vectors is important. Thus, if we differentiate the scalar and the vector products of two vectors, **A** and **B**, with respect to a scalar u, we get

$$d(\mathbf{A}\cdot\mathbf{B})/du = (d\mathbf{A}/du)\cdot\mathbf{B} + \mathbf{A}\cdot(d\mathbf{B}/du)$$
$$= \mathbf{A}\cdot d\mathbf{B}/du + \mathbf{B}\cdot(d\mathbf{A}/du) \qquad (1.16')$$

But,
$$d(\mathbf{A}\mathbf{x}\mathbf{B})/du = \mathbf{A}\mathbf{x}(d\mathbf{B}/du) + (d\mathbf{A}/du)\mathbf{x}\mathbf{B}$$
$$= \mathbf{A}\mathbf{x}(d\mathbf{B}/du) - \mathbf{B}\mathbf{x}(d\mathbf{A}/du) \qquad (1.17')$$

If a vector **A** is of constant length A, but changes direction only, then the differentiation of the scalar product $\mathbf{A}\cdot\mathbf{A}$, with respect to a scalar u gives

$$d(\mathbf{A}\cdot\mathbf{A})/du = dA^2/du = 2\mathbf{A}\cdot(d\mathbf{A}/du) = 0 \qquad (1.18')$$

Since neither the vector nor the derivative is to disappear, it follows from (1.18') that the derivative of a vector of constant length is at right angles to the vector itself. It is easy to see that if the length is constant, the end-point of the vector will move on a sphere. If the increment is infinitesimal, it is tangent to the sphere and thus perpendicular to the vector.

1.5 Space Derivative of a Scalar Quantity. The Concept of a Gradient Vector

Several scalar quantities, such as temperature, pressure, density, etc., vary in space and are continuous functions of space variables. We can express the space derivative of such a scalar by a vector which is called the gradient of that scalar quantity. To see how this is done, let us consider a pressure field around a pressure center, O. Since the field is continuous, we can locate surfaces of equal pressure, known as isobaric surfaces, at different distances from the center of the pressure. Let **r** be the position vector of a point P on one of these surfaces where the pressure is p (Fig. 1.6')

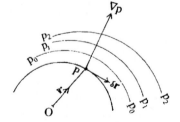

Fig. 1.6' Concept of the gradient of a scalar, p

Then, in Cartesian co-ordinates, we can write

$$d\mathbf{r} = dx\,\mathbf{i} + dy\,\mathbf{j} + dz\,\mathbf{k}$$
$$dp = (\partial p/\partial x)\,dx + (\partial p/\partial y)\,dy + (\partial p/\partial z)\,dz$$

From these expressions, it is evident that dp may be conceived as the scalar product of two vectors, d**r** and a vector

$$(\partial p/\partial x)\mathbf{i} + (\partial p/\partial y)\mathbf{j} + (\partial p/\partial z)\mathbf{k}$$

which is called the gradient of p. Therefore, we write

$$dp = (\text{grad } p)\cdot d\mathbf{r} \tag{1.19'}$$

When the isobaric surfaces are parallel to each other, then, according to rules of scalar products, the gradient of p vanishes along the isobaric surfaces, but becomes maximum at right angles to them.

1.6 Del Operator, ∇

We now introduce a vector differential operator, called the Del or Nabla operator, usually denoted by ∇, which has been found to be most useful in vector analysis. In the rectangular co-ordinate system, it is expressed as

$$\nabla = \partial/\partial x\,\mathbf{i} + \partial/\partial y\,\mathbf{j} + \partial/\partial z\,\mathbf{k} \tag{1.20'}$$

Since Del is a vector operator, it can be used in all kinds of scalar or vector operations. For example, in (1.19'), we can use it to write

$$dp = \nabla p \cdot d\mathbf{r}$$

We can also use the operator to write two integral theorems, viz., the Gauss's divergence theorem and the Stokes's vorticity theorem in following forms:

$$\text{Gauss's theorem,} \quad \iiint \nabla \cdot \mathbf{V}\,dv = \iint_S (\mathbf{V}\cdot d\mathbf{S}) \tag{1.21'}$$

where the triple integral on the left-hand side represents the volume(dv here represents a volume element) enclosed by the surface S represented by vector d**S** normal to it. $\nabla \cdot \mathbf{V}$ is called the divergence of the velocity vector **V**.

If the velocity vector has components u, v, w along the axes x,y,z respectively, we can write divergence as

$$\nabla \cdot \mathbf{V} = \partial u/\partial x + \partial v/\partial y + \partial w/\partial z$$

$$\text{Stokes's theorem,} \quad \oint \mathbf{V}\cdot d\mathbf{s} = \iint_S (\nabla \times \mathbf{V})\cdot d\mathbf{S} \tag{1.22'}$$

where S is the surface bounded by the line s. The line integral on the left-side of (1.22′) denotes circulation around the boundary line, while the vector product $\nabla \times \mathbf{V}$ on the right-side denotes the vorticity of the vector $\mathbf{V}$. Circulation and vorticity are thus related by the Stokes's theorem,

$$\text{Circulation} = \text{Vorticity} \times \text{Area}$$

The following are a few other examples of the use of the Del operator, which find important applications in meteorology. They are performed with a scalar ρ and vector $\mathbf{A}$ and $\mathbf{B}$. The results may be verified by substituting the components of the vectors.

$$\nabla \times \nabla \rho = 0$$
$$\nabla \cdot (\rho \mathbf{A}) = \nabla \rho \cdot \mathbf{A} + \rho \, \nabla \cdot \mathbf{A}$$
$$\nabla \times (\rho \mathbf{A}) = \nabla \rho \times \mathbf{A} + \rho \, \nabla \times \mathbf{A}$$
$$\nabla \cdot (\nabla \times \mathbf{A}) = 0$$
$$(\mathbf{A} \cdot \nabla) \mathbf{A} = (1/2) \nabla (\mathbf{A} \cdot \mathbf{A}) + (\nabla \times \mathbf{A}) \times \mathbf{A}$$
$$\nabla \times (\mathbf{A} \times \mathbf{B}) = \mathbf{A}(\nabla \cdot \mathbf{B}) - \mathbf{B}(\nabla \cdot \mathbf{A}) - (\mathbf{A} \cdot \nabla) \mathbf{B} + (\mathbf{B} \cdot \nabla) \mathbf{A}$$

1.7 Use of Del Operator in Different Co-ordinate Systems

1.7.1 Cartesian Co-ordinates (x, y, z)

If $\mathbf{r}$ be a position vector, $\mathbf{V}$ a velocity vector and α a scalar, then

$$\mathbf{r} = x\mathbf{i} + y\mathbf{j} + z\mathbf{k}$$
$$\mathbf{V} = u\mathbf{i} + v\mathbf{j} + w\mathbf{k}$$
$$\nabla \alpha = (\partial \alpha / \partial x) \mathbf{i} + (\partial \alpha / \partial y) \mathbf{j} + (\partial \alpha / \partial z) \mathbf{k}$$
$$\nabla \cdot \mathbf{V} = \partial u / \partial x + \partial v / \partial y + \partial w / \partial z$$
$$\nabla \times \mathbf{V} = (\partial w / \partial y - \partial v / \partial z) \mathbf{i} + (\partial u / \partial z - \partial w / \partial x) \mathbf{j} + (\partial v / \partial x - \partial u / \partial y) \mathbf{k}$$
$$\nabla \cdot \nabla \alpha = \nabla^2 \alpha = (\partial^2 / \partial x^2 + \partial^2 / \partial y^2 + \partial^2 / \partial z^2) \alpha$$

1.7.2 Spherical Co-ordinates (λ, ϕ, r)

Position vector, $\mathbf{r} = r\mathbf{k}$
Velocity vector, $\mathbf{V} = u\mathbf{i} + v\mathbf{j} + w\mathbf{k}$
where, in orthogonal curvilinear co-ordinates,

1.7 Use of Del Operator in Different Co-ordinate Systems

$dx = r \cos \phi \, d\lambda; \; dy = r d\phi; \; dz = dr$

$u = r \cos \phi \, d\lambda/dt; \; v = r \, d\phi/dt; \; w = dr/dt$

and,

Del operator, $\nabla = \mathbf{i}(1/r\cos\phi)\partial/\partial\lambda + \mathbf{j}(1/r)\partial/\partial\phi + \mathbf{k}\partial/\partial r$

$\nabla\alpha = \mathbf{i}(1/r\cos\phi)\partial\alpha/\partial\lambda + \mathbf{j}(1/r)\partial\alpha/\partial\phi + \mathbf{k}\partial\alpha/\partial r$

$\nabla\cdot\mathbf{V} = (1/r\cos\phi)[\partial u/\partial\lambda + \partial(v\cos\phi)/\partial\phi + 2w/r]$

$\nabla \times \mathbf{V} = (1/r^2\cos\phi)[r\cos\phi\{\partial w/\partial\phi - \partial(rv)/\partial r\}\mathbf{i} +$
$r\{\partial(u\,r\cos\phi)/\partial r - \partial w/\partial\lambda\}\mathbf{j} + \{\partial(rv)/\partial\lambda - \partial(u r\cos\phi)/\partial\phi\}\mathbf{k}$

$\nabla\cdot\nabla\alpha = \nabla^2\alpha = (1/r^2\cos\phi)[\partial\{(1/\cos\phi)\partial\alpha/\partial\lambda\}/\partial\lambda$
$+ \partial\{\cos\phi\,\partial\alpha/\partial\phi\}/\partial\phi$
$+ \partial\{r^2\cos\phi\,\partial\alpha/\partial r\}/\partial r]$

Appendix-1(B) Motion Under Earth's Gravitational Force

According to classical mechanics, the orbital velocity along a curve in plane polar co-ordinates is given by

$$v^2 = (dr/dt)^2 + r^2(d\theta/dt)^2; \; d/dt = (d\theta/dt)(d/dy) \qquad (1.23')$$

where r is the radial distance of a moving point on the curve from a fixed point and $d\theta/dt$ is its angular velocity (see Fig. 1.3′ in the text).

We may use (1.23′) to eliminate time t from the Eqs. (1.7) and (1.8) and obtain a relation between r and θ, which would give the path of the orbit. The procedure yields the following equation involving r and θ,

$$d\theta = 2A\,dr/[r^2\{v_0^2 - (2GM/r_0) + (2GM/r) - 4(A^2/r^2)\}^{1/2}] \qquad (1.24')$$

Let us put $u = 1/r$. (1.24′) is then transformed into

$$d\theta = -2A\,du/\{(v_0^2 - 2GM/r_0) + 2GMu - 4A^2u^2\}^{1/2} \qquad (1.25')$$

Integrating (1.25′), we get

$$\theta + \alpha = \cos^{-1}[(u - GM/4A^2)/\{(c/4A^2) + (GM/4A^2)^2\}^{1/2}] \qquad (1.26')$$

where $c = v_0^2 - 2GM/r_0$, and α is the phase difference.

Putting back r in (1.26′) and simplifying further, we get

$$r = (4A^2/GM)/[1 + (2AB/GM)\cos(\theta + \alpha)] \qquad (1.27')$$

where $B = [v_0^2 - 2GM/r_0 + (GM/2A)^2]^{1/2}$

or, $$r = k/[1+ \epsilon \cos(\theta + \alpha)] \tag{1.28'}$$
where $k = 4A/GM$, and $\epsilon = 2AB/GM$

Appendix-2 Adiabatic Propagation of Sound Waves

As in (3.2.1), if dU is expressed as functions of p and v, and since the changes are adiabatic (i.e., dQ = 0), we write

$$(\partial U/\partial p)_v dp + \{(\partial U/\partial v)_p + p\} dv = 0 \tag{2.1'}$$

or,
$$(\partial p/\partial v)_{ad} = -\{(\partial U/\partial v)_p + p\}/(\partial U/\partial p)_v$$
$$= -\{(\partial U/\partial T)_p(\partial T/\partial v)_p + p\}/\{(\partial U/\partial T)_v(\partial T/\partial p)_v\}$$
$$= -(\partial T/\partial v)_p\{p(\partial v/\partial T)_p + (\partial U/\partial T)_p\}/\{c_v(\partial T/\partial p)_v\} \tag{2.2'}$$

Now, since there is an equation of state (2.4.12) which connects the thermodynamic variables p, v and T, we may write

$$(\partial F/\partial p)dp + (\partial F/\partial v)dv + (\partial F/\partial T)dT = 0$$

Solving this equation for T, we get

$$dT = (\partial T/\partial v)_p dv + (\partial T/\partial p)_v dp$$

For an isothermal change, dT = 0, and the above reduces to

$$-(\partial p/\partial v)_T = (\partial T/\partial v)_p/(\partial T/\partial p)_v \tag{2.3'}$$

Inserting (2.3') in (2.2'), we get

$$-(\partial p/\partial v)_{ad} = -(\partial p/\partial v)_T\{p(\partial v/\partial T)_p + (\partial U/\partial T)_p\}/c_v \tag{2.4'}$$

From the First Law of thermodynamics, it follows that at constant pressure,

$$\delta Q = (\partial U/\partial T)_p dT + p(\partial v/\partial T)_p dT$$

or, dividing by dT, we get

$$\delta Q/dT = c_p = (\partial U/\partial T)_p + p(\partial v/\partial T)_p \tag{2.5'}$$

Substitution from (2.5') in (2.4') yields

$$-(\partial p/\partial v)_{ad} = -\gamma(\partial p/\partial v)_T \tag{2.6'}$$

Thus, the adiabatic velocity of sound $= (\gamma p/\rho)^{1/2}$ \quad (2.7')

Appendix-3 Some Selected Thermodynamic Diagrams

1. (T - S) diagram (also known as Tephigram) (Fig. 3.1′)
Co-ordinates:

Abscissa (x-axis)	T
Ordinate (y-axis)	$S\,(\ln \theta)$
Thermodynamic work:	$-c_p \int y\, dx$

2. Neuhoff diagram
Co-ordinates:

Abscissa (x-axis)	T
Ordinate (y-axis)	$-\ln p$
Thermodynamic work:	$-R \int y\, dx$

3. Stuve diagram
Co-ordinates:

Abscissa (x-axis)	T
Ordinate (y-axis)	$-p^{-\kappa}$
where	$\kappa = R/c_p$.
Thermodynamic work:	$c_p \int \ln (-y)\, dx$

4. Rossby diagram
Co-ordinates:

Abscissa (x – co-ordinate) – Specific humidity or humidity mixing-ratio
Ordinate (y – co-ordinate) – Potential temperature/Equivalent potential temperature

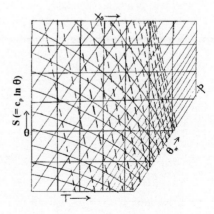

Fig. 3.1′ The T- S diagram

Appendix-4 Derivation of the Equation for Saturation Vapour Pressure Curve Taking into Account the Temperature Dependence of the Specific Heats (Joos and Freeman (1967))

Let us deduce the vapour pressure curve by considering equilibrium between the liquid and the vapour phases of water at a given pressure (p) and temperature(T). At equilibrium, the temperature is, of course, the same throughout the liquid-vapour system. Because we seek the equilibrium position at a given value of T and p, we must find the minimum value of G, the Gibbs potential or enthalpy ($= U - TS + pV$). If we take n_1 mols of vapour and n_2 mols of liquid, the total enthalpy of the system is

$$G = g_1 n_1 + g_2 n_2 \qquad (4.1')$$

where g_1 and g_2 refer to the enthalpy of 1 mol of the vapour and the liquid phase respectively.

The variation in each phase is to be taken under the conditions:

$$T = \text{constant}, p = \text{constant}, \text{ and also}, n_1 + n_2 = \text{constant}.$$

Since g_1, g_2 depend only upon T and p, they must remain constant, on account of the auxiliary conditions. We thus have for $\delta G = 0$

$$0 = \delta n_1 g_1 + \delta n_2 g_2 \qquad (4.2')$$

We also have, $\qquad 0 = \delta n_1 + \delta n_2 \qquad (4.3')$

The method of undetermined multipliers yields

$$g_1 + \lambda = 0, \; g_2 + \lambda = 0$$

or, $\qquad g_1 - g_2 = u_1 - u_2 + p(v_1 - v_2) + T(s_2 - s_1) = 0 \qquad (4.4')$

where p denotes the vapour pressure, s_2 and s_1 are the entropies of the vapour and the liquid phase respectively.

In computing the various terms in (4.4'), we assume that the vapour behaves like an ideal gas (this assumption may lead to considerable error if the vapour is near the point of condensation) and also note that $v_1 \gg v_2$.

To compute $(u_1 - u_2)$ in (4.4'), we make use of the relation (3.3.7) and write

$$u_1 = u_{1,0} + \int_0^T c_{v1} dT \qquad (4.5')$$

$$u_2 = u_{2,0} + \int_0^T c_{v2} dT \qquad (4.6')$$

The third term in (4.4') may be simplified to $p v_1$ in view of the fact that $v_1 \gg v_2$. Substituting for v_1 from the equation of state for vapour, we get

Appendix-4 Derivation of the Equation for Saturation Vapour Pressure Curve

$$p(v_1 - v_2) = R\,T = \int_0^T R\,dT \tag{4.7'}$$

Putting (4.5'), (4.6') and (4.7') together and using the Mayer's relation (3.3.6), we write for the first three terms

$$u_1 - u_2 + p(v_1 - v_2) = (u_{1,0} - u_{2,0}) + \int_0^T c_p dT - \int_0^T c_2 dT$$

$$= L(0) + \int_0^T c_p\,dT - \int_0^T c_2\,dT \tag{4.8'}$$

where we have written c_2 for c_{v2}, since in the case of a liquid there is little external work done by change of volume, and $L(0)$ for $(u_{1,0} - u_{2,0})$ which measures the difference in internal energy between the liquid and the vapour phase at temperature Absolute zero and represents the latent heat of vaporization at that temperature.

Now, to evaluate the term $T(s_2 - s_1)$, we turn to the differential equation

$$ds = (du + pdv)/T$$

and apply it first to the liquid.

In the case of the liquid, we can again neglect the volume expansion and write

$$ds_2 = c_2\,dT/T$$

Integrating between the limits 0 and T, we obtain

$$s_2 = \int_0^T c_2\,dT/T + s_{2,0} \tag{4.9'}$$

The term $s_{2,0}$, the entropy of the liquid at temperature Absolute zero, is zero for a homogenous fluid.

For the vapour phase, we express p in terms of v in the term $p\,dv$ and write the differential equation as

$$ds_1 = c_v\,dT/T + R\,dv/v$$

Indefinite integration yields

$$s_1 = \int c_v\,dT/T + R \ln v$$

$$= \int c_v\,dT/T + R \ln T + R \ln R - R \ln p + \text{Const} \tag{4.10'}$$

We now take the lower limit of integration at temperature Absolute zero and combine all terms not dependent upon T and p to give us a single constant $s_{1,0}$. This constant represents the entropy of the gas at $T = 0$ and $p = 1$. We further make the following substitutions in the second term on the right side of (4.10'):

$$R = c_p - c_v \text{ and } \ln T = \int dT/T$$

Thus, we get

$$s_1 = \int_0^T c_p \, dT/T - R \ln p + s_{1,0} \quad (4.11')$$

Now, the specific heat of an ideal gas does not vanish at $T = 0$, so a question of convergence arises in connection with the integral. To simplify matters, we can conceive of the specific heat as consisting of two parts: one part which does not vary with temperature and the other part which does and approaches zero rapidly as temperature decreases. We, therefore, write $c_p = c_{pc} + c_{pT}$, where c_{pc} is the part which remains constant and c_{pT} is that which varies with temperature. The latter part is absent totally for a monatomic gas.

Substituting for c_p and setting for $p = 1$ a very small lower limit in (4.11'), we get

$$s_1(T,1) - s_1(T_0,1) = \int_0^T c_{pT} dT/T + c_{pc} \ln T - c_{pc} \ln T_0 \quad (4.12')$$

In (4.12'), the lower limit of the integral has been put to zero because of the very rapid decline in the value of c_{pT} with temperature. Since the contribution of c_{pT} to the integral is negligible even at $T = 1°A$, we may interpret $s_{1,0}$ approximately as the entropy at $T = 1$ and $p = 1$. Strictly, it is the limiting value of the difference $s_1(T, 1) - c_{pc} \ln T_0$ as T_0 approaches zero.

Combining (4.8'), (4.9') and (4.11') and substituting in (4.4'), and solving for $\ln p$, we get the vapour pressure equation

$$\ln p = -L(0)/RT + (1/RT)\int_0^T c_2 \, dT - (1/RT)\int_0^T c_{pT} \, dT + (1/R)\int_0^T c_{pT} \, dT/T$$

$$+ (c_{pc}/R)\ln T - (1/R)\int_0^T c_2 dT/T + (s_{1,0} - s_{2,0} - c_{pc})/R \quad (4.13')$$

It is easy to see from (4.13') that if the variation of the specific heat with temperature be ignored, it will reduce to the Clausius-Clapeyron equation (4.8.5) with only a few changes of symbols. In (4.13'), we have used p for e_s, c_p for c_p', c_2 for c_w, R for R_v, and a different constant for A.

Appendix-5 Theoretical Derivation of Kelvin's Vapour Pressure Relation for e_r/e_s

We derive the formula following a procedure adopted by Wallace and Hobbs (1977):

Let us suppose that a volume of air resting on a plane surface of water in a closed space is supersaturated with water vapour and that a certain number of water vapour molecules combine or condense unto themselves spontaneously at constant pressure

Appendix-5 Theoretical Derivation of Kelvin's Vapour Pressure Relation for e_r/e_s

and temperature without the aid of an aerosol nucleus to form a viable water droplet of volume V and surface area A. If c_l and c_v be the chemical potentials of the liquid and the vapour phases of water respectively, and n the number of molecules per unit volume of the liquid, the condensation leads to a decrease in the Gibbs free energy of the system by $n V (c_v - c_l)$, where we define chemical potential for a particular phase of the material as a measure of Gibbs free energy per molecule of that phase at constant pressure and temperature. Now, the gain in Gibbs free energy in creating the surface area of the droplet is equal to A σ, where σ is the work required to create a unit area of the vapour-liquid interface and is usually called the surface energy of water. So, the net increase in Gibbs free energy of the system due to formation of the droplet is given by

$$\Delta G = A\sigma - n V(c_v - c_l) \qquad (5.1')$$

where $G(= U - TS + pV)$ denotes the Gibbs free energy of the system (3.7.6).

Now, the chemical potential of a particular phase of water changes when the vapour pressure of that phase changes reversibly at constant temperature. The changes are related by the following expression

$$dc = v\, de \qquad (5.2')$$

where e denotes vapour pressure and v the volume of one molecule of that phase.

Applying (5.2') to the case of a molecule each of the vapour and liquid phases of water, we obtain for the difference in chemical potential between the two phases, the expression

$$d(c_v - c_l) = (v_v - v_l)de \qquad (5.3')$$

where v_v and v_l denote the volume occupied by a molecule of the vapour and the liquid phase respectively.

Since $v_v \gg v_l$, (5.3') may be simplified to

$$d(c_v - c_l) = v_v de \qquad (5.4')$$

Substituting for v_v from the ideal gas law for one mol of water vapour, $ev_v = \kappa T$, where κ is Boltzmann's constant, in (5.4'), we get

$$d(c_v - c_l) = \kappa T(de/e) \qquad (5.5')$$

But it can be shown that when the vapour pressure e reaches the saturation value e_s which is the equilibrium vapour pressure over a plane surface of water, $c_v = c_l$. Integrating (5.5'), we, therefore, get

$$(c_v - c_l) = k\, T \ln(e/e_s) \qquad (5.6')$$

Substituting (5.6') in (5.1'), we get

$$\Delta G = A\sigma - nVkT\ln(e/e_s) \qquad (5.7')$$

For a droplet of radius R, (5.7') becomes

Fig. 5.1' Variation of ΔG with the radius R of the droplet in the case when $e/e_s < 0$, and when $e/e_s > 0$ (After Wallace and Hobbs, 1977, published by Academic Press, Inc)

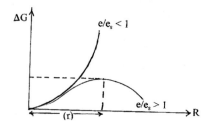

$$\Delta G = 4\pi R^2 \sigma - (4/3)\pi R^3 n\, k\, T \ln(e/e_s) \quad (5.8')$$

As long as $e < e_s$, $\ln(e/e_s)$ is negative and the right side of (5.8') is positive for all values of R, as shown in Fig. 5.1''.

This means that in this case when $e/e_s < 0$, ΔG is always positive and the system never attains thermodynamic equilibrium. In the case when $e > e_s$, the second term on the right side of (5.8' is negative, and ΔG first increases with R but decreases after R reaches a particular value, r.

Since $\partial(\Delta G)/\partial R = 0$ at $R = r$, we get from (5.8'),

$$\ln(e_r/e_s) = (2\sigma/r)(1/nkT) \quad (5.9')$$

This is Kelvin's formula (5.1).

Appendix-6 Values of Thermal Conductivity Constants for a Few Materials, Drawn from Sources, Including 'International Critical Tables'(1927), 'Smithsonian Physical Tables' (1934), 'Landholt-Bornstein' (1923–1936), 'McAdams' (1942) and others

Material	Density ρ (gm/cc)	Specific heat (Cal/gm/C)	Thermal Conductivity k(Cal/s/cm/C)	Thermal Diffusivity (k/ρc)
Soil, wetted (43% water)	1.67	0.53	0.0017	0.0019
Quartz sand (medium, fine and dry)	1.65	0.19	0.00063	0.0020
Quartz sand (8.3% moisture)	1.75	0.24	0.0014	0.0033
Sandy clay (15% moisture)	1.78	0.33	0.0022	0.0037

(*continued*)

Appendix-6 Values of Thermal Conductivity Constants for a Few Materials

Material	Density ρ (gm/cc)	Specific heat (Cal/gm/C)	Thermal Conductivity k(Cal/s/cm/C)	Thermal Diffusivity (k/ρc)
Soil, very dry	–	–	0.0004–8	0.002–3
Some wet soils	–	–	0.003–8	0.004–.010
Wet mud	1.50	0.60	0.0020	0.0022
Rocks and building materials:				
Brick masonry (at20C)	1.7	0.20	0.0015	0.0044
Concrete, av. Stone	2.3	0.20	0.0022	0.0048
Concrete, dams	2.47	0.22	0.0058	0.0107
Granite (at0C)	2.7	0.19	0.0065	0.0127
Limestone (at 0C)	2.7	0.22	0. 0048	0.0081
Marble	2.7	0.21	0.0055	0.0097
Sandstone	2.6	0.21	0.0062	0.0113
Traprock	–	–	–	0.0075
Rock material (av for earth)			–	0.010

(Room temperature (°C) assumed in all cases, except where specifically given).

Appendix-7 Physical Units and Dimensions

The following three Tables give the basic and derived S.I. units used in the text.

Table 7.1 Basic units

Variable	Unit	Symbol
Mass	Kilogramme	Kg
Length	Meter	m
Time	Second	s
Temperature	Kelvin	K
	Or, Absolute	A
	Or, Celsius	C

Table 7.2 Derived units

Variable	Name	Symbol (unit)
Velocity	–	$m\ s^{-1}$
Acceleration	–	$m\ s^{-2}$
Force	Newton	$N\ (Kg\ m\ s^{-2})$
Pressure	Pascal	$Pa\ (N\ m^{-2})$
Energy	Joule	$J(Nm)$
Power	Watt	$W\ (J\ s^{-1})$

Table 7.3 Decimal multiples and submultiples of SI units with prefixes and symbols

Multiple/Submultiple	Prefix	Symbol
10^{18}	Exa	E
10^{15}	Peta	P
10^{12}	Tera	T
10^{9}	Giga	G
10^{6}	Mega	M
10^{3}	Kilo	K
10^{2}	Hecto	H
10^{1}	Deka	Da
10^{-1}	deci	d
10^{-2}	centi	c
10^{-3}	milli	m
10^{-6}	micro	μ
10^{-9}	nano	n
10^{-12}	pico	p
10^{-15}	femto	f
10^{-18}	atto	a

Appendix-8 Some Useful Physical Constants and Parameters

Angular velocity of the earth (Ω)	7.292×10^{-5} radian s^{-1}
Mean radius of the earth	6.37×10^{6} m
Acceleration due to gravity (g) at msl at latitude 45°	$9.80616\ m\ s^{-2}$
Mass of the earth	5.988×10^{24} kg
Mean density of the earth	$5.50 \times 10^{3}\ kg\ m^{-3}$
Gravitation constant (G)	$6.673 \times 10^{-11}\ N\ m^{-2}\ kg^{-1}$
Universal Gas constant (R^*)	$8.314 \times 10^{3}\ J\ K^{-1}\ (kmol)^{-1}$
Mean molecular weight of dry air	$28.9\ kg\ (kmol)^{-1}$
Mean molecular weight of water vapour	$18.016\ kg\ (kmol)^{-1}$
Gas constant for dry air (R)	$287\ J\ K^{-1}\ kg^{-1}$
Mechanical equivalent of heat	$4.18\ J\ (Cal)^{-1}$
Specific heat of dry air at constant pressure (c_p)	$1004\ J\ K^{-1}\ kg^{-1}$

Appendix-8 Some Useful Physical Constants and Parameters

Specific heat of dry air at constant volume (c_v)	$717 \, \text{J K}^{-1} \, \text{kg}^{-1}$
Ratio of the specific heats (γ)	1.4
Latent heat of condensation at $0\,°\text{C}$	$2.5 \times 10^6 \, \text{J kg}^{-1}$
Freezing point of water	$273.16 \, \text{K}$
Standard sea level pressure	$101.325 \, \text{kPa}$
Standard sea level temperature	$288.15 \, \text{K}$
Standard sea level density	$1.225 \, \text{kg m}^{-3}$
Planck's constant (h)	$6.625 \times 10^{-34} \, \text{J s}$
Boltzmann's constant (k)	$1.380 \times 10^{-23} \, \text{J K}^{-1}$
Velocity of light (c)	$2.997 \times 10^8 \, \text{m s}^{-1}$
Avogadro number (N)	$6.0247 \times 10^{23} \, (\text{mol})^{-1}$
Stefan-Boltzmann constant (σ)	$5.6687 \times 10^{-8} \, \text{W m}^{-2} \, \text{K}^{-4}$
Wien's constant	$289.78 \times 10^{-5} \, \text{m K}$

References

Agafonova EG, Monin AS (1972) On the origin of the thermohaline circulation in the ocean. Okeanologia Issue No.6.
Aitken J (1923) Collected Scientific Papers (1880–1916) The University Press, Cambridge.
Arakawa A (1972) "Design of the UCLA General Circulation Model" Tech Rpt No 7, Dept of Meteorology, Univ of California, Los Angeles, California.
Arakawa A, Schubert WH (1974) Interaction of a cumulus cloud ensemble with the large-scale environment Part 1. J Atmos Sci 31:674–701.
Arenberg D (1939) Turbulence as the major factor in the growth of cloud drops. Bull Amer Meteor Soc 20:444–448.
Banks PM, Kockarts G (1973) Aeronomy. Academic Press, New York, Part A, p 3.
Bergeron T (1933) On the physics of cloud and precipitation. Mem Met Assoc, IUGG, Lisbon, 156–178.
Berlage HP (1966) The southern oscillation and world weather. Mededel Verhandel 88, Kon Ned Meteor Inst., p 152.
Bjerknes J (1966) A possible response of the atmospheric Hadley circulation to equatorial anomalies of ocean temperature. Tellus 18:820–829.
Bjerknes J (1969) Atmospheric teleconnections from the equatorial Pacific Mon Wea Rev 97:163–172.
Bjerknes V and collaborators (1933) Physikalische Hydrodynamik S-252, Berlin.
Blackadar AK (1957) Boundary layer wind maxima and their significance for the growth of the nocturnal inversion Bull Amer Meteor Soc, 38:283–290.
Bowen IS (1926) The ratio of heat losses by conduction and by evaporation from any water surface. Phys Rev 27:779–787.
Brunt D (1944) Physical and dynamical meteorology. Cambridge University Press, London, p 428.
Chang CP, Lim H (1988) Kelvin wave-CISK: A possible mechanism for the 30–50 day oscillations. J Atmos Sci 45:1709–1720.
Charney JG, Eliassen A (1964) On the growth of the hurricane depression. J Atmos Sci 21:68–75.
Chen L, Reiter ER, Feng Z (1985) The atmospheric heat source over the Tibetan plateau May-August 1979. Mon Wea Rev 113:1771–1790.
Couper-Johnston R (2000) El Nino – the weather phenomena that changed the world. Hodder and Stroughton, London.
Cwilong BM (1947) Sublimation centers in a Wilson chamber. Proc Roy Soc A 190:137–143.
Deacon EL (1949) Vertical diffusion in the lowest layers of the atmosphere. Quart J Roy Met Soc 75:89–103.
Defant A (1905) Gesetzmassigkeiten in der verteilung der verschiedenen tropfengrossen bei regenfallen. Akademie der Wissenschaften, Vienna, Sitzungsberichte, Mathematisch-Wissenschaftliche Klasse 2a, 114:585–646.

Defant A (1961) Physical Oceanography, vol 1. Pergamon Press, London.
Dobson GMB (1931) Ozone in the upper atmosphere and its relation to meteorology. Proc Roy Soc GB 26:373–383.
Dobson GMB, Brewer AW, Cwilong BM (1946) Meteorology of the stratosphere (The Bakerian Lecture). Proc Roy Soc A 185:144–175.
Ekman VM (1905) On the influence of the earth's rotation on ocean currents. Arkiv Math Astron Physik 2:1–54.
Ferrel W (1859) The motions of fluids and solids relative to the earth's surface. Math Monthly 1: 140, 210, 300, 366, 397.
Fultz D (1949) A preliminary report on experiments on thermally produced lateral mixing in a rotating hemispherical shell of liquid. J Met 56:17–33.
Fultz D (1950) Experimental studies related to atmospheric flow around obstacles. Pure Appl Geophys 17: 80–93.
Fultz D (1951) Experimental analogies to atmospheric motions Compendium of Meteorology 1235–1248.
Gerbier N, Berenger M (1961) Experimental studies of lee waves in the French Alps. Quart J Roy Meteor Soc 87:13–23.
Gill AE (1982) Atmosphere-Ocean-dynamics International Geophysics Series, vol 30. Academic Press Inc, New York, p 662.
Gutman G, Schwerdtfeger WS (1965) The role of latent and sensible heat for the development of a high pressure system over the subtropical Andes, in the summer. Meteor Rundsch 18:1–7.
Hadley G (1735) Concerning the cause of the general trade winds. Phil Trans Roy Soc London 39:58–62.
Hahn DG, Manabe S (1975) The role of mountains in the South Asian Monsoon Circulation. J Atmos Sci 32:1515–1541.
Halley E (1686) An historical account of the Trade Winds, and Monsoons, Observable in the seas between and near the Tropics, with an attempt to assign the phisical cause of the said Winds. Philos Trans R Soc London 16: 153–168.
Hanna S (1969) The formation of longitudinal sand dunes by large helical eddies in the atmosphere. J Atmos Sci 8:874–883.
Hendon HH, Salby ML (1994) The life cycle of the Madden-Julian Oscillation. J Atmos Sci 51:2225–2237.
Holton JR, Lindzen RS (1968) A note on Kelvin waves in the atmosphere. Mon Wea Rev 95:385–386
Holton JR (1979) An Introduction to Dynamic Meteorology, 2nd edn. Academic Press, New York, p 391.
Ingersoll LR, Zobel OJ, Ingersoll AC (1948) Heat conduction. McGraw-Hill Book Co, New York.
International Critical Tables (1927) McGraw-Hill Book Company, Inc, New York.
Johnson DR (1980) A generalized transport equation for use with meteorological co-ordinate systems. Mon Wea Rev 108:733–745.
Johnson DR, Townsend RD, Wei M-Y (1985) The thermally coupled response of the planetary scale circulation to the global distribution of heat sources and sinks. Tellus 37A:106–125.
Johnson DR, Yanai M, Schaack T (1987) Global and regional distribution of atmospheric heat sources and sinks during the GWE. In: Chang, Krishnamurti (eds) Monsoon Meteorology, Vol.9: 271–297.
Joos G, Freeman IM (collaborator) (1967) Theoretical Physics, 3rd edn. Published by Blackie & Sons Ltd, London.
Jordan CL (1958) Mean soundings for the West Indies area. J Met 15:91–97.
Joseph PV, Raman PL (1966) Existence of low level westerly jet stream over peninsular India during July. Ind J Met & Geophys 17:407–410.
Julian PR, Chervin RM (1978) A study of the southern oscillation and Walker circulation phenomenon. Mon Wea Rev 106:1433–1451.
Kelvin Lord (William Thompson) (1870) Saturation vapour pressure of liquids over a curved surface. Proc Roy Soc Edin Feb1870.

Kőhler H (1926) Zur thermodynamik der Kondensation an hydroskopischen kernen, Meddelanden fran statens, Met-hydr. Anst. Stockholm 13(8).
Krishnamurti TN, Kanamitsu M, Godbole R, Chang CB, Carr F, Chow JH (1976) Study of a Monsoon depression (II) – dynamical structure. J Meteor Soc of Japan 54:208–225.
Kuettner J (1959) The band structure of the atmosphere. Tellus 11:267–294.
Kuettner J (1967) Cloud streets: theory and observations. Aero Revue 42: (52–56) 109–112.
Landholt-Bornstein (1923–1936) Physikalische-Chemische Tabellen, Verlag Julius Springer, Berlin.
Lang KR (1999) The Sun. In: Beatty JK, Petersen CC, Chaikin A (eds) The new solar system, 4th edn. Sky Publishing Corporation, Cambridge MA p 23–38.
Leith CE (1978) Objective methods of weather prediction. Ann Rev Fluid Mechan 10.
Lim H, Chang C-P (1983) On the dynamics of midlatitude-tropical interactions and the winter monsoon. In: Chang and Krishnamurti (eds) Monsoon meteorology Chap12 405–434.
Lindzen RS, Batten ES, Kim JW (1968) Oscillations in the atmosphere with tops. Mon Wea Rev 96:133–140.
Lindzen RS, Holton JR (1968) A theory of quasi-biennial oscillation. J Atmos Sci 25:1095–1107.
London J, Sasamori T (1971) Radiative energy budget of the atmosphere. Space Res 11:639–649.
Long RR (1951) A theoretical and experimental study of the motion of certain atmospheric vortices. J Met 8:207–221.
Lorenz EN (1960) Energy and numerical weather prediction. Tellus 12:364–373.
Lorenz EN (1969) The nature and theory of the general circulation of the atmosphere. World Meteorological Organization, Geneva, Switzerland.
Luo H, Yanai M (1983) The large-scale circulation and heat sources over the Tibetan Plateau and surrounding areas durin the early summer months of 1979. I. Precipitation and kinematic analysis. Mon Wea Rev 111:922–944.
Luo H, Yanai M (1984) The large-scale circulation and heat sources over the Tibetan Plateau and surrounding areas during the early summer of 1979 II. Heat and moisture budgets. Mon Wea Rev 112:966–989.
McAdams WH (1942) Heat transmission, 2nd edn. McGraw-Hill Book Company, Inc. New York.
Madden RA (1986) Seasonal variations of the 40–50 day oscillation in the tropics. J Atmos Sci 43:3138–3158.
Madden RA, Julian PR (1971) Detection of a 40–50 oscillation in the zonal wind in the tropical Pacific. J Atmos Sci 28:702–708.
Madden RA, Julian PR (1972) Description of global-scale circulation cells in the tropics with a 40–50 day period. J Atmos Sci 29:1109–1123.
Matsuno T. (1966) Quasi-geostrophic motions in the equatorial area. J Meteoro Soc of Japan 44:25–42.
Meetham AR (1936) Conference on Atmospheric Ozone (Oxford), Quart Jour Roy Met Soc Suppt 62:59.
Mintz Y (1951) The observed zonal circulation of the atmosphere. Bull Amer Met Soc 35:208–214.
Mintz Y (1968) Very long-term global integrations of the primitive equations of atmospheric motion: an experiment in climate simulation. Met Monogragh 8:20–36, Amer Meteor Soc Boston Mass.
Mitra SK (1952) The Upper Atmosphere, 2nd edn. The Asiatic Society, Calcutta-16, India.
Monin AS (1975) The role of the oceans in climate models. In "The Physical basis of climate and climate modelling", GARP Pub. Series no.16, WMO.
Murakami T (1987) Effects of the Tibetan plateau. In: Chang and Krishnamurti (eds) Monsoon meteorology. Chap 8, 235–270.
Nitta T (1983) Observational study of heat sources over the eastern Tibetan plateau during the summer monsoon. J Meteor Soc of Japan 61:590–605.
Normand CWB (1921) Thermodynamics of the wet and dry-bulb hygrometer Memoirs of the India Meteorological Dept. 23: p.1.
Ooyama K (1969) Numerical simulation of the life cycle of tropical cyclones. J Atmos Sci 26:3–40.

Palmen E (1951) Mean meridional circulation cells over the globe in winter. Quart J Roy Meteor Soc 77:337.

Palmen E, Alaka MA (1952) Angular momentum budget. Tellus 4:324.

Palmen E, Newton CW (1969) Atmospheric circulation systems. Academic Press, New York.

Peixoto JP, Crisi AR (1965) Sci Rep Massachusetts Inst. of Technology, USA.

Petterssen S (1956) Weather analysis and forecasting, 2nd edn. vol 1. Motion and motion systems McGraw-Hill Book Co, Inc, New York.

Phillips NA (1956) The general circulation of the atmosphere: A numerical experiment. Quart J Roy Met Soc 82:123–164.

Phillips NA (1963) Geostrophic motion. Rev Geophys 1:123–176.

Ramage CS, Raman CRV (1972) Meteorological Atlas of the International Indian Ocean Expedition, vol. 2 (upper air) National Science Foundation, Washington, D.C.

Rao GV, Erdogan S (1989) The atmospheric heat source over the Bolivian plateau for a mean January. Bound-Layer Meteor 46:13–33.

Raschke E, Vonder Haar TH, Bandeen WR, Pasternak M (1973) The annual radiation balance of the earth-atmosphere system during 1969–70 from Nimbus 3 measurements. J Atmos Sci 30:341–364.

Reed RJ, Rodgers DG (1962) The circulation of the tropical stratosphere in the years 1954–1960. J Atmos Sci 19:127–135.

Richardson LF (1920) Stability and the criterion of turbulence Proc Roy Soc A 97:359.

Richardson LF (1922) Weather prediction by numerical process, Cambridge University Press, London and New York.

Riehl H (1978) Introduction to the atmosphere, 3rd edn. McGraw-Hill, Tokyo.

Riehl H, Alaka MA, Jordan CL, Renard RJ (1954) The Jet stream (ChapterVII), Met. Monograph, 2, no.7 Published by Amer Meteor Soc.

Riehl H, Malkus JS (1958) On the heat balance of the equatorial trough zone. Geophysica 6:503–538.

Rossby CG (1937) On the mutual adjustment of pressure and velocity distributions in certain simple current systems, I J Mar Res 1:15–28.

Rossby CG (1938a) On the mutual adjustment of pressure and velocity distributions in certain simple current systems, II. J Mar Res 2:239–263.

Rossby CG (1938b) On the temperature changes in the stratosphere resulting from shrinking and stretching. Beitr Phys Freien Atmos 24:53–59.

Rossby CG (1940) Planetary flow patterns in the atmosphere. Q J R Met Soc 66(suppl):68–97.

Rossby CG (1947) On the distributions of angular velocity in gaseous envelopes under influence of large-scale horizontal mixing processes. Bull Amer Meteor Soc 28:53–68.

Saha KR, Saha AK (1939) Is life possible in other planets? Science and Culture, vol 4. Indian Science News Association, Calcutta (India) pp 445–458.

Saha KR (1968) On the instantaneous distribution of vertical velocity in the monsoon field and structure of the monsoon circulation. Tellus 20:601–620.

Saha MN, Srivastava BN (1931) A Treatise on Heat The Indian Press Pvt (Ltd) Allahabad (India), revised reprinted edn, 1969.

Saha S, Nadiga S, Thiaw C, Wang J, Wang W, Zhang Q, van den Dool HM, Pan H.-L, Moorthi S, Behringer D, Stokes D, Pena M, Lord S, White G, Ebisuzaki W, Peng P, Arkin P, Xie PP (2006) The NCEP climate forecast system. J Climate, 19:3483–3517.

Sandstrom JW (1910) In: V Bjerknes and collaborators (eds) Dynamic meteorologyand hydrology, vol 2. Carnegie Institution, Washington, DC.

Sardeshmukh PD (1993) The baroclinic χ problem and its applications to the diagnosis of atmospheric heating rates. J Atmos Sci 50:1099–1112.

Schaack TK, Johnson DR (1994) January and July global distribution of atmospheric heating for 1986, 1987, and 1988. J Climate 7:1270–1285.

Schaack TK, Johnson DR, Wei M-Y (1990) The three-dimensional distribution of atmospheric heating during the GWE. Tellus 42A:305–327.

Schaefer VJ (1946) Production of ice crystals in a cloud of super cooled water droplets. Science 104:457–459 Also, Bull Amer Met Soc 29:175–182(1948).

Schmidt W (1908) Zur Erklarung der gesetzmassig verteilung der Tropfengrossen bei Regenfallen, Meteorologische Zeitschrift, Bd.25, S. 498 Braunschweig.
Schott G (1931) Der Peru strom und seine nordlichen Nachbargebiete in normaler und anormaler Ausbildung Ann d Hydrogr u Mar Meteorol, Bd .59:161–169, 200–213.
Sellers WD (1965) Physical climatology. University of Chicago Press, Chicago, p 84.
Shukla J (1978) CISK-barotropic-baroclinic instability and the growth of monsoon depressions. J Atmos Sci 35:495–508.
Simpson GC (1928) Further studies in Terrestrial Radiation. Mem. R Meteor Soc London 3:21.
Simpson GC (1929) The distribution of Terrestrial Radiation. Mem R Meteor Soc London 3:23.
Simpson GC (1941) On the formation of cloud and rain. Quart J Roy Meteor Soc 67:99–133.
Squires P (1958) The microstructure and colloidal stability of warm clouds, Part - 1 the relationship between structure and stability. Tellus, X: 256–261.
Smagorinsky J (1963) General circulation experiments with the primitive equations: I. The basic experiment. Mon Wea Rev 91:99–164.
Smithsonian Physical Tables (1934) Smithsonian Institution, Washington.
Starr VP (1968) Physics of negative viscosity phenomena. McGraw-Hill Book Co, New York.
Stickley AR (1940) An evaluation of the Bergeron-Findeisen precipitation theory. Mon Wea Rev 68:272–280.
Sverdrup HU, Johnson MW, Fleming RH (1942) The Oceans. Prentice-Hall, Inc., Englewood Cliffs, NJ.
Taylor GI (1915) Eddy motion in the atmosphere. Phil Trans Roy Soc A 215:1–26.
Taylor GI (1921) Experiments with rotating fluids. Proc Roy Soc (A) 100:114–121.
Taylor GF (1941) Aeronautical meteorology. Pitman Publishing Corporation, New York.
Thomson J (1857) On the grand currents of atmospheric circulation, British Association Meeting, Dublin (unpublished). See (1892), Phil Trans Roy Soc London (A) 183:653–684.
Thorpe AJ, Guymer TH (1977) The nocturnal jet. Quart J Roy Met Soc 103:633–653.
Twomey S, Wojciechowski (1969) Observations of the geographical variation of cloud nuclei. J Atmos Sci 26:684–688.
U.S.National Academy of Sciences (1975) Report of the panel on 'Understanding Climate Change'.
US Navy Weather Research Facility Arctic Forecast Guide, 1962.
Van den Dool HM (1990) Time-mean precipitation and vertical motion patterns over the United States. Tellus 42A:51–64.
Vettin F (1857) Ueber den aufsteigenden Luftstrom, die Entstehung des Hagels und der Wirbel-Stürme. Ann Phys (Leipzig) 102:246–255.
Vettin F (1884) Experimentelle Darstellung von Luftbewegungen unter dem Einfluss von Temperatur-unter schieden und Rotations-Impulsen. Meteor Z 1:227–230, 271–276.
Walker GT (1923) Correlations in seasonal variations of weather VIII. Mem Indian Meteor Dept 24:75–131.
Walker GT (1924) World weather IX. Mem Indian Meteor Dept 24:275–332.
Walker GT (1928) World weather III. Mem Roy Meteor Soc II 17:97–106.
Wallace JM, Hobbs PV (1977) Atmospheric science – an introductory survey. Academic Press, New York, p 467.
Wallace JM, Kousky VE (1968) Observational evidence of Kelvin waves in the tropical stratosphere. J Atmos Sci 25:900–907.
Wei M-Y, Johnson DR, Townsend RD (1983) Seasonal distributions of diabatic heating during the first GARP Global Experiment. Tellus 35A:241–255.
Wiin-Nielsen A (1967) On the annual variation and spectral distribution of atmospheric energy. Tellus 19:540–558.
Wilcke JC (1785) Forsok til Uplysning Om Luft-hvirflar och Sky-drag K. Vet Acad Nya Hand 6(2):290–307.
Winston JS, Gruber A, Gray TT, Varnadore MS, Earnest CE, Mannelli LP (1979) Earth-atmosphere radiation budget analyses derived from NOAA satellite data June 1974-February 1978, vol.2. Meteorological Satellite Laboratory, U. S. Department of Commerce, Washington, DC.

Woodcock AH (1942) Soaring over the open sea. Sci Monthly 55:226–232.
Xie PP, Arkin PA (1996) Analyses of global monthly precipitation using gauge observations, satellite estimates and numerical model predictions. J Climate 9:840–858.
Yeh T-C, Gao Y-X (1979) Meteorology of the Tibetan Plateau. Scientific Publications Agency, Beijing (in Chinese).

Author Index

Agafonova, E.G., 134–135
Aitken, J., 60–61, 67
Alaka, M.A., 322–323
Arakawa, A., 289, 309
Arenberg, D., 72–73
Aristotle, 3
Arkin, P.A., 77

Bandeen, W.R., 137, 141, 142
Banks, P.M., 21
Berenger, M., 244
Bergeron, T., 73, 75–76
Berlage, H.P., 268–269
Bjerknes, J., 190, 196, 266, 270
Bjerknes, V., 72, 188, 275
Bowen, I.S., 124
Brewer, A.W., 106
Brunt, D., 32, 49, 86–87, 99–100, 120–121

Carr, F., 309
Chang, C.B., 309
Chang, C.-P., xx, 265, 271
Charney, J.G., 275–277, 293, 309
Chen, L., 138, 145, 147, 148
Chervin, R.M., 269
Chow, J.H., 309
Copernicus, 3
Couper-Johnston, R., 268, 272
Cwilong, B.M., 67–68

Deacon, E.L., 213–214
Defant, A., 69, 134
Dobson, G.M.B., 67, 106

Earnest, C.E., 137, 142, 143–144
Ekman, V.M., 208, 212–213, 215

Eliassen, A., 293, 309
Erdogan, S., 138, 145

Feng, Z.Q., 148
Ferrel, W., 312
Fleming, R.H., 132
Freeman, I.M., 23, 344
Fultz, D., 327–330

Galileo, 3, 94
Gao, Y.-X., 138, 145, 146, 148
Gerbier, N., 244
Gill, A.E., 208, 225, 244, 251, 271
Godbole, R., 309
Gray, T.T., 137, 143, 144
Gruber, A., 137, 143, 144
Gutman, G., 138
Guymer, T.H., 221

Hadley, G., 311
Hahn, D.G., 331
Halley, E., 311
Hanna, S., 220
Hendon, H.H., 264
Hobbs, P.V., 346, 348
Holton, J.R., 181–182, 214, 220, 254, 257, 290, 295, 297, 299

Ingersoll, A.C., 122
Ingersoll, L.R., 122

Ji, M., 273
Johnson, D.R., 138, 149, 150
Johnson, M.W., 132
Joos, G., 23, 344
Jordan, C.L., 57, 58

Joseph, P.V., 220
Julian, P.R., 262, 263, 269

Kanamitsu, M., 309
Kelvin Lord, 61
Kepler, 3–4, 6–8
Kockarts, G., 21
Kohler, H., 60
Kousky, V.E., 243, 260
Krishnamurti, T.N., 271, 309
Kuettner, J., 220

Landholt-Bornstein, 348
Lang, K.R., 90, 94, 95
Leetmaa, A., 273
Lim, H., 265, 271
Lindzen, R.S., 257, 297
London, J., 139
Long, R.R., 327
Lord, S., 291
Lorenz, E.N., 304, 324, 325, 327
Luo, H., 138, 145, 147, 148

McAdams, W.H., 348
Madden, R.A., 262, 263, 264, 265
Malkus, J.S., 327
Manabe, S., 331
Mannelli, L.P., 137, 143
Matsuno, T., 254, 271
Meetham, A.R., 106
Mintz, Y., 313, 314, 331
Mitra, S.K., 103, 105, 106
Monin, A.S., 130, 134, 135
Moorthi, S., 291
Murakami, T., 138, 271

Nadiga, S., 291
Newton, 7, 33
Nitta, T., 138, 145, 147, 148
Normand, C.W.B., 46–47

Ooyama, K., 58

Palmen, E., 313, 321–323, 327
Pan, H.-L., 291
Pasternak, M., 137, 141, 142
Pena, M., 291
Peng, P., 291
Petterssen, S., 182
Phillips, N.A., 328–331
Ptolemy, 3

Raman, cry, 183
Raman, P.L., 220

Rao, G.V., 138, 145
Raschke, E., 137, 141, 142
Reed, R.J., 261
Reiter, E.R., 148
Renard, R.J., 325
Richardson, L.F., 208, 213, 275, 285
Riehl, H., 114, 325, 327
Rodgers, D.G., 261
Rossby, C.G., 244, 246, 250, 312–313

Saha, A.K., 11
Saha, K.R., 11, 220
Saha, M.N., 11, 27, 41, 42, 81, 87, 92, 109
Saha, S., 291
Salby, M.L., 264
Sandstrom, J.W., 182
Sardeshmukh, P.D., 202
Sasamori, T., 139
Schaack, T.K., 138
Schaefer, V.M., 67
Schmidt, W., 69, 72, 211
Schott, G., 267
Schubert, W.H., 309
Schwerdtfeger, W.S., 138
Sellers, W.D., 326
Shukla, J., 309
Simpson, G.C., 66, 119, 137, 139, 140, 141
Smagorinsky, J., 331
Squires, P., 73
Srivastava, B.N., 11, 27, 41, 42, 81, 87, 92, 109
Starr, V.P., 317, 324
Stickley, A.R., 69, 76
Stokes, D., 291
Sverdrup, H.U., 132

Taylor, G.F., 65
Taylor, G.I., 122, 327
Thiaw, C., 291
Thomson, J., 38, 63, 312, 316
Thorpe, A.J., 221
Twomey, S., 64

Van den Dool, H.M., 202, 291
Varnadore, M.S., 137, 143, 144
Vettin, F., 327
Vonder Harr, T.H., 137, 141, 142

Walker, G.T., 266, 268, 272
Wallace, J.M., 243, 260, 348
Wang, J., 291
Wang, W., 291

Wei, M.-Y., 138
White, G., 291
Wiin-Nielsen, A., 324
Wilcke, J.C., 327
Winston, J.S., 137, 142, 143, 144
Wojciechowski, T.A., 64
Woodcock, A.H., 220

Xie, P.P., 77, 291

Yanai, M., 138, 145, 147–148, 149, 150
Yeh, T.C., 138, 145, 146, 148

Zhang, Q., 291
Zobel, O.J., 122

Subject Index

Absolute acceleration, 159
Absolute angular momentum, 189, 311, 319–324
Absolute motion, 158–159, 188
Absolute temperature, 11, 17, 27, 54, 84, 86, 302
Absolute velocity, 159, 160, 191
Absolute vorticity, 191, 193, 194, 195, 197, 198, 237–238, 282, 294–295, 307–308
Absorption of radiation, 81, 107
Acceleration, 5, 6, 12, 14, 32, 66, 130, 149, 158–162, 175–176, 207, 246, 249, 258, 261, 276, 333
Acoustic or sound waves, 33, 94, 227, 230, 232, 233–234, 239, 242, 277, 285, 290
Ageostrophic wind, 216, 278
Air cooling, 125
Albedo, 113, 114, 117, 137, 141, 148
Angular momentum balance, 320–324
Angular velocity, 4, 5, 7, 159, 162, 170, 176, 188, 245, 321, 328
Annular experiments, 328–331
Atmospheric absorption, 107–108
Atmospheric boundary layer, 222–225

Baroclinic atmosphere, 189
Baroclinic instability, 293, 295–300, 302–306
Baroclinic model, 282–284, 288, 295, 299, 300, 331
Barotropic atmosphere, 189, 193, 198, 218–219, 227, 238
Barotropic instability, 306–308
Barotropic model, 276, 282, 331
Blackbody radiation, 83
Boundary layer, 23, 42, 55, 58, 170, 207–225, 264, 291, 329
Bowen's ratio, 124

Boyle's law, 24–25
Buoyancy flux, 134–135
Buoyancy oscillations, 31–33, 239, 240–241

Carbon–nitrogen chain reaction, 92
Carnot cycle, 37, 39, 42, 51
Carnot engine, 36–38, 42
Charles' and Gay-Lussac's law, 25
Chromosphere, 96, 97
Circulation, 42, 78, 93–94, 106, 124, 134–135, 141, 143, 187–206, 217–221, 225, 253, 260, 262–267, 270–271, 291, 301, 303, 306, 311–331
Circulation theorem, 188–190, 196
Clausius-Clapeyron equation, 51–54
Cloud classification
  types, 68, 69–71, 72
Cloud condensation nuclei, 60–61, 63–64
Cloud formation in atmosphere, 62–64
Coastal upwelling, 224–225
Composition of atmosphere, 10–12
Conditional instability, 55–56, 293, 308–309
Conditional instability of second kind (CISK), 293, 308–309
Conduction of heat, 18, 33, 36, 39, 42, 79, 81, 106, 121, 122, 125, 129, 147, 329
Convection, 18, 33, 42–43, 58, 79–80, 81, 93, 94, 121, 123, 138, 220, 221, 264–265, 270, 309, 317
Convergence, 57–58, 78, 168, 175, 192, 197, 204, 206, 218–219, 224–225, 235, 264, 265, 287, 301, 308, 309, 316, 321, 323
Coriolis force, 161–162, 170, 176, 177–178, 180, 192, 208, 216, 220, 250, 277, 293
Corona, 90, 95, 96–98, 101
Critical velocity, 12
Cyclostrophic motion, 179–180

363

Deformation, 202–203, 204, 205–206, 247, 251–252, 271
Density, 12, 14, 15, 21–22, 44, 45–46, 63, 71, 89–90, 91, 96, 97, 106, 121, 129–130, 131, 134–135, 148, 155, 159, 164, 168, 169, 174, 183, 189, 198, 212, 223, 227, 231–232, 235, 237, 241, 263, 277, 285, 302, 327–328
Desert coolers, 124–125
Dew-point, 44–45, 47, 113
Diabatic heating, 138, 168, 175, 201, 256, 279, 317, 331
Differential properties of wind field, 202–205, 206
Dispersion of waves, 230–231
Divergence, 58, 138, 149, 168, 170, 187–206, 224, 225, 234, 235, 245, 258–259, 265, 270–271, 277–280, 282–283, 286, 301, 309, 323
Drop-size distribution in clouds fog and rain, 63, 64–65
Dry adiabatic lapse rate, 31, 48, 50, 55, 56, 121–122
Dynamical instability, 293–309
Dynamical models, 275–291

Eddy flux of sensible heat, 324
Eddy flux of water vapour, 326
Eddy transports, 212, 322, 324–327
Eddy viscosity, 207, 210, 211, 216
Ekman drift, 222–224
Ekman layer, 208, 213, 214–217
Ekman pumping, 224–225
Ekman spiral, 215, 222
Ekman stress, 222
Electrical attraction, 72
Electromagnetic radiation, 80, 88, 94, 102
El Niño, 266–268, 270, 272, 273
El Niño/La Niña, 266–268
El Niño-southern oscillation (ENSO), 266–273
Energy balance equation, 145–148
Energy transformations, 247–250, 299–300, 309
Entropy, 33–35, 39, 40, 41, 42, 49, 50, 56, 286
Equation of continuity, 155, 167–168, 174, 200, 286, 321
Equations of state, 23–26, 45
Equator, 4–5, 15, 17, 42, 93, 95, 98, 105, 109, 110, 116, 134, 137, 141–144, 151, 190, 240, 253–273, 282, 311–312, 314, 316–317, 320, 324, 327
Equatorial oscillation, 253–273
Equatorial radius, 4, 5
Equatorial waves, 253–273

Equivalent potential temperature, 47–48, 50–51, 56–57, 58, 293
Evaporation, 10, 34, 41–42, 46, 49, 51, 68, 73, 78, 115, 123–125, 130, 134–135, 147–148, 324, 326
Evaporative cooling, 47, 124–125
Evaporative heat flux, 123–125, 145
Exchange co-efficient, 116, 211–212, 216

First law of thermodynamics, 27–28, 34, 50, 138, 155, 168
Free atmosphere, 22, 208, 212, 218
Free energy, 40, 41
Free enthalpy, 41
Frictionally-controlled boundary layer, 212–217
Frictional wind, 155, 157–158, 212–217, 321

Gas constant for dry air, 45
Gas constant for water vapour, 52
Gas laws, 23–26, 52, 183, 184, 255, 289, 302
General circulation, 141, 143, 260, 311–331
Generalized system of co-ordinates, 199, 203
Geocentric view, 3
Geopotential, 6, 157, 159, 164, 184, 276, 279, 280–281, 301
Geopotential height, 6, 184, 276, 291
Geopotential surface, 6
Geostrophic adjustment, 246–247, 250, 251
Geostrophic wind, 170, 177–178, 181–182, 184, 185, 202, 215, 216, 217, 221, 278–280, 294, 329
Gibbs' potential, 41
Gradient vector, 338–339
Gradient wind, 176–177, 178, 181–182
Gravitation, 3–8, 9, 21, 98, 159, 208, 239, 246–247, 250, 302, 308
Gravitational constant, 4
Gravitational force, 4–6, 159, 250
Gravity, 4–6, 12–14, 27, 45, 65, 66, 130, 134, 149, 155, 157, 161, 208, 227–228, 232, 234–237, 239–252, 253–254, 257–259, 276–278, 290, 298, 320, 328
Gravity waves, 227, 232, 234, 237, 239, 240, 241, 243, 244–252, 253, 254, 257–258, 261, 262, 276–278, 279, 290
Greenhouse effect, 117–119, 132
Group velocity, 230–231, 243, 251, 257, 262

Hadley circulation, 266, 271, 312, 313, 316, 328
Heat, 10, 15–18, 20–21, 27–42, 43–44, 46–55, 74, 79–81, 82, 87, 93–97, 99–100, 102, 104, 115–136, 137–151, 155, 168, 175,

189, 199, 201, 213, 220–221, 256, 259, 260, 262, 265, 266, 271–272, 279, 300, 302, 304, 309, 311, 313, 317, 324–328, 330–331
Heat balance, 36, 44, 115–136, 137–151, 155
Heat balance over Tibetan plateau, 138, 145, 146, 147–148, 151, 317–318
Heat capacity, 16, 18, 130, 266
Heat sources and sinks, 42, 137–151, 271, 317
Heat wave, 115–136, 137–151
Heliocentric theory, 3, 4
Helioseismology, 93–94
Helmholtz potential, 40
Humidity-mixing-ratio, 44, 45, 49, 56, 57
Hydrodynamical attraction, 69–72
Hydrological cycle, 43, 59, 326
Hydrostatic approximation, 14, 15, 21, 31, 50, 130, 184, 237, 245, 276, 279, 322, 328
Hygrometers, 46–48

Ice chamber, 63, 67
Ice-crystal effect, 74–75, 76
Ice nuclei, 76
Ideal gas law, 26, 52, 183, 184, 255, 289, 302
Inertial instability, 293, 294–295
Inertial motion, 178–179
Internal gravity waves, 237, 239–244, 253, 257, 262
Isobaric co-ordinate system, 173, 174, 196, 255, 285
Isogons, 182

Jetstream, 185, 192, 220, 271

Karman constant, 213
Kelvin scale of temperature, 17
Kelvin wave, 240, 253–254, 255–257, 258, 259–262, 264, 271–272
Kepler's laws, 3, 7–8
Kinetic energy of atmosphere, 319–320
Kinetic energy balance, 319–320
Kinetic theory of gases, 27
Kirchhoff's law of radiation, 82–83, 84

Laboratory simulation of general circulation, 327–330
Lapse rate of temperature, 14, 15, 18, 31, 48, 50, 56, 120, 121–122, 214
Law of central forces, 6–8

Madden-Julian oscillation (MJO), 253, 262–265
Magnetic storms, 100–101, 106
Mass continuity equation, 138, 149–151, 218

Maxwell-Boltzmann's law, 27
Melting point of ice, 53
Mesosphere, 19–20, 102
Meteorological instruments, 23
Mixing-length hypothesis, 211–212
Moist adiabatic lapse rate, 48–49
Molecular velocity, 11–12
Molecular viscosity, 66, 207, 209, 210
Molecular weight, 11, 25, 26, 29, 123
Mountain lee waves, 244

Natural co-ordinates, 173, 175–180, 191–192
Net radiation, 141–145, 147
Neutral stability curve, 298–300
Neutrinos, 92, 98
Nocturnal jet, 179, 214, 220–221
Nondivergent models, 279–281
Normand diagram, 47
Numerical experiment on general circulation, 331
Numerical weather prediction (NWP), 275–291, 331

Oceanic boundary layer, 222–224
Optical properties of ocean water, 131–132
Oscillations in atmosphere, 227–252
Ozone, 10, 20, 22, 83, 100, 102–106, 108, 117, 138
Ozone hole, 103–104
Ozone and weather, 105–106
Ozonosphere, 20, 102–106

Perturbation technique, 227, 231–232, 241, 254, 295
Phase difference, 229
Phases of water, 53–55
Phase velocity, 228, 232
Photosphere, 90, 93, 94–95, 96, 102
Photosynthesis, 9, 135–136
Physical constants and parameters, 350–351
Planck's law of radiation, 84–85
Planetary motion, 3–4, 6–7
Poicaré waves, 250–251
Polar auroras, 100–101
Potential temperature, 30–31, 32, 47–48, 50–51, 56–58, 106, 149–150, 162, 193–195, 208, 213, 233, 241, 293, 303–304
Potential vorticity, 193–195, 245–246, 251–252, 281–282
Pressure
    mean sea level pressure, 15, 16, 18, 119, 316
    vertical variation of pressure, 14–15
Prevost's theory of heat exchanges, 81

Primitive equation model, 276, 284–290, 309, 331
Properties of atmosphere, 9–26
Proton–proton chain reaction, 91–92

Quasi-balanced winds, 173–185
Quasi-biennial oscillation (QBO), 240, 253, 260–262
Quasi-geostrophic approximation, 278–279
Quasi-geostrophic models, 276, 278–279, 288, 290, 295, 300, 304–306, 309, 331
Quasi-geostrophic theory, 328, 329

Radiation, 10, 20, 22–23, 42, 79–88, 89–98, 99–114, 115–119, 121, 129–134, 137–145, 147–148, 221, 249, 262, 264
Radiative cooling, 74, 147–148
Radiative heat flux, 129–134, 138
Radiative heating, 115, 116, 119–120, 147
Rainfall from cold clouds, 75–76
Rainfall from warm clouds, 75–76
Reflection of radiation, 107–108, 114, 131–132
Refraction of radiation, 82, 131–132
Relative acceleration, 159–162
Relative humidity, 44, 47, 62
Relative motion, 159–162, 165, 190
Relative velocity, 159, 161, 191, 321
Relative vorticity, 191, 192, 194–195, 197, 238, 245, 277, 278
Reversing factor, 96
Reynold's number, 74, 207–208
Richardson's criterion, 208, 213–214
Rossby diagram, 238, 254, 259, 313
Rossby-gravity waves, 254, 257–260, 261, 262
Rossby number, 178, 328–329, 330
Rossby radius of deformation, 247, 251–252, 271
Rossby waves, 227, 232, 237–239, 272, 282
Rotation, 4–5, 14, 80, 93, 159, 161, 180, 187–191, 197, 203–206, 221, 239, 241, 249, 251, 253, 280, 311, 328–330

Salinity, 129–130, 134
Satellite radiation data, 141–145
Saturation, 34, 44, 47–49, 50, 51–53, 54–58, 61, 63, 73–75, 123, 125
Saturation vapour pressure, 34, 44, 47, 49, 51–53, 54–55, 61–63, 73–75
Scalar, 287, 319
Scale analysis, 169–171, 173, 276–277, 280, 306
Scattering of radiation, 107–108
Secondary circulation, 217–221, 301
Second law of thermodynamics, 36–39, 42, 52

Sensible heat flux, 120–123, 124, 145–146, 147–148
Shallow water equations, 245–246
Shearing stress, 157–158, 207–208, 210, 212–213
Simple pendulum, 228–229
Snell's law, 132
Solar atmosphere, 89, 93, 94, 95–97
Solar constant, 89, 108–109, 116, 141
Solar spectrum, 99–100, 104, 108, 109
Solar wind, 89, 97–98, 100–101, 102
Solenoids, 189–190, 196
Southern oscillation (SO), 268–270
Specific heat at constant pressure, 28–29, 46
Specific heat at constant volume, 28, 302
Specific heat of gases, 28–29
Specific heat of water vapour, 49, 52, 53
Specific humidity, 44, 78, 326
Spectral distribution of radiant energy, 86–88
Spherical co-ordinate system, 164–167
Spin-down effect, 217–221
Static stability, 31–33, 56, 174, 201, 279, 288, 294–295, 298–299, 330–331
Stefan-Boltzmann law of radiation, 84, 85–86, 117, 118
Stokes' law, 65, 187, 190, 197, 199, 205
Stratosphere, 19–20, 23–24, 86, 100, 102, 103, 104, 105–106, 134, 140, 240, 255, 259–260, 262, 264, 313–314, 316
Streamlines, 180–183, 191–192, 204, 206, 281, 317–318
Structure of sun
  convective layer, 93–94
  core, 91–92
  interior, 90–91
  radiative layer, 92–93
Sun, 3–8, 9–11, 18, 20–22, 69, 74, 81–83, 86, 89–98, 99–100, 102, 108, 109–111, 117, 125, 127, 132, 135–136, 143, 148, 151
Sunspots, 89, 93–94, 95, 97, 109
Supersaturation, 60–64
Surface layer, 5, 9, 18, 94, 130–134, 207, 212–214, 216, 221, 293
Symmetric standing waves, 93

Temperature
  vertical variation of temperature, 18–21
Terrestrial radiation, 140, 141
Thermal conductivity, 129
Thermal Rossby number, 329–330
Thermal wind, 183–185, 220, 261, 297, 298–301, 306, 311, 326, 329–330
Thermocline, 133–134, 225, 237, 272, 273
Thermodynamic diagram, 56–58

Subject Index 367

Thermodynamic energy equation, 155, 168, 174–175, 200, 201, 233, 241, 255, 277, 278–279, 281, 283–284, 286–287, 288–289
Thermodynamic equilibrium, 40–41
Thermodynamics, 12, 27–42, 43–58, 81, 138, 155, 168, 233
Thermohaline circulation, 134–135
Third law of thermodynamics, 41
Trajectories, 180–182
Transition layer, 96, 212, 214–217
Triple point, 53–55
Triple vector product, 160
Troposphere, 19, 20, 22, 23, 56–57, 86, 102, 106, 134, 201, 260, 262–265, 308, 313–314, 326
Turbulence, 18, 36, 39, 42, 43, 72–73, 121–123, 208, 217, 221, 240, 244
Turbulent layer, 93, 221
Turbulent motion in atmosphere, 208–211

Universal gas constant, 25–26, 41

Upper atmosphere, 98, 99, 100–102, 104, 108
Upwelling in ocean, 224–225

Vapour pressure, 34, 44–45, 47–49, 51–53, 54–55, 61–63, 73–75, 123, 148
Vector, 4–5, 14, 17, 78, 156–163, 165–167, 170, 173, 175–176, 180, 185, 187–190, 198, 209, 211, 213–214, 216, 222–223, 225, 242, 271, 277, 319
Vertical eddy flux, 237
Vertical motion in atmosphere, 200–202
Virtual temperature, 45–46
Vorticity, 93, 187–206, 218–219, 237–239, 245–246, 251–252, 278–284, 294–295, 300–301, 306–308

Walker circulation, 199, 270–271
Water vapour in atmosphere, 43–58
Waves in atmosphere, 88, 227–252, 253, 328
Waves in ocean, 227–252
Wien's displacement law, 84, 85, 86
Windchill effect, 125